Sound FX

Unlocking the Creative Potential of Recording Studio Effects

Alexander U. Case

ELSEVIER

AMSTERDAM • BOSTON • HEIDELBERG • LONDON
NEW YORK • OXFORD • PARIS • SAN DIEGO
SAN FRANCISCO • SINGAPORE • SYDNEY • TOKYO
Focal Press is an Imprint of Elsevier

Focal
Press

Acquisitions Editor: Catharine Steers
Publishing Services Manager: George Morrison
Senior Project Manager: Brandy Lilly
Assistant Editor: David Bowers
Marketing Manager: Marcel Koppes
Cover Design: Greg Harris
Interior Design: Joanne Blank

Focal Press is an imprint of Elsevier
30 Corporate Drive, Suite 400, Burlington, MA 01803, USA
Linacre House, Jordan Hill, Oxford OX2 8DP, UK

 Recognizing the importance of preserving what has been written, Elsevier prints its books on acid-free paper whenever possible.

Library of Congress Cataloging-in-Publication Data
Case, Alex U.
  Sound FX : unlocking the creative potential of recording studio effects /
Alex U. Case. – 1st ed.
     p. cm.
  Includes bibliographical references and index.
  ISBN-13: 978-0-240-52032-2 (pbk. : alk. paper)
  ISBN-10: 0-240-52032-7 (pbk. : alk. paper)   1. Sound–Recording and reproducing.   2. Sound studios.   3. Music.   I. Title.
  TK7881.4.C38 2007
  781.49028'5–dc22

                                                          2007005937

British Library Cataloguing-in-Publication Data
A catalogue record for this book is available from the British Library.

ISBN: 978-0-240-52032-2

For information on all Focal Press publications
visit our website at www.books.elsevier.com

07 08 09 10 11 5 4 3 2 1

Printed in the United States of America

For Dolores and Joe

who taught me how to learn
and showed me I could teach.

Contents

Contents

Acknowledgements

This book is a collaboration that would not have been possible or fun without the tireless patience and deep talent of many folks at Focal Press: Emma Baxter started this whole thing. Catharine Steers and David Bowers handled in-person and electronic interactions across continents with grace and skill. Brandy Lilly and Jodie Allen wielded the fine tooth comb necessary to convert this writer's efforts into a textbook fit for publication. Stephanie Barrett and her team created the groovy cover.

Nicholas LaPenn created the more than 100 line drawings in this text. His talent with things visual is clear. Because he is also a recording engineer, he brings a combination of skills that makes this book pleasing to the eye, yet always technically informative and accurate. Mr. LaPenn sought additional help from Philip Dignard when I put before him a hand-drawn little sketch labeled, "The Linear Piano," which they would make into a highlight of this text: Figure 1.16.

The fine art opening each section of the book resonates thanks to the profound talent of art director Elizabeth Meryman. Her ability to draw a connection between the specific material in this text and all of art history humbles me. I am honored this book is enriched by the work of these talented artists: Salvador Dali, Alan Rath, Man Ray, and Catharine Widgery.

Less obvious, but no less significant influences include:

Charles Hodge, who was kind enough to influence the hearing acuity of his younger brother in positive ways, first by playing Led Zeppelin loud enough in his room that I could hear it in mine, and second, by remaining an audio enthusiast even when he has a more grown up career in another field. My brother has always had a better stereo (and surround system) than I. I am coming to terms with this.

Stanley Nitzburg, for enthusiastic support of and tireless interest in his daughter's husband's avocation. His keen observations related to audio and acoustics always advance my understanding and increase my pleasure in my chosen field.

Patricia Nitzburg, for enjoying an uncountable number of concerts and concert halls with her author-in-law. I always hear better when I get to hear what she hears.

Colleagues have helped me grow intellectually and artistically. I wish to particularly recognize:

Will Moylan, John Shirley and Bill Carman in the Sound Recording Technology Program at the University of Massachusetts Lowell. I am proud to be part of your team.

The many Sound Recording Technology undergraduate and graduate students with a gift for asking the right questions.

Tony Hoover, a consultant in acoustics with deep knowledge of and passion for music. He is as inspiring in the field of acoustics as he is with a guitar in the studio.

The many outstanding faculty and students at Berklee College of Music who first helped identify the possibilities for this text.

Jim Anderson of New York University, a world class engineer whose generosity of insight, innovation, and bravery in his craft are modestly cloaked in the mild-mannered appearance of just a regular-multi-Grammy-winning-guy.

Chris Jaffe and Yasushi Shimizu, whose careers in technical fields have made music sound more glorious, and who shared with me a bit of their expertise.

Dave Moulton, with whom I've had the pleasure of being a co-author, and co-conspirator on matters ranging from audio, to education, to wine.

Rick Scott at Parsons Audio in Wellesley, Massachusetts, who patiently keeps me well-informed and well-stocked on all things FX.

My friends who enquired after page-counts and deadlines, and deferred meals and margaritas in support of the writing of this book.

My bride, Amy, has participated in and contributed to the creation of this text every step of the way. Soliciting her input on all matters — text and lyrics, art and illustrations — is as natural for me as it is for a dog to pad about in a circle before lying down. If I could, I would keep her in my pocket.

Introduction

The reader is welcome to jump immediately to the chapter dedicated to whichever effect is on their mind. However, the value of all chapters in Sections 2, 3 and 4 is best realized by reading first the material in Section 1. Master the concepts, both fundamental and advanced, in Section 1, and you are then prepared to crack open any of the Amplitude Effects of Section 2 or the Time Effects of Section 3. Presented in a logical order, Amplitude Effects and Time Effects needn't be read in strict sequence.

The discussion in Section 4 integrates these effects in important applications, from a general mixdown, to the all-important snare drum. That is best saved until after each effect family has been mastered individually, so the preferred approach is to work through Section 4 last. Readers with even an intermediate understanding of audio will find they can start this book anywhere. Of course, those readers have already thumbed ahead to page 131 and may never read this sentence.

The most important music of our time is recorded music. The recording studio is its principle musical instrument. The recording engineers and music producers who create the music we love know how to use signal processing equipment to capture the work of artists, preserving realism or altering things wildly, as appropriate. While the talented, persistent, self-taught engineer can create sound recordings of artistic merit, more productive use of the studio is achieved through study, experience and collaboration. This book defines the technical basis of the most important signal processing effects used in the modern recording studio, highlights the key drivers of sound quality associated with each, shares common production techniques used by recording engineers with significant experience in the field, references many of the touchstone recordings of our time, and equips the reader with the knowledge needed to comfortably use effects devices correctly, and, more importantly, to apply these tools creatively.

Equalization is likely the most frequently used effect of all, reverb the most apparent, delay the most diverse, distortion the most seductive, volume the most underappreciated, expansion the most underutilized, pitch shifting the most abused, and compression the most misunderstood. All effects, in the hands of a talented, informed and experienced engineer, are rich with production possibilities.

Alex Case
Portsmouth, New Hampshire, 2007

Section 1
Sound—Signals, Systems, and Sensation

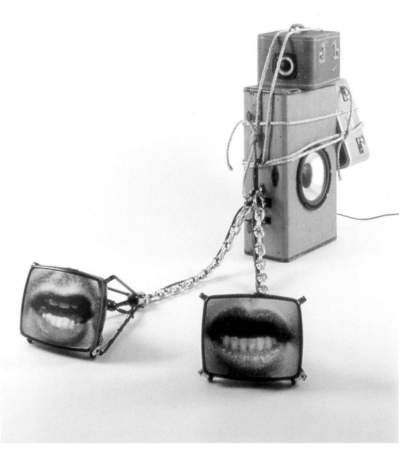

Family, 1994. Reprinted with permission from the artist, Alan Rath, and Haines Gallery, San Francisco.

Audio Waveform

"Catch a wave
and you're sittin' on top of the world."
— "CATCH A WAVE," THE BEACH BOYS, *SURFER GIRL* (CAPITAL RECORDS, 1963)

Ask a cartoonist to draw sound and they'll likely come up with something quite similar to Figure 1.1. These sketches of sound have technical merit.

When a guitar is strummed, a drum struck, or a trombone blown, we know sound will follow. The motion of the soundboard of the guitar, the vibration of the heads of the drum, and the resonance of the air within the plumbing of the trombone ultimately drive the air nearest our eardrums into action. We hear the air vibrate near us due to a chain of events that started perhaps some distance away at any such musical instrument or sound source. It is a separate matter, but we likely hope the sound made is music.

1.1 Medium

The air between a musical instrument and a listener is a springy gas. When squeezed together, it pushes back apart. If pulled apart, it snaps back together. Picture air as a three-dimensional network of interconnected springs, as in Figure 1.2. Any push or tug at just one point causes the whole system to jiggle in reaction. A continuous vibration of any one particle leads to a corresponding continuous vibration of the whole system. Motion of one element causes it to compress and stretch neighboring springs, which in turn push and pull against other springs further down the line.

Sound in air is a pressure wave with compressions (increases in air pressure) and rarefactions (reductions in air pressure) analogous to the squeezing together and stretching apart of elements of this vibrating spring system. Particles of air push and pull on one another very much as if connected by springs. Displace a bit of the air in one location, such as on-stage, and it causes a chain reaction throughout the space to the audience. As long as

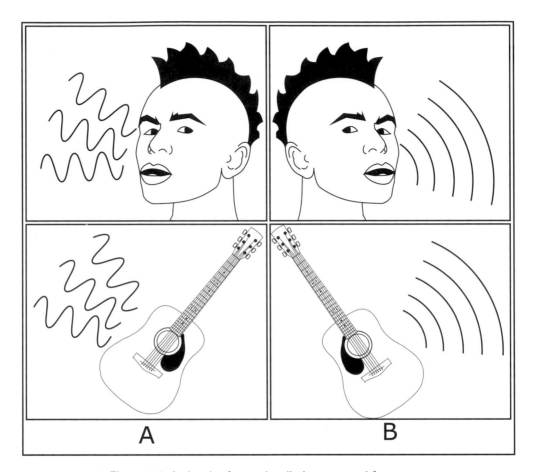

▲ *Figure 1.1 A sketch of sound radiating outward from a source.*

the source and receiver are near enough to each other, air motion at the instrument's location will eventually cause, however faintly, a bit of air motion at the listener's location.

Slight increases in pressure occur when air particles are squeezed closer together. A loudspeaker cone, kick drum head, or piano soundboard moving toward the listener will do this. Decreases in pressure occur when air particles are pulled apart — the loudspeaker cone, kick drum head, or piano soundboard moving away from the listener.

1.2 Amplitude versus Time

The physiology and neurology associated with the human hearing system search constantly for changes in air pressure. Passing through the ear

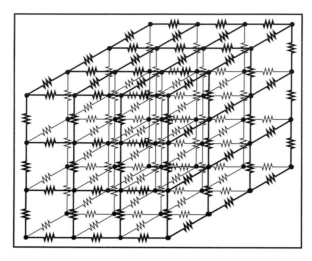

▲ *Figure 1.2 Air is a springy medium, like a three-dimensional network of springs.*

canal, changes in sound pressure push and pull on the ear drum, triggering a chain reaction that ultimately leads to the perception of sound. The pressure of a sound wave is the same type of air pressure associated with pumping air into a tire: PSI (pounds per square inch) in the some parts of the world, or kPa (kilopascals) elsewhere. Micropascals (μPa) is the preferred order-of-magnitude expression of air pressure for sound that humans can healthily hear.

A common way to graph sound plots air pressure as the vertical axis and time as the horizontal axis. Such a graph describes sound at a single, fixed location in space. As sound occurs, the air pressure at that point increases and decreases several times per second. The familiar plots of sound, in textbooks and comic books, accurately portray this concept.

Figure 1.1(a) shows sound as a squiggly line radiating from the sound source. Zoom in on the squiggly line and it might look like a sine wave as shown in Figure 1.3, or the more general waveform of Figure 1.4. A line is drawn to zig and zag, up and down, describing air pressure as it is occurs over time. The higher parts of the curve represent instances of increased pressure (compression), and the lower parts represent decreased pressure (rarefaction). A straight, horizontal line describes no change in pressure (i.e., no sound).

A lack of sound does not mean there is no air pressure, rather only that there is no *change* in air pressure. When the air pressure is unchanging, our eardrums aren't moving. In other words, we have nothing to hear.

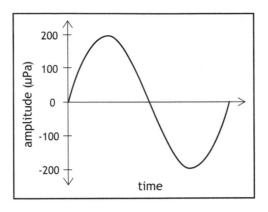

▲ *Figure 1.3 A pure tone — a sine wave.*

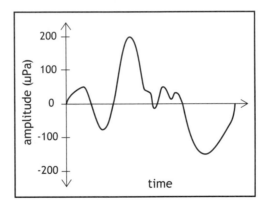

▲ *Figure 1.4 A general waveform.*

The horizontal axis in these figures meets the vertical axis, not at a pressure of zero micropascals, but at the ambient atmospheric pressure around you today. The negative amplitudes in these curves represent a reduction in air pressure below ambient, not negative pressure. The air pressure is always positive; sound represents varying degrees of positive. The precise value of pressure where the x-axis meets the y-axis is not an audio concern; it is a matter for those who track weather. The ambient pressure published in weather reports is the centerline for the pressure oscillations of our music.

1.2.1 AMPLITUDE CONFUSIONS

Discussing the amplitude of a signal, it is natural to want to assign it a numerical value. Reducing the amplitude of a signal to a single number gets a little tricky. What is the amplitude of the signal shown in Figure 1.3? The plot shows that at its highest point, it reaches an amplitude of +200 μPa. Similarly, the lowest pressure shown is −200 μPa. One might correctly describe the signal as having a *peak amplitude* of 200 μPa or a *peak-to-peak amplitude* of 400 μPa. Because this signal is perfectly sinusoidal, the peak amplitude or the peak-to-peak amplitude fully describes the general amplitude of the signal, even though it is constantly changing. The amplitude of the waveform in between the peaks follows the known pattern of a pure tone.

The slightly more complicated waveform of Figure 1.4 unravels this amplitude notation methodology. Its positive peak is still 200 μPa, but its negative peak is −150 μPa, with several intermediate positive and negative peaks in between. If this signal is a musical waveform, it will surely keep changing shape, with local maxima and minima that change as the song plays. There is no single consistent positive or negative peak. As most audio signals lack the perfect symmetry of a sine wave, a better way to express the amplitude of an audio waveform is required.

Perhaps the *average* amplitude would be helpful. This approach is frustrated by the fact that audio spends about as much time above zero as below. In the case of the sine wave (see Figure 1.3), the average amplitude is exactly zero. No matter what the peak amplitude is (it may be raised or lowered by any amount), the average amplitude remains zero.

In search of a number that describes the amplitude and does not average zero, it might be tempting just to ignore the negative half of the wave. Averaging only the positive portion, a nonzero figure can at last be calculated. This remains problematic. The negative portion of the cycle also contributes to the perception of amplitude. Turning up the volume while music is being played causes the negative portion of the waveform to become more negative still. More extreme amplitudes, positive or negative, may be interpreted as louder. The more extreme air pressure changes lead to more extreme motion of the eardrum. Humans are impressed by amplitude whether a pressure reduction pulls the eardrum outward or a pressure increase pushes the eardrum inward. It's amplitude either way. So the negative swings in air pressure must contribute to any numerical expression of amplitude as much as the positive ones, and, therefore, should not be ignored.

Musical signals, though lacking the perfect symmetry of a sine wave, share this tendency to average zero. The springy air, in reaction to the driving action of a loudspeaker, compresses and stretches. Each pressure increase is followed by a pressure decrease. At the end of the song, the air returns to ambient pressure, the loudspeaker cone returns to its original position, and the eardrum returns to its starting point.

One way to allow the negative portion of the oscillating wave to contribute to the amplitude calculation is to average the absolute value of the amplitude. Make all negative amplitudes positive, keep all positive values positive, and find the running average. The resulting expression for amplitude can track the perception of amplitude reasonably well. *VU meters* do exactly this, averaging the absolute value of the amplitude observed over the preceding 300 milliseconds.

There is further room for improvement: *root mean square* (RMS). Measuring RMS amplitude properly allows both negative and positive parts of the wave to influence the resulting number for amplitude. RMS might best be understood by working through the acronym in reverse. *Square* the amplitudes to be measured, so that a positive value always results. Take the *mean* (a.k.a. average) value of the amplitudes observed. Finally take the square *root* of the result to undo the fact that the contributing amplitudes were all squared before being averaged.

RMS amplitude is more convenient for scientists and equipment designers, as it is this type of average amplitude that must be used in calculations of energy, power, heat, etc. Audio engineering rarely needs such precision. The simpler absolute value average of the VU meter is almost always a sufficient indicator of amplitude.

1.2.2 TIME IMPLICATIONS

The amplitude versus time plot reveals fundamental information about audio waveforms. A pure-tone sine wave (see Figure 1.3) consists of a simple, never-changing pattern of oscillation. Measure the length of time associated with each cycle to determine the waveform's *period*. Count the number of times it cycles each second for a determination of its *frequency*. Period is the time it takes for exactly one cycle to occur, with units of seconds per dimensionless cycle, or simply seconds. Frequency describes the number of cycles that occur in exactly one second, with units of dimensionless cycles per second. Therefore, units for frequency live entirely in the denominator (per second, or /s) and have been given the alternative unit of hertz (Hz).

Note that counting the number of cycles per second (frequency) is the opposite of counting the number of seconds per cycle (period). Mathematically, they are reciprocals:

$$f = \frac{1}{T}$$

(1.1)

and

$$T = \frac{1}{f}$$

(1.2)

where f = frequency, and T = period.

1.3 Amplitude versus Distance

The springiness of air ensures that any localized changes in air pressure near a sound source will cause a chain reaction of air pressure changes, above and below the current air pressure, all around that source. Even a slight disturbance of air pressure will ripple outward. In order to describe the state of air pressure along some distance, a different pair of axes is needed: air pressure versus location or air pressure versus distance.

At a fixed instant in time, a plot is made of the air pressure as a function of its location in space. Figure 1.5 shows such a snapshot. Returning to Figure 1.1(a), where an illustrator strategically failed to label any axes, one can conclude that the curves radiating outward from the sound source might be amplitude versus time or amplitude versus distance. The rings of Figure 1.1(b) present sound in a slightly different way. This familiar sketch of sound is a snapshot of amplitude versus distance, showing just the positive peaks of a propagating wave, or just the negative excursions, or just the zero crossings. Called *isobars*, the rings of sound radiating outward from the sound source indicate the spatial distribution of points of equivalent pressure. This is a helpful image for audio engineers; it works in comics too.

The amplitude versus distance expression of sound leads to another fundamental property of waveforms: *wavelength*, which is the distance traveled during exactly one cycle. Drive 55 miles per hour for one hour, and the distance covered is exactly 55 miles. Distance traveled can be calculated through the multiplication of speed by time. The speed of sound in air (under normal temperature and pressure) is 344 m/s. The always-friendly

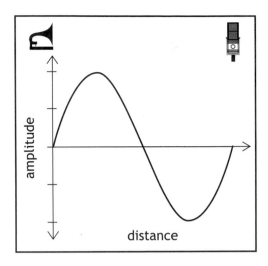

▲ *Figure 1.5 A snapshot in time shows amplitude over a distance from sound source to receiver.*

metric system does fail us a bit here, as the speed of sound in feet per second is about 1,130 ft/s. For rock and roll, it is often acceptable to round this down to an even 1,000 ft/s.

To calculate the wavelength, then, multiply this speed-of-sound figure by the appropriate amount of time. Recalling that the time it takes a wave to complete exactly one cycle is, by definition, its period:

$$\lambda = cT \qquad (1.3)$$

where λ = wavelength, c = speed of sound, and T = period.

Expressing wavelength as a function of frequency (f) requires substitution of frequency for period. Using Equation 1.2:

$$\lambda = \frac{c}{f} \qquad (1.4)$$

Precise calculations are straightforward, but it is worth noting that wavelengths can be juggled in one's head in the heat of a recording session without resorting to pencil, paper, or calculator, provided the speed of sound sticks to the fair approximation of 1,000 feet per second.

A representative middle frequency is a 1-kHz sine wave. Using Equation 1.4,

$$\lambda = \frac{c}{f}$$

$$\lambda_{1,000} = \frac{1,000 \, \text{ft/s}}{1,000 \, \text{Hz}} \tag{1.5}$$

Recalling the units underlying hertz are cycles per second (/s),

$$\lambda_{1,000} = \frac{1,000 \, \text{ft/s}}{1,000/\text{s}} \tag{1.6}$$

which leads to the final result for the convenient, approximate wavelength for a 1,000-Hz sine wave:

$$\lambda_{1,000} = 1 \, \text{ft} \tag{1.7}$$

This middle frequency, 1,000 Hz, which has a period of 1 ms, conveniently has a wavelength of approximately 1 ft. This alignment of "ones" — 1 kHz, 1 ms, 1 ft — is a useful point of reference that an engineer can bring to every recording session (Figure 1.6).

1.4 Amplitude versus Frequency

Plots of amplitude versus time and amplitude versus distance are helpful and will be used throughout this text. An important third way of describing signals must also be understood. When music is enjoyed, listeners are

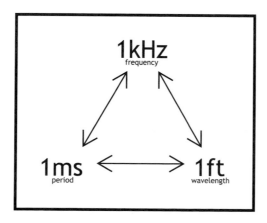

▲ *Figure 1.6 The alignment of ones.*

certainly aware that there are changes in amplitude over time at whatever location they currently occupy. Without a computer screen in front of them offering the information visually, listeners don't consciously pay attention to the fine details shown in the amplitude versus time plots.

The physiology of human hearing in fact analyzes sound as a function of frequency. The mammalian hearing system breaks sound up along the frequency axis. A pure tone is perceived as spectrally narrow and activates only a localized portion of the hearing anatomy. More complex sounds containing a range of frequencies, such as music, stimulate a broader portion of the hearing. Separating sound into different frequency ranges allows for the evaluation and enjoyment of sound across a spectral range. Humans simultaneously process the low-frequency sounds of a bass guitar in parallel with the higher-frequency sounds of a cymbal, all the while sorting out the complex detail of a vocal occupying a range of frequencies in between.

Amplitude versus frequency (Figure 1.7) is therefore an important graphical representation of sound. This plot must make assumptions about space and time. Generally, location is fixed, creating a plot that represents the sound at one place only, perhaps the comfortable couch the listener uses when listening to their favorite music. In addition, time must be constrained to some finite duration.

The right hand side of Figure 1.7 shows the amount of amplitude in a signal as a function of frequency. Is it for the last second of the signal? The preceding minute? The entire song? These time increments are all perfectly valid. An engineer might want to know the spectral content of the signal for any window of time.

In fact, such a display can be updated continuously, as the audio occurs. *Real-time analyzers* (RTAs) do exactly this. When they are set to a "fast" setting, they describe the signal that just occurred over the last 100 milliseconds or so in the intuitive terms of amplitude versus frequency. When they are set to "slow," that window in time expands to about 1,000 milliseconds.

The amplitude versus frequency plot therefore represents the signal at a fixed location and for a specific duration, identifying the distribution across frequency during that part of the signal.

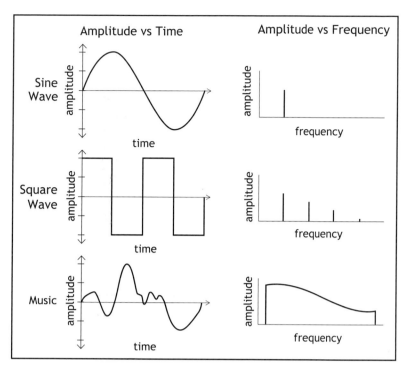

▲ *Figure 1.7 The frequency content of a signal.*

1.5 Complex Waves

While vocals, guitars, and didgeridoo waveforms are of more musical interest, simple building-block waveforms are worthy of study (Figure 1.8). The purest tone is a sine wave. This specific pattern of amplitude, repeating without fail, contains just a single frequency. The sine wave is plotted (Figure 1.8a) using:

$$Y(t) = A_{peak} \sin(2\pi f t) \qquad (1.8)$$

where $Y(t)$ is the amplitude (y-axis) as a function of time t (in seconds), A_{peak} is the peak amplitude (likely in volts or units of pressure), and f is the frequency of the sine wave (in hertz). For this and all waveform equations, readers more comfortable with degrees of phase instead of radians should simply replace 2π radians with 360 degrees within the argument of the sin function. Therefore, Equation 1.8 would become:

$$Y(t) = A_{peak} \sin(360 f t) \qquad (1.9)$$

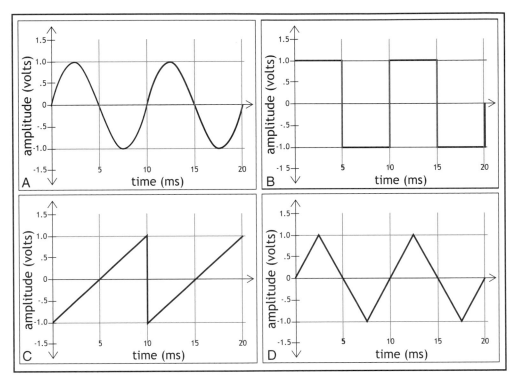

▲ *Figure 1.8 100-Hz waveforms: (a) sine wave, (b) square wave, (c) sawtooth wave, and (d) triangle wave.*

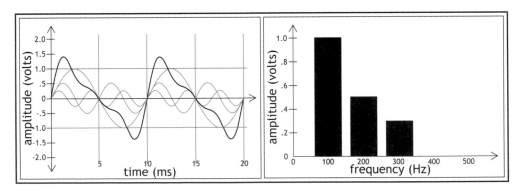

▲ *Figure 1.9 A 100-Hz complex wave with three harmonics.*

Sine waves may be combined (Figure 1.9) to create more complex waves. Beginning with a fundamental frequency of 100 Hz, a second and third harmonic are added, each with unique amplitude. The resulting waveform is simply the continuous algebraic sum of the amplitude of each of these three sine waves as time goes by:

$$Y(t) = A_1 \sin(2\pi ft) + A_2 \sin(2\pi 2ft) + A_3 \sin(2\pi 3ft) \tag{1.10}$$

where A_n is the amplitude of the nth harmonic. In Figure 1.9,

$$Y(t) = \sin(2\pi ft) + 0.5 \sin(2\pi 2ft) + 0.25 \sin(2\pi 3ft) \tag{1.11}$$

and $f = 100$ Hz. Any number of sine wave components may be part of this mathematical exercise. Specific recipes of sine waves should be noted, such as those of the square wave, the triangle wave, and the sawtooth wave.

1.5.1 SQUARE WAVES

A square wave is described by the following equation:

$$Y(t) = \frac{4A_{\text{peak}}}{\pi} \sum_{n=1}^{\infty} \left\{ \frac{1}{2n-1} \sin[2\pi(2n-1)ft] \right\} \tag{1.12}$$

It is an infinite sum of precisely these harmonic components. The harmonics must fall at exactly these frequencies — these specific multiples of the fundamental. If any frequency is shifted, even a little, the wave becomes nonsquare.

As important as the frequency is, each harmonic must also have the specified amplitude. In this case, the amplitude of each harmonic, numbered n, decreases inversely with n. That is, the nth harmonic is scaled by the factor $1/n$. The term out front, $4/\pi$, serves to give the full bandwidth square wave a convenient peak amplitude of unity. If you increase or decrease the amplitude of any or several of the contributing harmonics, the waveform becomes less square.

It is important to note that the square wave contains only odd harmonics. Harmonics that are even multiples of the fundamental frequency are simply not part of the recipe. The presence of any amount of any even multiple of the fundamental frequency would make the resulting wave less square. Equation 1.12 has become a little clumsy in an effort to force the harmonics to always be *odd* multiples of the fundamental frequency. The term, $(2n - 1)$, which appears twice in the equation, enables the series to step through values of n and create only odd multiples of f. If the series is restated using only odd numbers, m, it might be easier to follow:

$$Y(t) = \frac{4A_{\text{peak}}}{\pi} \sum_{m=1,3,5,\dots}^{\infty} \frac{1}{m} \sin(2\pi mft) \tag{1.13}$$

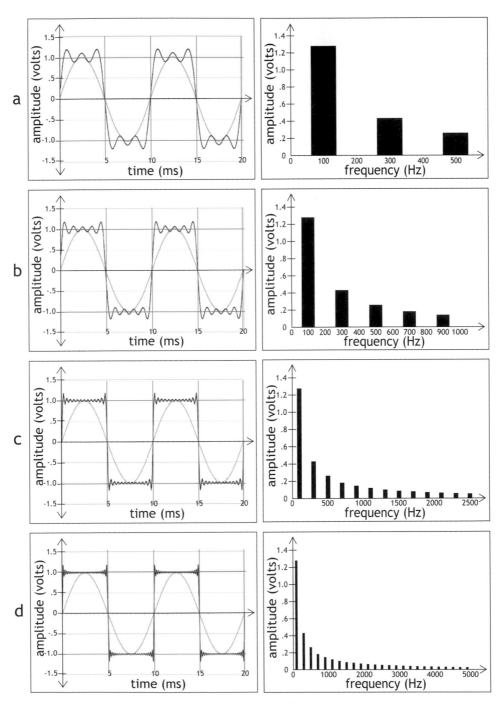

▲ *Figure 1.10 100-Hz square wave through the addition of harmonics up to (a) 500 Hz (3 harmonics), (b) 1,000 Hz (5 harmonics), (c) 2,500 Hz (13 harmonics), and 5,000 Hz (25 harmonics).*

A square wave with a fundamental frequency of 100 Hz and a peak amplitude of 1 volt (Figure 1.8b) uses Equation 1.12 or 1.13 to create:

$$Y_{100}(t) = \frac{4}{\pi}\left[\sin(2\pi 100t) + \frac{1}{3}\sin(2\pi 300t) + \frac{1}{5}\sin(2\pi 500t) + \frac{1}{7}\sin(2\pi 700t) + \ldots\right] \qquad (1.14)$$

The significance of the harmonics is shown in Figure 1.10. The contribution of evermore upper harmonics, in strict adherence to the amplitudes and frequencies specified, is clear through visual inspection. The waveform becomes increasingly more square as the bandwidth reaches upward and the number of harmonics included in the summation grows.

This makes clear the need for wide-bandwidth audio systems when square waves (think MIDI, SMPTE, and digital audio) are to be recorded and transmitted. A cable that rolls off the high frequencies of the signal within will attenuate the necessary harmonics that make up a square wave, in effect making a square wave less square. A perfectly square wave is achieved only through the rather impractical inclusion of an infinite number of the prescribed harmonics.

1.5.2 SAWTOOTH WAVES

The sawtooth wave might be considered a variation on the square wave theme. Retain both odd and even harmonics, continue to diminish the amplitude of each nth harmonic by $1/n$, rescale the overall amplitude to preserve unity peak amplitude, and a sawtooth wave with positive slope results (Figure 1.8c):

$$Y(t) = -\frac{2}{\pi}\sum_{n=1}^{\infty}\frac{1}{n}\sin(2\pi nft) \qquad (1.15)$$

A 100-Hz sawtooth wave is built up through increasing bandwidth in Figure 1.11. Careful calculation through up to $n = 50$ is shown, but the proper sawtooth does not occur until $n = \infty$.

The traditional beginning of a sine wave is that instant where the amplitude is crossing up through zero toward positive amplitude. The minus sign in Equation 1.15 dictates that all sine wave components of a sawtooth initially head in the negative direction instead. For this reason, comparison is made in Figure 1.11 to a negative sine wave, a sine wave multiplied by –1.

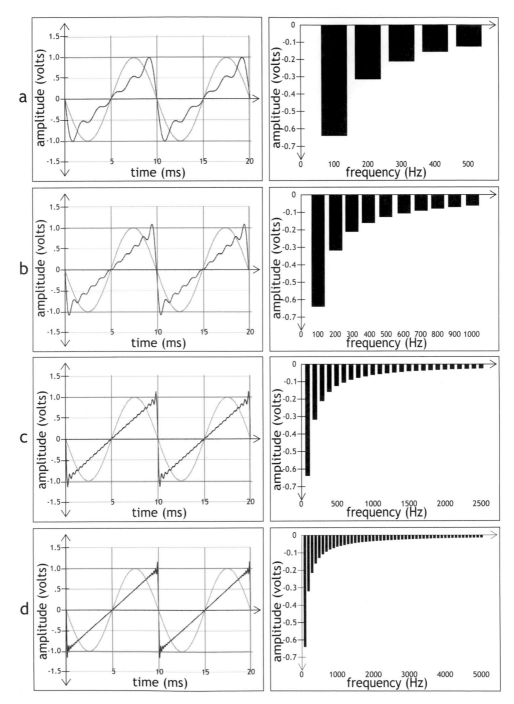

▲ Figure 1.11 100-Hz sawtooth wave through the addition of harmonics up to (a) 500 Hz (5 harmonics), (b) 1,000 Hz (10 harmonics), (c) 2,500 Hz (25 harmonics), and 5,000 Hz (50 harmonics).

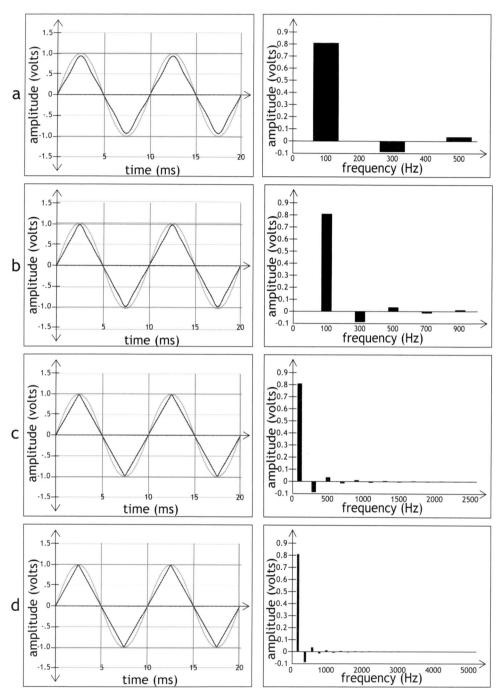

▲ *Figure 1.12 100-Hz triangle wave through the addition of harmonics up to (a) 500 Hz (3 harmonics), (b) 1,000 Hz (5 harmonics), (c) 2,500 Hz (13 harmonics), and 5,000 Hz (25 harmonics).*

The sum of all sine waves in the equation (an infinite number of precisely scaled multiples of fundamental frequency f) causes the net wave to leap to −1 before steadily rising toward +1. The instant when the sawtooth wave reaches +1 is also the fortuitous instant when each and every component harmonic happens to be beginning a negative cycle anew. This symphony of sine waves crossing upward through zero but multiplied by −1 causes the net amplitude to snap to −1 again. The pattern repeats.

1.5.3 TRIANGLE WAVES

The triangle wave (Figure 1.8d) comes from a different set of carefully scaled odd harmonics:

$$Y(t) = \frac{8A_{peak}}{\pi^2} \sum_{n=1}^{\infty} \frac{-1^{(n-1)}}{(2n-1)^2} [2\pi(2n-1)ft] \tag{1.16}$$

In addition to the requisite scaling to achieve unity peak amplitude and the use of the term $(2n - 1)$ to generate odd harmonics, notice the additional need for an alternating polarity among the harmonic components. The term, $-1^{(n-1)}$ causes the harmonics to switch sign with each increment of n. The polarity of every other harmonic is positive, while the polarity of each harmonic in between is negative. The summation of these particular components, some adding to the total while others subtract, leads to a triangle wave.

Figure 1.12 demonstrates the significance of adding additional harmonics to the fundamental sine wave. As the amplitude of successive harmonics falls proportional to $1/n^2$, this complex wave is more dependent on lower harmonics than the square and sawtooth waves. This is evident in two ways. Note the towering significance of the lower harmonics on the right-hand side of Figure 1.12. Note also how the wave obtains its characteristic sharpness and comes quite close to resembling the full bandwidth shape with just 13 harmonics.

Very much as multitrack music is built from a mix of component production elements such as drums, bass, keys, and vocals, individual pitched waveforms that make up each multitrack element are themselves made up of a specific mix of sinusoidal components. It is our job to make art from these humble ingredients.

1.6 Decibel

It is difficult to do anything in audio and not encounter the decibel. As discussed below, the decibel offers a precise calculation that quantifies properties of an audio signal in a very useful form. The fact is one may never trouble to dig out these equations and perform a decibel calculation during the course of a recording session. But the hardware designers and software code jockeys who create the effects processors and recording devices that fill the studio certainly do. If an audio engineer is to speak comfortably and accurately about decibels, it helps to know a little of the math that makes it possible. Those who are bored or frustrated by the math should at least know that someone went to a lot of trouble to find a way to express the level of the signal in a way analogous to the expression of pitch. The decibel offers a perceptually meaningful description of amplitude, one that the ears and brain can make sense of.

The decibel appears in some form on almost every faceplate and every user interface of every signal processor in the recording studio. Understanding the meaning of quantities in decibels is essential to understanding sound effects. There is an equation that absolutely defines the decibel (dB):

$$dB = 10 \times \log_{10}\left(\frac{power_A}{power_B}\right) \qquad (1.17)$$

The English translation of that equation goes something like, "The decibel is ten times the logarithm of the ratio of two powers." This straightforward statement is rich with meaning.

The equation for the decibel has two features built-in. First is the *logarithm*. The mathematical properties of this function are considered in detail shortly, but it is important to understand the motivation for digging up the logarithmic, or log, function in the first place. The log is part of the decibel equation to make the math more convenient. It makes the vast range of amplitudes typical of audio much easier to deal with.

The second key element of the decibel equation is the *ratio* of powers within parentheses. The decibel equation uses a ratio so as to be consistent with the human perception of power and related quantities. The equation attempts to create a number that describes the amplitude of a musical waveform. For the decibel to be useful, the resulting number needs to have some connection to the human perception of this property of sound.

These two feature — log and ratio — make the decibel a versatile and useful way to express the amplitude of our musical waveforms.

1.6.1 LOGARITHM

The log represents nothing more than a reshuffling of how the numbers are expressed. The two following equations are both true and say very nearly the same thing:

$$10^y = X \tag{1.18}$$

$$\log_{10}X = y \tag{1.19}$$

Equation 1.18 is relatively straightforward. Ten raised to the power y gives the result X. For example:

$$10^3 = 10 \times 10 \times 10 = 1,000 \tag{1.20}$$

The logarithm enables us to undo the calculation mathematically. Starting with the answer from above, 1,000, the log function leads back to 3.

$$\log_{10}(1,000) = 3 \tag{1.21}$$

Said another way, Equation 1.19 answers the question, "What power of 10 will give us this number, X?" To take the log of 1,000 as in Equation 1.21 is to ask, "What power of 10 gives us 1,000?" The answer is 3: $10^3 = 1,000$, so $\log_{10}(1,000) = 3$.

What power of 10 will give us 1,000,000? With an eye for powers of 10, or perhaps with the help of a calculator, it is easily confirmed that the $\log_{10}(1,000,000) = 6$. Ten raised to the sixth power gives us one million, as Equation 1.9 would describe it. Now calculate the power of 10 that gives 100 trillion: $\log_{10}(100,000,000,000,000) = 14$.

Herein lies the motivation for logarithms in audio. They make big numbers — potentially very big numbers — much smaller: 100,000,000,000,000 becomes 14. It converts governmental budgets into football scores.

The log function is an acquired taste. Those with little or no exposure to them will likely find the logarithm awkward at first.

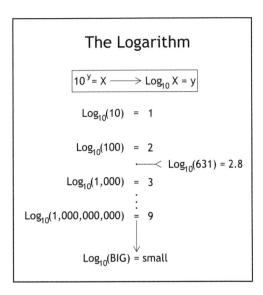

▲ *Figure 1.13 The logarithm mathematically connects a potentially very large number to a much smaller number.*

There's an interesting twist, so follow along in Figure 1.13. The log function calculates the power of 10 needed to create a number — any number greater than zero. So while the $\log_{10}(100) = 2$ and the $\log_{10}(1,000) = 3$, the log function can also find values in between. For example, the $\log_{10}(631)$ is about 2.8. In other words,

$$10^{2.8} = 631 \tag{1.22}$$

To know this, a calculator, a computer, a slide rule, a class geek, or some tables full of logarithm answers are required. These aren't calculations easily done in the head, with the help of counting on fingers and toes.

The logarithm mathematically connects a potentially very large number, and it can be any number greater than zero, to a much smaller number. This is useful because the range of amplitudes humans can hear is truly vast. The smallest sound pressure that average healthy humans can hear, rounded off for convenience, is about 20 micropascals. Compare that amount to the amplitude of air pressure associated with the onset of physical pain in our hearing system. (Please note: The risk of hearing damage starts well before the pain begins, so listen safely and wisely. Please do not risk hearing damage.). Pain starts happening at about 63,000,000 micropascals. The difference between detection and pain in the human experience of air pressure is many millions of micropascals. Listening to a conversation at normal levels might occur with amplitudes of around 20,000 micropascals. It is reasonable to monitor a pop mix at about 630,000 micropascals. One might occasionally crank it to more than 6,000,000 micropascals. Even at this level, the neighbors aren't complaining, the drummer wants it louder, and, if it doesn't go on for too long, there isn't much risk of hearing damage yet. Jet engines and power plants are much, much louder still, on the order of hundreds of millions of micropascals.

The problem with these numbers is clear: they are too big to be useful in the studio. "Yeah, let me push the snare up about 84, pull the strings down about 6,117, and see if the mix sits right at 1,792,000." This is too awkward; the decibel to the rescue. The mathematical log function exhibits the following helpful property:

$$\log_{10}(BIG) = small \tag{1.23}$$

The log of a big number results in a much smaller number. Human hearing is capable of interpreting a vast range of amplitudes. The numbers used to describe amplitude become unwieldy if not routinely subjected to the logarithm, so it is a fundamental part of the decibel equation.

1.6.2 RATIOS

Also built into the decibel equation is a ratio (consult again Equation 1.17). Mathematically, a ratio strategically reduces two numbers to a single, informative number.

Consider pitch. Musical harmony is built on ratios. The octave, for example, represents a doubling of pitch, a ratio of 2 : 1. The orchestra tunes (generally) to A440. Also identified as A4, this is the first A above middle C. A440 is a musical note whose fundamental frequency is exactly 440 Hz. Double that frequency to 880 Hz to create a note exactly one octave higher. To lower the pitch by exactly one octave, reverse the ratio (e.g., 1 : 2, or mathematically, $^1/_2 = 0.5$). That is, cut the frequency in half. An octave below A440 has a fundamental frequency of 220 Hz.

All the musical intervals represent ratios. The numerical value of the ratio represents a scaling factor that, when multiplied by the starting frequency, finds the new frequency needed to reach the desired musical interval. As the intervals deviate from the octave, the ratios are no longer built on simple whole numbers, and the exact ratios depend on the type of tuning used. For example, a perfect fifth is a ratio of 3 : 2 ($^3/_2 = 1.5$) in just intonation.

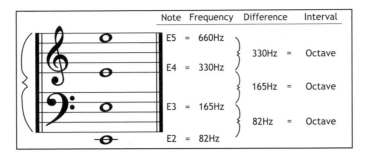

Note	Frequency	Difference	Interval
E5 = 660Hz			
		330Hz =	Octave
E4 = 330Hz			
		165Hz =	Octave
E3 = 165Hz			
		82Hz =	Octave
E2 = 82Hz			

▲ *Figure 1.14 Frequency changes of the octave.*

In equal-tempered tuning, a perfect fifth is achieved by multiplying by about 1.498. A perfect fourth represents the ratio of $4:3$ ($^4/_3 = 1.3333$) in just intonation, but the slightly different 1.3348 for equal temperament. No matter which form of tuning is selected, it is always a ratio that rules. In order to create a note that is a specified musical interval away, multiply the starting pitch by the value of the appropriate ratio.

It doesn't make musical sense to think in terms of the actual number of hertz between two notes. A big band arranger won't ask the trumpet player to play 217 Hz above the tenor sax. Instead, ratios are used. Score the horn a minor third above, or an octave above, and the trumpeter can oblige.

Looking at Figure 1.14, consider each note's fundamental pitches mathematically, not musically. Start two E's below middle C, labeled E2. It has a fundamental frequency of about $82^1/_2$ Hz — a meaningless observation for the performing musician, perhaps, but an important one for the recording musician. Up one octave, the pitch is exactly twice the starting pitch. That's the very definition of an octave. One octave up leads to E3, with a fundamental pitch of about 165 Hz (2×82.5). One more octave up is E4, the first E above middle C. E4 has a fundamental pitch of about 330 Hz. E5 is yet another octave above and has a pitch of about 660 Hz. Four pitches, one musical value. They are all labeled "E," and they all sound very similar, musically speaking. In harmony, E at any octave performs very nearly identical functions.

What's the difference between one E natural and the next E up? An octave. But there's a subtle illusion going on here. Using Figure 1.14, watch as the absolute numbers fail and the ratio takes over. The "distance" (as measured in hertz) from E2 to E3 is 82 Hz. This 82-Hz difference has meaning to the human hearing system; it's an octave. Starting at E3 and going up to E4 traverses an octave again. However, measured in hertz, this octave represents a difference of 165 Hz. E4 to E5 is an octave, worth 330 Hz.

An octave equals 82 Hz in one instance, 165 Hz in another situation, and then 330 Hz in the third case. The octave cannot be expressed in hertz unless the starting pitch is known. It can always be expressed as a ratio: 2 : 1.

The musical significance of the octave is well known, offering the most consonant (i.e., least dissonant) pairing of two notes of different pitches. Experienced musicians also attach specific sensory meaning to many (probably all) of the other intervals: the buzz of the perfect fifth, the warmth of the major third, and the bittersweet mood of the minor seventh. Such complex, advanced human feelings about the pitch differences between two notes reduce almost insultingly to some pretty straightforward math. To go up an octave, multiply by 2. Done. It doesn't matter what the starting pitch is.

Minor headache: Beyond the octave, the numbers aren't so neat. To go up a perfect fifth in the most common form of tuning in pop music, equal temperament, simply multiply by about 1.49830708. To go up a major third (in equal-tempered tuning) multiply by the unwieldy (and rounded off) 1.25992105. The numbers are rather unappealing. But the fundamental principle is comfortingly straightforward. Don't add a certain number of hertz to go up by a certain musical amount. Instead, multiply by the numerical value of the appropriate ratio.

The idea of the ratio is built into our musical pitch-labeling scheme. Notes are described on the familiar musical staff and labeled with the familiar short, repeating alphabet from A to G. Peek at the numbers and something peculiar is revealed. If we plot the musical staff using linear mathematics, in which all the lines and spaces of our traditional notation system are spaced an equal number of hertz apart rather than simply an equal distance apart, we get the rather strange looking staff shown in Figure 1.15. The traditional notation scheme masks the actual quantities involved — for good reason. The *musical* relevance of the notes is captured in the notation system. The relationship between C and G is always the same: it's a perfect fifth at any octave at any location on the staff. Therefore, it is shown that way on paper. It isn't musically important how many hertz apart two notes are, but it is certainly important how many lines and spaces apart they are, as arrangers well know.

The keyboard of the piano presents the same illusion, physically. Figure 1.16 shows a piano in which the number of hertz between the notes determines the physical size and location of the keys, which are unplayable

▲ *Figure 1.15 The linear staff.*

and unmusical. The layout of a proper keyboard repeats a pattern based on the musical meaning of the notes, not the linear value of the frequencies of the notes.

Ratios are a part of music. On sheet music and on the keyboard, the ratio is a proven, convenient way to take physical properties and rearrange them in a way that is consistent with their musical meaning. As with pitch (frequency), so it is with amplitude (voltage or pressure). That is, human perception "consumes" musical pitch in a relative way. It is the ratio relationship between notes that creates musical harmony, not their value in hertz. The human perception of amplitude behaves similarly, so a way to quantify the amplitude of audio signals that has musical meaning is needed. The decibel, built in part on the ratio, accomplishes this.

Research has shown that in order to double the apparent loudness of a signal, the power must increase approximately tenfold (Figure 1.17) Starting with a power of 1 watt, doubling the apparent loudness leads to 10 watts. Doubling the loudness from 1 watt required an increase of 9 watts. Repeat this exercise starting at a different power. Beginning with 10 watts, doubling the loudness requires that the power be scaled up ten times to a new value of about 100 watts. This doubling in loudness requires an increase of 90 watts. The next doubling, to 1,000 watts, comes courtesy of a 900-watt addition of power. In all cases, the perceptual impact was the same: the signal became roughly twice as loud. This is the power amplitude analogy of the octave. Here, we are talking about loudness, not pitch. But just as the perception of pitch is driven by a ratio (multiply by 2 to go up an octave), so is the perception of power (multiply by 10 to double the apparent loudness).

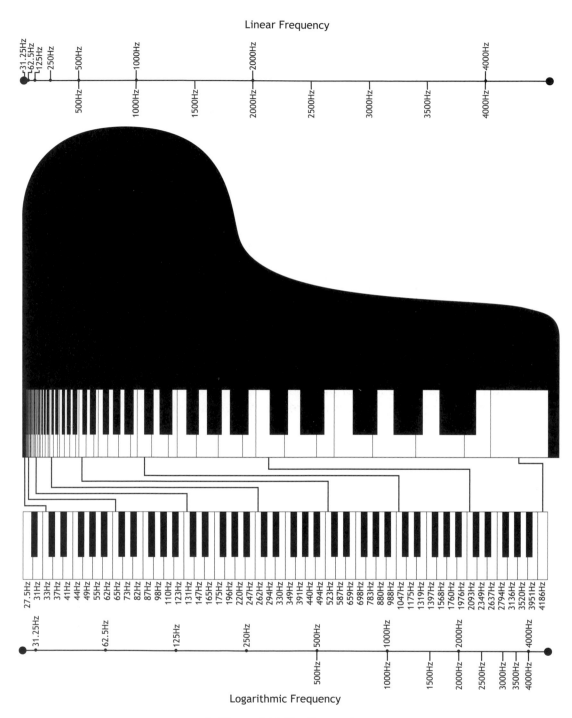

▲ *Figure 1.16 The linear piano.*

Power (watts)	Difference (watts)		(dB)
1,000w			
	>900w	=	10dB
100w			
	> 90w	=	10dB
10w			
	> 9w	=	10dB
1w			

▲ *Figure 1.17 The power changes of perceived doublings.*

The equation for the decibel, therefore, has a ratio built in. As Figure 1.17 reveals, the decibel difference between each of the power settings is always the same: 10 decibels. For each equivalent perceptual change, the actual number of watts needed is different, depending on the initial power setting. The decibel equation brings consistency in numbers to a consistent sensory event. The amount of change required to double the loudness — to have the same perceptual impact on our listening systems — is always 10 dB.

Decibels provide engineers the amplitude equivalent of the musical pitch labeling scheme. They convert the physical quantity into a numerical expression that is highly consistent with the perception of that quantity. Through experience, audio engineers develop a very specific idea in their mind about what a 3 dB or 6 dB increase in level sounds like. Experienced musicians are able to start at one pitch and find by ear any other pitch, be it up an octave, or down a fifth, or any other interval away. Music schools offer ear training to teach this ability for pitch. Audio schools offer audio ear training to accomplish the same thing, in the amplitude domain.

1.6.3 REFERENCES

A close look at the decibel equation reveals that it is a single number expression for two numbers. That is, 30 dB represents a comparison of one number to another. It does not make sense to say that a power of 1,000 watts equals 30 dB. To use 1,000 watts in the decibel equation (Equation 1.17), a second wattage must be put into the ratio. Starting with a reference of 1 watt, it is correct to calculate that 1,000 watts is 30 decibels higher than 1 watt:

$$10 \times \log_{10}\left(\frac{1{,}000}{1}\right) = 30 \qquad (1.24)$$

The decibel is meaningless without mentioning two numbers. It is always necessary to compare two numbers with the decibel. Often, an engineer wishes to make statements that compare amplitude to the current value, as in, "Turn the snare up 3 dB." That is shorthand for, "Make the amplitude of the snare 3 decibels louder than the current amplitude." If one were to resort to the equation, the current amplitude is used in the bottom of the ratio (the denominator), and the top of the ratio (the numerator) gets the amplitude of the new, louder snare that is desired. Of course, this equation is never used during a session. The faders on a computer screen or on a mixer are labeled in decibels already. Someone else already did the calculations for us.

If a signal isn't being compared to its current value, then it is compared to some reference value. A good starting point might be a reference of 1 watt; 10 watts is 10 dB above this reference, and 100 watts is 20 dB above the same reference. So the correct way to express decibels here is something like, "100 watts is 20 decibels above the reference of 1 watt."

It gets tiring, always expressing a value in decibels above or below some reference value. Here's the time saver: If the reference is 1 watt, express it as dBW (pronounced, "dee bee double you"). The "W" tacked on to the end identifies the reference as exactly 1 watt. This shortens the statement to, "100 watts is 20 dBW." Done. The reference, which is required for the decibel statement to be meaningful, is attached to the dB abbreviation.

Note that the statement is not, "100 watts is 20 dB." That is incorrect. A single number is not expressible as a decibel. There must be some value stated as a point of comparison. Using "dBW" instead of just "dB" is the subtle addition that gives these statements meaning.

So while Equation 1.17 is the general equation for the decibel, a more specific equation using a reference of 1 watt is helpful:

$$dBW = 10 \times \log_{10}\left(\frac{power}{1\,watt}\right) \tag{1.25}$$

Other subequations exist, with different reference values. For example, sometimes 1 watt is too big to be a useful reference power. Use the much smaller milliwatt (0.001 watt) instead. If the power reference is one milliwatt, the suffix attached to dB is a lower case "m," for milli:

$$dBm = 10 \times \log_{10}\left(\frac{power}{0.001\,watt}\right) \tag{1.26}$$

A little physics lets us leave the power domain and create expressions based on the quantities we see more often in the studio: sound pressure and voltage. Sound pressure decibel expressions use the threshold of hearing (20 micropascals) as the reference pressure, and we tack on the suffix "SPL" to express *sound pressure level* in terms that have perceptual meaning:

$$dBSPL = 20 \times \log_{10}\left(\frac{pressure}{20\,\mu Pa}\right) \tag{1.27}$$

Note that the equation changes a little. Instead of multiplying the logarithm by 10, sound pressure statements require multiplication by 20. This is a result of the physical relationship between acoustic power and sound pressure. Power is proportional to pressure squared. The sound power terms within Equation 1.17 become sound pressure squared instead. It is a property of logarithms that this power of two within the logarithm may be converted to multiplication by two outside the logarithm. Hence the 10 becomes 20 for decibel calculations related to pressure. Likewise, we can use decibels to describe the voltage in our gear. Electrical power is proportional to voltage squared, so again we use a 20 instead of a 10 in the decibel calculations for voltage.

In the voltage domain, a few references must be dealt with:

$$dBu = 20 \times \log_{10}\left(\frac{voltage}{0.775\,volt}\right) \tag{1.28}$$

$$dBV = 20 \times \log_{10}\left(\frac{voltage}{1\,volt}\right) \tag{1.29}$$

It is a quirk of history that the unwieldy reference of 0.775 volt was chosen. The interested reader can use Ohm's Law to apply a standard 1 milliwatt of power across a load of 600 ohms (which was a standard in the early telecommunications industry, not audio). A voltage of 0.775 V results. Even as the idea of a 600-ohm load lost its significance in the modern professional audio industry, the quirky standard voltage remains. Someone, tired of the clumsiness of that number, chose an easier to remember reference: 1 volt. Good idea. Confusing result. Too many standards. It sort of misses the point of a "standard," doesn't it? The output voltages specified in the back of the manual for any signal-processing device might be expressed in dBu, for example, +4 dBu. Or it might show up in dBV, like −10 dBV. In both cases, the manual is just indicating the nominal output level, relative to a chosen industry reference voltage.

1.6.4 ZERO DECIBELS

The meaning of zero decibels must be considered. When a meter of a signal processor reports a value of 0 dB, it does not mean that no signal is present. It in fact means that the amplitude is identical to the reference.

Working in dBu, consider an input that is 0.775 volt. Expressing this input in dBu using Equation 1.28, we get:

$$dBu_{0.775} = 20 \times \log_{10}\left(\frac{0.775}{0.775}\right) \tag{1.30}$$

$$dBu_{0.775} = 20 \times \log_{10}(1) \tag{1.31}$$

Importantly,

$$\log_{10}(1) = 0 \tag{1.32}$$

Therefore, the special case of having an input identical to the reference voltage leads to:

$$dBu_{0.775} = 20 \times 0 \tag{1.33}$$

$$dBu_{0.775} = 0 \text{ dBu} \tag{1.34}$$

When the signal hits zero, be it 0 dBu, 0 dBV, or any other decibel reference, the amplitude equals the reference.

1.6.5 NEGATIVE DECIBELS

The log of one is zero (see Equation 1.32). The log of a value greater than unity is a positive number; it's greater than zero. The log of a value less than unity is a negative number; it is less than zero.

So when a signal is greater than the reference value being used (0.775 volt, using dBu for example), the decibel calculation produces a positive result. Any positive expression of decibels indicates a signal is higher in amplitude than the reference.

When a signal is less than the reference, the decibel calculation gives a negative result: +3 dBu indicates a signal that is 3 decibels greater than 0.775 volt, and −3 dBu indicates a signal that is 3 decibels less than 0.775 volt.

Discussing amplitude with units of micropascals or volts, while technically correct and quantitatively useful, is not productive in the recording studio. A different expression of amplitude is preferred. The decibel gives audio engineers a way to communicate matters related to amplitude that is perceptually meaningful and musically useful. So don't turn it up 2 volts, turn it up 6 dB!

1.7 Dynamic Range

Musical dynamics are so important to a composition and performance that they are notated on every score and governed closely by every bandleader, orchestra conductor, and music director. Making clever use of loud parts and soft parts is a fundamental part of performance, composition, and arranging in all styles of music from classical to jazz and folk to rock. In the studio, we must concern ourselves with a different, but related sort of dynamics (Figure 1.18): *audio dynamic range*.

Exploring the upper limit of audio dynamic range comes naturally to most musicians and engineers. The music is often turned up until it distorts. It is as if the instruction manual required it. The volume setting on everything from guitar amps to home stereos is often found to push the limits, flirting with distortion. It is a basic property of all audio equipment: turn it up too loud and distortion results.

At the other amplitude extreme lives a different audio challenge. If the musical signal is too quiet, the inherent noise of the audio equipment itself

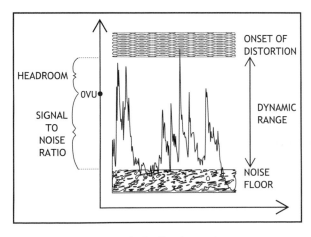

▲ *Figure 1.18 Audio dynamic range.*

becomes audible and possibly distracting. Cassette tapes with their characteristic hiss and LPs with their crackles and rumble demonstrate the challenge that a noise floor presents. In fact, all audio equipment has a noise floor — equalizers, compressors, microphones, and even patch cables. Yes, even a cable made of pure gold, manufactured in zero gravity during the winter solstice of a leap year, costing half a year's salary, will still have a noise floor, however faint.

A constant part of the recording craft then is using all these pieces of audio equipment in that safe amplitude zone, above the noise floor but below the point at which distortion begins. That safe zone is, in fact, the audio dynamic range. It defines and, using decibels, quantifies the range of usable amplitude between the noise a piece of gear makes and the level at which the piece of equipment starts to distort.

The target in between these two extremes is typically labeled zero VU. 0 VU is a nominal level for a signal through a piece of equipment. At 0 VU, the music gets through well above the self-noise of the equipment, but safely under the point where it starts to distort.

If we recorded pure sine waves for a living, we would raise the signal amplitude up just to the point of distortion, back off a smidge in level, and hit record. Thankfully, musical waveforms are nothing like sine waves. The amplitude of a musical waveform races wildly up and down due to both the character of the particular musical instrument and the way it is being played. Musical instruments lack the amplitude predictability of a sinusoid.

Some signals are more predictable than others. Electric guitars amps cranked to the limit have very little dynamic range. Many guitar sessions find the engineer recording the way Nigel does, with the amp set to 11. Many (but not all) guitar amps sound best when they are cranked to within inches of their lives. This leaves no room for audio peaks to get through at a higher level. The meters on the mixing console and the multitrack recorder simply zip up to 0 VU at the downbeat, and then barely move until the end of the song. Chugga chugga crunch Ch-Chugga. Chugga chugga crunch Ch-Chugga. The meters do not budge until the guitarist stops playing. Crunchy rhythm rock-and-roll guitars are a case study in limited dynamic range.

Percussion, on the other hand, can be a complicated pattern of hard hits and delicate taps. Such an instrument is a challenge to record well, presenting extremely wide and difficult to predict dynamic range.

Every instrument offers its own complicated dynamics. The musical dynamic range of the instrument must somehow be made to fit within the audio dynamic range of the studio's equipment. Otherwise, the listeners are going to hear distortion, noise, or both.

Accommodating the unpredictability of all musical events, we record at a level well below the point where distortion begins. The amplitude "distance" (expressed in decibels) between the target operating level and the onset of distortion is called *headroom*. This provides the engineer a safety cushion, absorbing the musical dynamics of the instrument recorded without exceeding the audio dynamic range of the gear used to capture the recording.

The relative level of the noise floor compared to 0 VU, again expressed in decibels, is the signal-to-noise ratio. It quantifies the level of the noise, relative to the nominal signal level. The trick, of course, is to send the audio signal through at a level well above the noise floor so that listeners will not even hear that hiss, hum, grit, and gunk that might be lurking low in the piece of equipment.

Making effective use of dynamic range influences how audio engineers record to any format, from analog tape to digital hard disk. It also governs the levels used when sending audio through a compressor, delay, reverb, or any other type of audio equipment. It is a constant tradeoff of noise at low amplitudes versus distortion at high amplitudes.

1.8 Sound Misconceptions

With a more thorough understanding of the audio waveform, common errors in understanding and judgment can be avoided.

1.8.1 MISTAKING THE MESSAGE FOR THE MEDIUM

The transmission of sound from a source to a receiver does not require the delivery of air particles from the sound-making object to the listener. Sound *waves* propagate from the source to the receiver. The actual carrying medium — the air — does not. The springiness of air ensures that as the sound source drives the air, its influence spreads outward. The influence of air should not be confused with actual one-way motion of air. When a loudspeaker cone sends music towards a listener, it does not do so by sending a steady breeze of air into the person's face. The pressure wave propagates; the air essentially stays put.

During the sound event, the air molecules stay in the same general space, moving about in a cloud as molecules are wont to do. As the pressure wave passes, these constantly moving gas particles are squeezed together and pulled apart slightly by the approach and retreat of neighboring air molecules or the vibration of the sound-producing instrument. After the sound wave has departed, the air molecules return to their original approximate locations.

Humans listen to the message carried by air. There is no reason to be particularly interested in any specific air particles themselves.

1.8.2 DON'T PICTURE THESE SKETCHES

The actual motion of the air particles associated with sound is in the same direction as the propagating sound wave itself. A wave propagating left to right causes the brief, organized jiggle, left and right, of air particles as the compression/rarefaction cycle occurs.

When the motion of the particles is in the same direction as the propagation of the wave, it is classified a *longitudinal* wave. Sound does not belong to the slightly more intuitive family of waves known as *transverse* waves. Transverse waves have particle motion perpendicular to the wave motion. Making waves in water or snapping a rope demonstrates transverse waves. Toss the proverbial pebble in the allegorical still pond, and the vertical disruption of water ripples horizontally outward. Up and down motion of the end of a rope triggers a wave of up and down motion that is transmitted through the length of the rope. This seemingly insignificant fact — sound waves are longitudinal — causes some headaches when describing sound and has lead many an audio enthusiast into confusion.

Consider sound propagating from a loudspeaker toward a listener. Is the sound curving back and forth, up and down, as it heads horizontally from the speaker to the ears? No, it most definitely is not. For the sound heading from a loudspeaker to a listener in the same horizontal plane, there is nothing up and down about it. When sound propagates horizontally, the most interesting motion of the air particles is horizontal too. It might be sketched as a wave curving up and down, but the air particles in fact move only side to side.

We resort to drawing up and down lines because it is too difficult to draw it literally (Figure 1.19). The clouds of air particles gathering into patterns

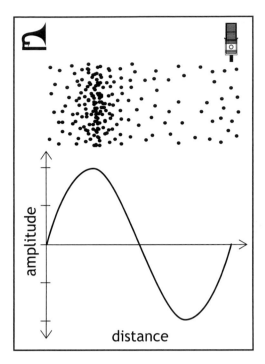

▲ *Figure 1.19 A pressure wave between source and receiver.*

of increased and decreased density is correct, but tedious to draw and clumsy to study.

So the pressure versus time plots (see Figures 1.1a, 1.3, and 1.4) are the preferred depiction of sound over time. One must keep in mind, however, that these graphs are a description of sound, but not a literal illustration of sound.

Signal Flow

2

"And you may ask yourself, 'How do I work this?' "
— "ONCE IN A LIFETIME," TALKING HEADS, *REMAIN IN LIGHT* (SIRE, 1980)

Ever been lost in a big, old city? The roads seem to wander randomly. Few streets intersect at right angles. Individual streets change names, become one way, or simply dead end without warning. I have a recurring nightmare that, as a tourist in a city, I reach an intersection while driving down a one-way street and innocently encounter all three signs of doom at once: no left turn, no right turn, and do not enter. There is no flashing red light to warn me as I approach this dreaded intersection. Naturally there is no sign identifying the street I am on, nor the street I have reached. The cars, with me stuck in the middle, just line up, trapped by traffic. My blood pressure rises. My appointments expire. I wake up in a sweat, vowing to walk, not drive, and to always ask for directions.

2.1 Types of Sessions

Without some fundamental understanding of how a studio is connected, anyone can eventually find themselves at the audio equivalent of this intersection: feedback loops scream through the monitors, no fader seems capable of turning down the vocal, drums rattle away in the headphones but aren't getting to the multitrack — I could go on. Believe me, I could go on.

At the center of this mess is the mixing console, in the form of an independent piece of hardware or part of the digital audio workstation. In the hands of a qualified engineer, the mixing console manages the flow of all audio signals, combining them as desired, and offers the engineer the controls needed to get each bit of audio to its appropriate destination safely and smoothly. The untrained user can expect to get lost, encounter fender benders, and quite possibly be paralyzed by gridlock.

The ultimate function of the console is to control, manipulate, process, combine, and route all the various audio signals racing in and out of the different pieces of equipment in the studio or synth rack; it provides the appropriate signal path for the recording task at hand.

Consider mixdown. The signal flow goal of mixing is to combine several tracks of music that have been oh-so-carefully recorded on a multitrack into two tracks of music (for stereo productions, more for surround) that friends, radio stations, and the music-buying public can enjoy. They all have stereos, so we "convert" the multitrack recording into stereo: 24 or more tracks in, 2 tracks out. The mixer is the device that does this.

Clearly, there's a lot more to mixing than just combining the 24-plus tracks into a nice sounding 2-track mix. For example, one might wish to add reverb, equalization (EQ), compression, and a little turbo-auto-pan-flange-wah-distortion™ (patent pending). This book seeks to elevate the reader's ability to use these effects and more. Before tackling the technical and creative issues of all of these effects, one must first understand how the audio signals flow from one device to the next.

It is the mixing console's job to provide the signal flow structure that enables all these devices to be hooked up correctly. It ensures that all the appropriate signals get to their destinations without running into anything. A primary function of the console is revealed. The mixer must be able to hook up any audio output to any audio input (Figure 2.1).

In connecting any of these outputs to any of these inputs, the console is asked to make a nearly infinite number of options possible. Mixdown was alluded to briefly above, but engineers do more than mix. This signal routing device has to be able to configure the gear for recording a bunch of signals to the multitrack simultaneously, as occurs at a big band recording session. It should also be able to make the necessary signal flow adjustments required to permit an overdub on the multitrack, recording a lead vocal while listening to the drums, bass, guitars, and brass previously recorded to the multitrack. Additionally, one might need to record or broadcast live in stereo, surround, or both. Fortunately, all sessions fall into one of the following categories.

2.1.1 BASICS

If the music is to be created by live musicians, the multitrack recording project begins with the *basics* session. In the beginning, nothing is yet

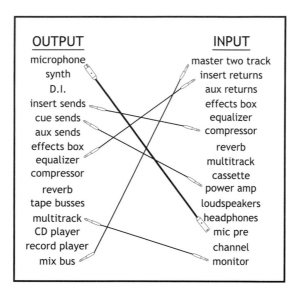

▲ *Figure 2.1 The mixer can connect any output to any input.*

recorded onto multitrack. Any number of musicians are in the studio's rooms and isolation booths playing, and the engineer is charged with the task of recording the project's first tracks onto the multitrack recorder.

Most band-based productions go like this: The entire band plays the song together. The engineer records each musical instrument onto separate tracks. It is fully expected that the singer will want to redo the vocal as an overdub later (*overdubs* are discussed below). The same goes for the guitarist. Even though the plan is to replace these performances with new recordings later, the recordist still tracks everything. Everything performed is recorded, needed or not.

Sometimes the most desirable performance — the "keeper" take — is the one that happens during basics. Try as they might, in one overdub session after another, the performers may not be able to top it in terms of performance value. At basics, there is less pressure on these "scratch" performances. Scratch tracks are performed and recorded as a guide to the rest of the band. These performers just sing/play along so the band can keep track of practical matters (e.g., which verse they are on), accelerating and crescendoing in expressive ways. A more careful track will be recorded later — that day, in a few weeks, or possibly several months later. Such freedom often leads to creativity and chance taking, key components of a great musical performance. So it is a good idea to record the singer and guitarist during the basics session.

With the intent to capture many, if not most, tracks as overdubs later anyway, the audio mission of the basics session is often reduced to getting the most compelling drum and bass performance onto the multitrack. Sometimes even the bass part gets deferred into an overdub. So for basics, the engineer records the entire band playing all at once so that they can get the drummer's part tracked.

Check out the setup sheet (Figure 2.2) for a very simple basics session. It is just a trio — drums, electric bass, guitar, and vocals — and yet there are at least 15 microphones going to at least 10 tracks ("at least" because it is easy and not unusual to place even more microphones on these same instruments. For example, one might create a more interesting guitar tone through the combination of several different kinds of microphones in different locations around the guitar amp.). If the studio has the capability, it is tempting to use even more tracks (e.g., record the electric bass using a Direct Inject (DI) box on a separate track from the bass cabinet microphone).

The console is in the center of all this, as shown in Figure 2.3. It routes all those microphone signals to the multitrack so they can be recorded. It routes them to the monitors so the audio can be heard. It routes those same signals to the headphones so the band members can hear each other, the producer, and the engineer. It also sends and receives audio to and from any number of signal processing effects: compressors, equalizers, reverbs, etc.

2.1.2 OVERDUBS

During an *overdub* session, a performer records a new track while listening to the tracks recorded so far. An electric guitar is overdubbed to the drums and bass recorded at an earlier basics session. A keyboard overdub adds another layer to the drums, bass and guitar arrangement that has been recorded. Note that an overdub describes a session where additional material is recorded so that it fits musically with a previously recorded set of tracks. Either blank tracks are utilized for the new tracks or any unwanted tracks are erased to make room for the new track. Overdubbing refers to laying a new performance element onto an existing set of performances. While it often is the case, overdubbing does not require anything be erased while recording. The new track does not have to go over an old track, it might be recorded onto its own new track.

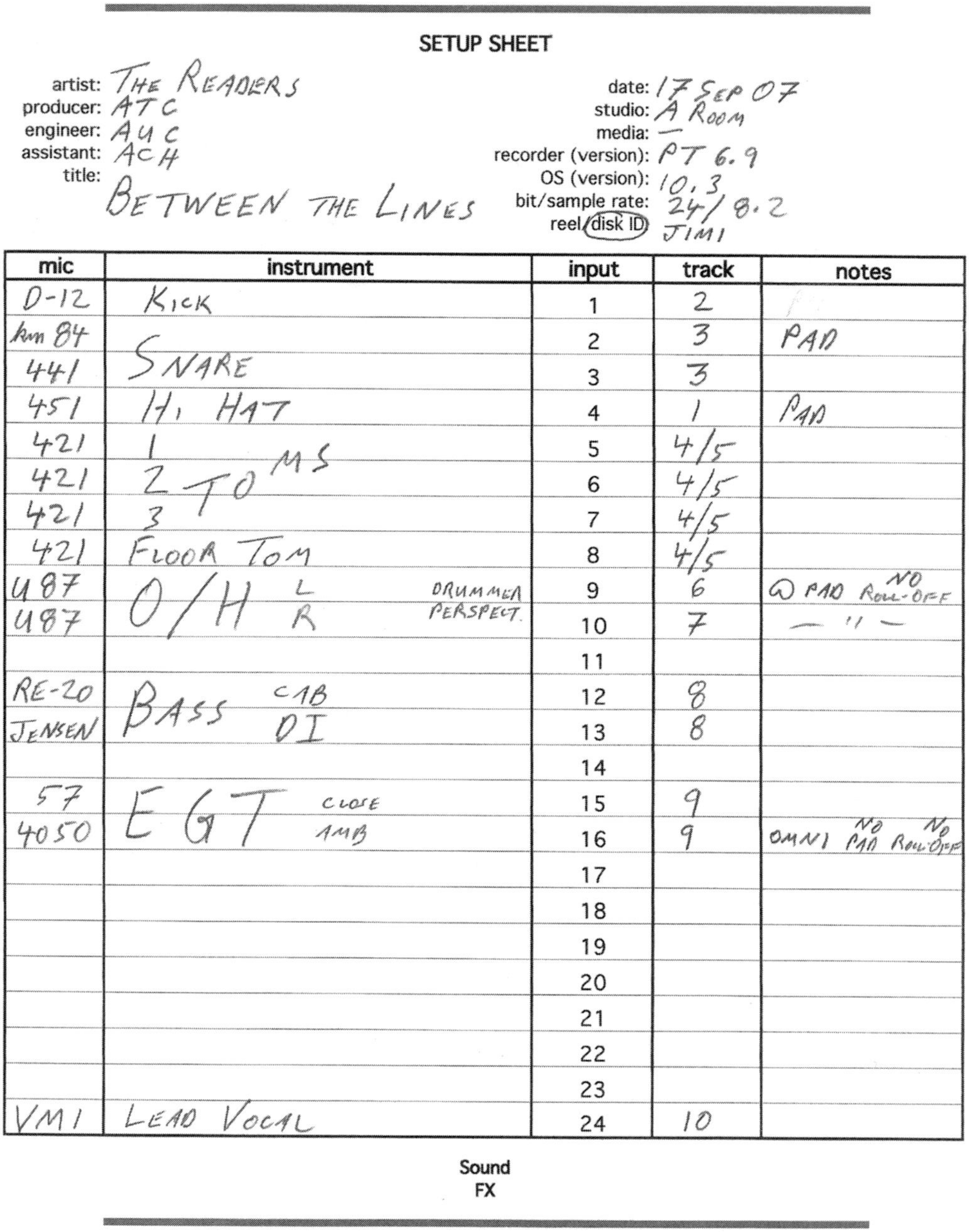

SETUP SHEET

artist: *THE READERS*
producer: *ATC*
engineer: *AUC*
assistant: *ACH*
title: *BETWEEN THE LINES*

date: *17 SEP 07*
studio: *A ROOM*
media: *—*
recorder (version): *PT 6.9*
OS (version): *10.3*
bit/sample rate: *24 / 8.2*
reel/(disk ID): *JIMI*

mic	instrument	input	track	notes
D-12	KICK	1	2	
km 84		2	3	PAD
441	SNARE	3	3	
451	HI HAT	4	1	PAD
421	1	5	4/5	
421	2 TOMS	6	4/5	
421	3	7	4/5	
421	FLOOR TOM	8	4/5	
U87	O/H L DRUMMER	9	6	@ PAD NO ROLL-OFF
U87	R PERSPECT.	10	7	— '' —
		11		
RE-20	BASS CAB	12	8	
JENSEN	DI	13	8	
		14		
57	E GT CLOSE	15	9	
4050	AMB	16	9	OMNI NO PAD NO ROLL-OFF
		17		
		18		
		19		
		20		
		21		
		22		
		23		
VM1	LEAD VOCAL	24	10	

Sound
FX

▲ *Figure 2.2 Setup sheet for a simple basics session.*

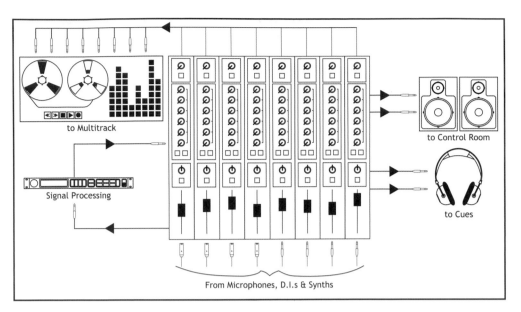

▲ *Figure 2.3 Signal flow for a basics session.*

At the overdub session, there are often fewer musicians playing, fewer microphones in action, and possibly fewer band members around. It is often a much calmer experience. During basics there is the unspoken, but strongly implied, pressure that no one can mess up, or the whole take will have to be stopped and the song restarted from the top. The crowd in the studio is overwhelming — the whole band is there. The crowd in the control room is watching. The lights, meters, microphones, and cables surround the musicians, giving them that "in the lab, under a microscope" feeling. Performance anxiety often fills the studio of a basics session. Overdubs, on the other hand, are as uncomplicated as one singer, one microphone, a producer, and an engineer. Dim the lights. Relax. Do a few practice runs. Any musical mistakes at this point are known only to this intimate group; no one else will hear them. They will be erased. If the performer doesn't like what they did, they can just stop. No worries. Try again. No hurries. Ah, low blood pressure, relatively speaking.

During overdubs, the console routes the microphones to the multitrack recorder. The console is also used to create the rough mix of the live microphones with all of the tracks already recorded on the multitrack. It sends the mix to the control room monitors. Simultaneously, it creates a separate mix for the headphones; the recording artists generally require

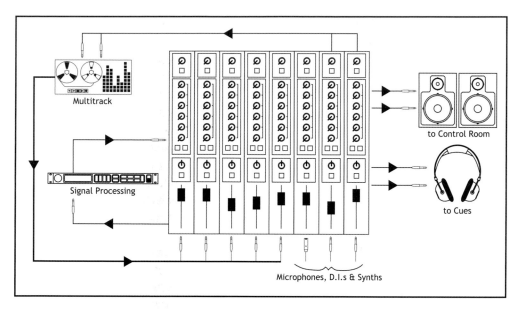

▲ *Figure 2.4 Signal flow for an overdub session.*

headphone mixes tailored to suit their needs as performers that is different from the mix needed by the engineer and producer in the control room. In addition, one never misses an opportunity to patch in a compressor and/or some other effects. Figure 2.4 lays out the console in overdub mode. The overdub session is likely easier on the performer and the engineer, but the console works every bit as hard as it did during the more complicated basics session.

2.1.3 MIXDOWN

After basics and overdubs, any number of multitrack elements have been recorded. They are at last combined, with any effects desired, into a smaller number of tracks suitable for distribution to the consumer (stereo or surround) in the *mixdown* session.

At mixdown, the engineer and producer use their musical and technical abilities to the max, coaxing the most satisfying loudspeaker performance out of everything the band recorded. There is no limit to what might be attempted. There is no limit to the amount of gear that might be needed. On a big-budget pop mix, it is common for nearly every track (and there are at least 24, probably many more) to get equalized and compressed. Most tracks probably get a dose of reverb and/or some additional effects as well. A few hundred patch cables are used. Perhaps several tens, probably

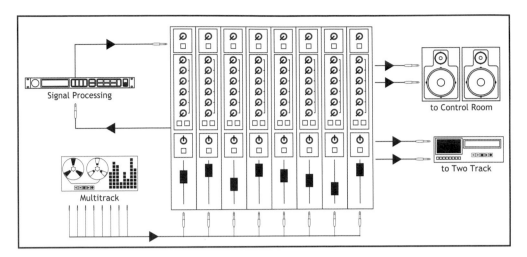

▲ *Figure 2.5 Signal flow for a mixdown session.*

hundreds of thousands of dollars worth of outboard signal processing is used. Mixing automation is required (see Chapter 15), and an enormous console capable of handing so many tracks and effects is desired. During earlier recording and overdubbing sessions, the engineer might have thought, "This is sounding like a hit." It's not until mixdown when they'll really feel it. It's not until the gear-intense, track-by-track assembly of the tune that they'll think, "This sounds like a record!"

The mixing console accommodates this need for multiple tracks and countless effects, as shown in Figure 2.5.

2.1.4 LIVE TO TWO

For many recording sessions, one bypasses the multitrack recorder entirely, recording a live performance of any number of musicians straight to the two-track master machine, or sending it live to a stereo broadcast or the sound reinforcement loudspeakers. A live-to-two session is the rather intimidating combination of all elements of a basics and a mixdown session. Performance anxiety haunts the performers, the producer, and the engineer.

The console's role is actually quite straightforward (Figure 2.6): microphones in, stereo mix out. Of course the engineer may want to patch in any number of signal processors. Then, the resulting stereo feed goes to the control room monitors, the house loudspeakers, the headphones, the two-track master recorder, and/or the transmitter.

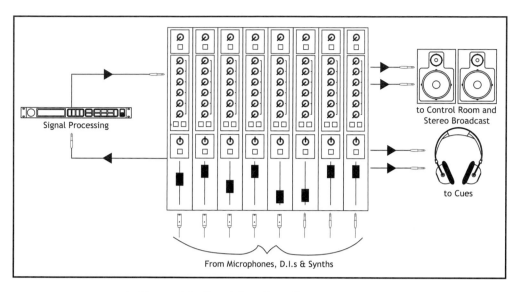

▲ *Figure 2.6 Signal flow for a live-to-two session.*

2.2 Console Signal Flow

These four types of sessions define the full range of signal flow requirements of the most capable mixer. Yet despite having distilled the possibilities into these key categories, the console demands to be approached with some organization. Broadly, the inexperienced engineer can expect to be frustrated by two inherent features of the device: complexity of flow and quantity of controls.

Complexity is built into the console because it is capable of providing the signal flow structure for truly any kind of recording session one might encounter. The push of any button on the console might radically change the signal flow configuration of the device. In this studio full of equipment, the button might change what's hooked up to what. A fader that used to control the snare microphone going to track 16 of the multitrack might instantly be switched into controlling the baritone sax level in the mix feeding the studio monitors. It gets messy fast.

The sheer quantity of controls on the work surface of the mixer is an inevitable headache because the console is capable of routing so many different kinds of outputs to so many different kinds of inputs. Twenty-four tracks used to be something of a norm for multitrack projects. Most contemporary productions exceed this. What about the number of microphones and signal processors? Well, let's just say that in the hands and ears of a great engineer, more can be better. The result is consoles that fill the room — or two or three large computer monitors — with knobs, faders, and switches. The control room starts to look like the cockpit of the space shuttle, with a mind-numbing collection of controls, lights, and meters. These two factors, complexity and quantity, conspire to make the console a confusing and intimidating device to use. It need not be.

2.2.1 CHANNEL PATH

In the end, a mixer is not doing anything especially tricky. The mixer just creates the signal flow necessary to get the outputs associated with today's session to the appropriate inputs. The console becomes confusing and intimidating when the signal routing flexibility of the console takes over and the engineer loses control over what the console is doing. It's frustrating to do an overdub when the console is in a live-to-two configuration. The darn thing won't permit the engineer to monitor what's on the multitrack recorder. If, on the other hand, the console is expecting to mixdown, but the session plans to record basics, one is likely to experience that helpless feeling of not being able to hear a single microphone that's been setup and plugged in. The band keeps playing, but the control room remains silent. It doesn't take too many of these experiences before console phobia sets in. A loss of confidence maturing into an outright fear of using certain consoles is a natural reaction. Through total knowledge of signal flow, this can be overcome.

The key to understanding the signal flow of all consoles is to break the multitrack recording process — whether mixing, overdubbing, or anything else — into two distinct signal flow stages. First is the *channel path*. Also called the *record path*, it is the part of the console used to get a microphone signal (or synth output) to the multitrack recorder. It usually has a microphone preamp at its input, and some numbered multitrack busses at its output. In between lies a fader and maybe some equalization, compression, echo sends, cue sends, and other handy features associated with capturing a great sound (Figure 2.7a).

2.2.2 MONITOR PATH

The second distinct audio path is the *monitor path*. Also called the *mix path*, it is the part of the console used to actually hear the sounds being recorded. It typically begins with the multitrack returns and ends at the mix bus. Along the way, the monitor path has a fader and possibly another collection

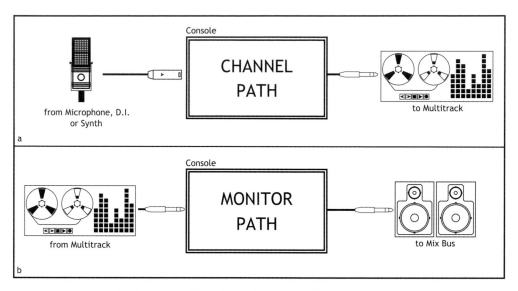

Console

CHANNEL
PATH

from Microphone, D.I.
or Synth

to Multitrack

a

Console

MONITOR
PATH

from Multitrack

to Mix Bus

b

▲ *Figure 2.7 The channel path and the monitor path.*

of signal processing circuitry like equalization, compression, and more (Figure 2.7b).

Making a crisp, logical distinction between channel path and monitor path enables the engineer to make sense of the plethora of controls sitting in front of them on the console. Engineers new to the field should try to hang on to these two different signal paths conceptually as this will help them understand how the signal flow structure changes within the mixer when going from basics to overdubs to mixdown to live-to-two. Mentally divide the console real estate into channel sections and monitor sections so that it is always clear which fader is a channel fader and which is a monitor fader.

2.2.3 SPLIT CONSOLE

Console manufacturers offer two channel/monitor layouts. One way to arrange the channel paths and monitor paths is to separate them physically from each other. Put all the channel paths on, for example, the left side of the mixer, and the monitor paths on the right, as in Figure 2.8(a). Working on this type of console is fairly straightforward. See the snare drum signal overloading the multitrack recorder? This is a recording problem, so one must reach for the record path. Head to the left side of the board and grab the channel fader governing the snare microphone. What if the levels to multitrack look good, but the guitar is drowning out the vocal? This is a

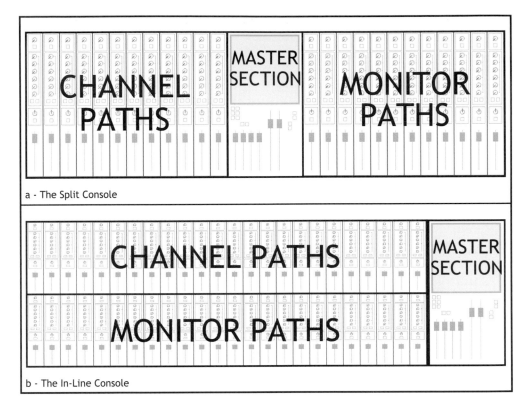

a - The Split Console

b - The In-Line Console

▲ *Figure 2.8 Console configurations.*

monitoring problem. Reach over to the right side of the console and fix it with the monitor faders. Sitting in front of 48 faders is less confusing when one knows the 24 on the left are controlling microphone levels to multitrack (channel faders) and the 24 on the right are controlling mix levels to the loudspeakers (monitor faders). So it's not too confusing that there are two faders labeled, "Lead Vocal." The one on the left is the microphone level to the multitrack recorder; the one on the right is the multitrack return to the monitor mix.

The digital audio workstation is a split console that integrates outboard analog and digital components in the studio with digital processing within the computer. The same studio building blocks are needed: channel path to multitrack recorder to monitor path. The channel paths live partly outside and partly inside the digital audio workstation, using equipment outside the computer and the signal processing available within the computer for recording signals to the multitrack recorder. An analog front end consisting of a microphone preamplifier and any optional analog effects desired (compression and equalization are typical) feeds the analog-to-digital converters. Once digital, the audio signal may be further processed using the digital signal processing abilities of the digital audio workstation as the signal is recorded to multitrack.

After the multitrack, the signals are monitored on separate signal paths (monitor paths, naturally) within the digital audio workstation. The record path offered the chance to add any desired effects as the signal was recorded to the multitrack. The separate monitor paths give the engineer the chance to manipulate the signals after the multitrack, as it feeds the control room speakers, the artist's headphones, etc.

With built-in software, the structure of the mixing console within a digital audio workstation is incredibly flexible, generally being custom-tailored to each recording session. If the artist wishes to add a tambourine to one of the songs, the engineer clicks a few commands and adds an additional track to the multitrack plus all the associated mixer components. In hardware, the number of inputs is determined when the mixer is built and purchased. In software, one acquires an upper limit of inputs (the computer lacks resources to accommodate more than some number of inputs), but can add or delete inputs freely as the production dictates as long as the track count remains below the limit.

2.2.4 IN-LINE CONSOLE

A clever, but often confusing, enhancement to the hardware-based console is the in-line configuration. Here, the channel and monitor paths are no longer separated into different modules physically located on opposite ends of the mixer. In fact, they are combined into a single module (see Figure 2.8b).

Experience has shown that an engineer's focus, and therefore the signal processing, tends to be oriented toward either the channel path or the monitor path, but not both. During tracking, the engineer is dedicating ears, brains, heart, and equipment to the record path, trying to get the best possible sounds onto the multitrack. The monitoring part of the console is certainly being used. The music being recorded couldn't be heard otherwise. But the monitor section is creating a "rough mix," giving the engineer, producer, and musicians an honest aural image of what is being recorded. The real work is happening on the channel side of things. The monitor path accurately reports the results of that work. Adding elaborate signal processing on the monitor path only adds confusion at best, and misleading lies at worst. For example, adding a "smiley face" equalization curve — boosting the lows and the highs so that a graphic EQ would seem to smile (see Chapter 5) — on the monitor path of the vocal could hide the fact that a boxy, thin, and muffled signal is what's actually being recorded onto the multitrack.

It turns out that for all sessions — tracking, overdubbing, mixing, and live-to-two — engineers only really need signal processing once, in the channel or the monitor path. The basics session has a channel path orientation as discussed above. Mixing and live-to-two sessions are almost entirely focused on the final stereo mix, so the engineer and the equipment become more monitor path centric.

This presents an opportunity to improve the console. If the normal course of a session rarely requires signal processing on both the monitor path and the channel path, then why not cut out half the signal processors? If half the equalizers, filters, compressors, aux sends, etc. are removed, the manufacturer can offer the console at a lower price, or spend the freed resources on a higher-quality version of the signal processors that remain, or a little bit of both. As an added bonus, the console gets a little smaller, and a lot of those knobs and switches disappear, reducing costs and confusion further still. This motivates the creation of the in-line console.

On an in-line console, the channel path and the monitor path are combined into a single module so that they can share some equipment. Switches lie next to most pieces of the console letting the engineer decide, piece by piece, whether a given feature is needed in the channel path or the monitor path. A single equalizer, for example, can be switched into the record path during an overdub and then into the monitor path during mixdown. The same logic holds for compressors, expanders, aux sends, and any other signal processing in the console. Of course, some equipment is required for both the channel path and the monitor path, like faders and pan pots. So there is always a channel fader and a separate monitor fader. The in-line console, then, is a clever collection of only the equipment needed, when it is needed, and where it is needed.

In-Line Headaches

An unavoidable result of streamlining the console into an in-line configuration is the following kind of confusion. A single module, which now consists of two distinct signal paths, might have two very different audio sounds within it. Consider a simple vocal overdub. A given module might easily have a vocal microphone on its channel fader but some other signal, like a guitar track, on its monitor fader. The vocal track is actually monitored on some other module and there is no channel for the guitar because it was overdubbed in an earlier session.

What if the levels to tape look good, but the guitar is drowning out the vocal? This is a monitoring problem. The solution is to turn down the monitor fader for the guitar. But where is it? The in-line console has monitor faders on both "sides" of the console. Unlike the split design, an in-line console presents the engineer with the ability to both record and monitor signals on every module across the entire console. Each module has a monitor path. Therefore, each module might have a previously recorded track under the control of one of its faders. Each module also has a channel path. Therefore, each module might have a live microphone signal running through its channel fader too.

To use an in-line console, the engineer must be able to answer the following question in a split second: Which of the perhaps 100 faders in front of me controls the guitar signal returning from the multitrack recorder? Know where the guitar's monitor path is at all times, and don't be bothered if the channel fader sharing that module has nothing to do with the guitar track. The monitor strip may say, "Guitar," but the engineer knows that the channel on the same module contains the vocal being recorded. It is essential to know how to turn down the guitar's monitor fader without fear of accidentally pulling down the level of the vocal going to the multitrack recorder.

One must maintain track sheets, setup sheets, and other session documentation. These pieces of paper can be as important as the tape/hard disk that stores the music. However, rather than just relying on these notes, it helps to maintain a mental inventory of where every microphone, track, and effects unit is patched into the mixer. Much to the frustration of the assistant engineer who needs to watch and document what's going on, and the producer who would like to figure out what's going on, many engineers don't even bother labeling the strip or any equipment for an overdub session or even a mix session. The entire session setup and track sheet is in their heads. Engineers who feel they have enough mental memory for this should try it. It helps one get their mind fully into the project. It forces the engineer to be as focused on the song as the musicians are. They have lines, changes, solos, and lyrics to keep track of. The engineer can be expected to keep up with the microphones, reverbs, and tracks.

This comes with practice. When one knows the layout of the console this intimately, the overlapping of microphones and tracks that must occur on an in-line console is not so confusing. Sure the split console offers some geographic separation of microphone signals from tape signals, which makes it a little easier to remember what's where. Through practice, all engineers learn to keep up with all the details in a session anyway. The in-line console becomes a perfectly comfortable place to work.

Getting Your Ducks in a Row

If an engineer dials in the perfect equalization and compression for the snare drum during a basics session but fails to notice that the processing is inserted into the monitor path instead of the channel path, the engineer is in for a surprise. When the band, the producer, and the engineer listen to the snare track, at a future overdub session, they will find that the powerful snare was a monitoring creation only and was not preserved in the multitrack recording of the snare drum. It evaporated on the last playback of the last session. Hopefully, the engineer documented the settings of all signal processing equipment anyway, but it would have been more helpful to place the signal processing chain in front of the multitrack machine, not after. That is, these effects likely should have been channel path effects, not monitor path effects.

Through experience, engineers learn the best place for signal processing on any given session. Equalization, compression, reverb, and the feeds to the headphones — each has a logical choice for its source: the channel path or monitor path. It varies by type of session. Once an engineer has lived through a variety of sessions, these decisions become instinctive. The mission — and it will take some time to accumulate the necessary experience — is to know how to piece together channel paths, monitor paths, and any desired signal processing for any type of session. Then the signal flow flexibility of any mixer, split or in-line, is no longer intimidating. By staying oriented to the channel portion of the signal and the monitor portion of the signal, one can use either type of console to accomplish the work of any session. The undistracted engineer can focus instead on helping make music.

What's That Switch Do?

Even your gluttonous author will admit that there is such a thing as too much. When excellent engineers with deep experience recording gorgeous tracks are invited to work for the first time in a large, world-class studio, and sit in front of a 96-channel, in-line console for the first time, they will have trouble doing what they know how to do (recording the sweet tracks) while they are bothered by what they may not know how to do (use this enormous console with, gulp, more than 10,000 knobs and switches). Good news: That vast control surface is primarily just one smaller control group (a regular in-line module) repeated over, and over, and over again. When an engineer learns how to use a single module — its channel path and its monitor path — they then know how to use the whole collection of 96 modules.

2.3 Outboard Signal Flow

The signal flow challenge of the recording studio reaches beyond the console. After microphone selection and placement refinements, audio engineers generally turn to signal-processing devices next, using some combination of filters, equalizers, compressors, gates, delays, reverbs, and multi-effects processors to improve or reshape the audio signal. This leads to a critical signal flow issue: How are effects devices and plug-ins incorporated into an already convoluted signal path through the console or workstation?

2.3.1 PARALLEL AND SERIAL PROCESSING

Philosophically, there are two approaches to adding effects to a mix. Consider first the use of reverb on a vocal track (discussed in detail in Chapter 11). The right dose of reverb might support a vocal track that was recorded in a highly absorptive room with a close microphone. It is not merely a matter of support, however. A touch of just the right kind of reverb (it is known in the studio world as "magic dust") can enable the vocal to soar into pop music heaven, creating a convincing emotional presence for a voice fighting its way out of a pair of loudspeakers. Quite a neat trick, really. The distinguishing characteristic of this type of signal processing is that it is *added* to the signal, it does not *replace* the signal.

This structure is illustrated in Figure 2.9. The dry (i.e., without reverb, or more generally, without any kind of effect) signal continues on its merry way through the console as if the reverb were never added. The reverb itself is a parallel signal path, beginning with some amount of the dry vocal, going through the reverb processor, and returning elsewhere on the console to be combined to taste with the vocal and the rest of the mix.

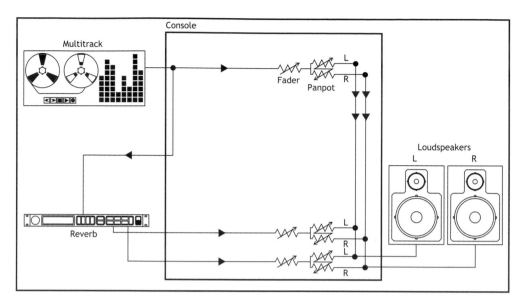

▲ *Figure 2.9 Parallel processing.*

Consider, in contrast, the use of equalization (detailed in Chapter 5). A typical application of equalization is to make a spectrally mediocre or problematic track beautiful. A dull acoustic guitar, courtesy of some boost around 10 or 12 kHz, is made to shimmer and sparkle. A shrill vocal gets a carefully placed roll-off somewhere around 3 and 7 kHz to become more listenable. This type of signal processing changes an undesirable sound into a new-and-improved version. In the opinion of the engineer who is turning the knobs, it "fixes" the sound. The engineer does not want to hear the problematic, unprocessed version anymore, just the improved, equalized one. Therefore, the processed sound *replaces* the old sound.

To do this, the signal processing is placed in series with the signal flow, as shown in Figure 2.10. Adding shimmer to a guitar is not so useful if the murky guitar sound is still in the mix too. And the point of equalizing the vocal track was to make the painful edginess of the sound go away. The equalizer is dropped in series with the signal flow, between the multitrack machine and the console, for example, so that only the processed sound is heard. For the equalizer, it is murky or shrill sound in, and gorgeous, hi-fidelity sound out. Equalizing, compressing, de-essing, wah-wah, distortion, and such are all typically done serially so that listeners hear the affected signal and none of the unaffected signal.

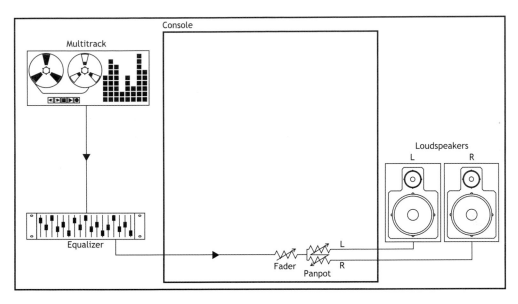

▲ *Figure 2.10 Serial processing.*

Parallel processing, like reverb, adds to the sound. Serial processing, like EQ, replaces the sound.

2.3.2 EFFECTS SEND

Not surprisingly, these two flow structures — parallel and serial — require different signal flow approaches in the studio. For parallel processing, some amount of a given track is sent to an effects unit for processing. Enter the effects send. Also known as an echo send or aux send (short for auxiliary), it is a simple way to tap into a signal within the console and send some amount of that signal to some destination. Probably available on every module of the console or digital audio workstation, the effects send is really just another fader. Neither a channel fader nor a monitor fader, the aux send fader determines the level of the signal being sent to the signal processor. Reverb, delay, and such are typically done as parallel effects and therefore rely on aux sends (Figure 2.11).

There is more to the echo send than meets the eye, however. It is not just an "effects fader." An important benefit of having an effects send level knob on every module on the console is that, in theory, the engineer could send some amount of every component of the project to a single effects unit. To give a perfectly legitimate example, if the studio invested several thousand

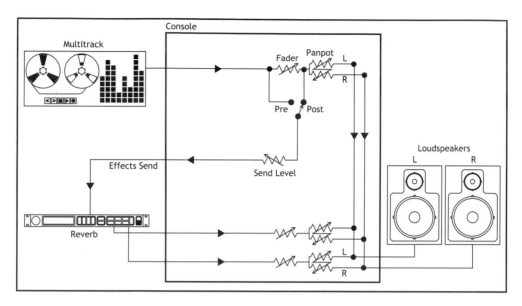

▲ *Figure 2.11 Effects send.*

dollars on a single, super-high-quality, sounds-great-on-everything sort of reverb, then the recording engineer is probably going to want to use it on several tracks. Unless the studio has several very high-quality (i.e., very expensive) reverbs, it is not practical to waste it on just the snare, or just the piano, or just the vocal. The aux send makes it easy to share a single effects unit across a number of tracks. Turn up the aux send level on the piano track a little to add a small amount of this reverb to the piano. Turn up the aux send level on the vocal a lot to add a generous amount of the same reverb to the vocal. In fact, the aux send levels across the entire console can be used to create a separate mix of all the music being sent to an outboard device. It is a mix the engineer does not usually listen to; it is the mix the reverb "listens" to when generating its sound.

As if there already was not enough for the engineer to do during a session, it is informative to review the faders that are in action: The channel faders are controlling the levels of the various signals going to the multitrack, the monitor faders are controlling the levels of all the different tracks being heard in the control room, and the effect sends are controlling the specific levels of all the different components of music going to the reverb. Three different sets of faders have to be carefully adjusted to make musical sense for their own specific purposes.

So far, so good. Hang-on though, as there are two more subtleties to be explored. First, as one is rarely satisfied with just one kind of effect in a multitrack production, the engineer would probably like to employ a number of different signal processors within a single project. Each one of these effects devices might need its own aux send. That is, the engineer might have one device that has a lush and long reverb dialed in; another that is adding a rich, thick chorus effect; and perhaps a third unit that is generating an eighth note echo with two or three fading repetitions. The lead vocal might be sent in varying amounts to the reverb, chorus, and delay, while the piano gets just a touch of reverb, and the background vocals get a heavy dose of chorus and echo and a hint of reverb. The console must have more than one aux send to do this; in this particular case, three aux sends are required on each and every module of the console.

The solution, functionally, is that simple. More aux sends. It is an important feature to look for on consoles and digital audio workstations as the number of aux sends determines the number of different parallel effects devices one can use simultaneously during any session.

Beyond this ability to build up several different effects submixes, aux sends offer the engineer a second, very important advancement in signal flow capability: cue mixes. Generally sent to headphones in the studio or, in the case of live sound, fold-back monitors on the stage, an aux send is used to create a mix the musicians use to hear themselves and each other. As the parameters are the same as an effects send (the ability to create a mix from all the channels and monitors on the console), the cue mix can rely on the same technology.

With this deeper understanding of signal flow, faders should be reviewed. Channel faders control any tracks being recorded to the multitrack, monitor faders build up the control room mix, aux send number one might control the mix feeding the headphones, aux send number two might control the levels going to a long hall reverb program, aux send number three might adjust the levels of any signals going to a thick chorus patch, and aux send number four feeds a few elements to a delay unit. Six different mixes carefully created and maintained throughout a single overdub session. The recording engineer must balance all of these mixes quickly and simultaneously. Oh, and by the way, it is not enough for the right signals to get to the right places. It all must make technical and musical sense. The levels to the multitrack need to be just right for the recording medium — tape or hard disk, analog or digital. The monitor mix needs to sound thrilling, no matter how elaborate the multitrack project has become. The headphone mix needs to sound inspiring — the recording artists have some creating to do, after all. The effects need to be appropriately balanced, as too much or not enough of any signal going to any effects unit may cause the mix to lose impact. This is some high-resolution multitasking. It is much more manageable if the console is a comfortable place to work. Experience through session work in combination with careful study of this chapter will lead to technical and creative success.

Prefader Send

With all these faders performing different functions on the console, it is helpful to revisit the monitor fader to see how it fits into the signal flow. Compare the monitor mix in the control room to the cue mix in the headphones. The singer might want a vocal-heavy mix (also known as "more of me") in their headphones as they sing, with extra vocal reverb for inspiration and no distracting guitar fills. No problem. Use the aux send dedicated to the headphones to create the mix the artist wants. The engineer and the producer in the control room, however, have different priorities. They don't want a vocal-heavy mix; they need to hear the vocal in an appropriate musical context with the other tracks. Moreover, extra reverb on the vocal would make it difficult to evaluate the vocal performance being recorded as it would perhaps mask any pitch, timing, or diction problems. Of course the guitarist on the couch in the rear of the control room (playing video games) definitely wants to hear those guitar fills. Clearly, the cue mix and the control room mix need to be two independent mixes. Using aux sends for the cue mix and monitor faders for the control room mix, the engineer creates two separate mixes.

Other activities occur in the control room during a simple vocal take. For example, the engineer might want to turn up the piano and pull down the guitar in order to experiment with some alternative approaches to the arrangement. Or perhaps the vocal pitch sounds uncertain. The problem may be the 12-string guitar, not the singer. So the 12-string is temporarily attenuated in the control room to enable the producer to evaluate the singer's pitch versus the piano. All these fader moves in the control room need to happen in a way that doesn't affect the mix in the headphones — an obvious distraction for the performer. The performers are in there own space, trying to meet a lot of technical and musical demands in way that is artistically compelling. They should not be expected to do this while the engineer is riding faders here and there throughout the take.

The mix in the headphones needs to be totally independent of the mix in the control room, hence the pre/post switch, included in Figure 2.11. A useful feature of many aux sends is that they can grab the signal before (i.e., pre) or after (i.e., post) the fader. Clearly, it is desirable for the headphone mix to be sourced prefader so that it will play along independently, unchanged by any of these control room activities.

Postfader Send

The usefulness of a postfader send is revealed when one looks in some detail at the aux send's other primary function: the effects send. Consider a very simple two-track folk music mixdown: fader one controls the vocal track and fader two controls the guitar track (required by the folk standards bureau to be an acoustic guitar). The well-recorded tracks are made to sound even better by the oh-so-careful addition of some room ambience to support and enhance the vocal while a touch of plate reverb adds fullness and width to the guitar (see Chapter 11 for more on the many uses of reverb). After a few hours — more likely five minutes — of tweaking the mix, the record label representatives arrive and remind the mixer that, "It's the vocal stupid." Oops; the engineer is so in love with the rich, open, and sparkly acoustic guitar sound that the vocal track was a little neglected. The label's issue is that listeners will not be able to reliably make out all the words. It is pretty hard to sell a folk record when the vocals are unintelligible, so it has to be fixed. Not too tricky though. Just turn up the vocal.

Here is the rub: While pushing up the vocal fader will change the relative loudness of the vocal over the guitar and therefore make it easier to follow the lyrics, it also changes the relative balance of the vocal versus its own reverb. Turning up the vocal leaves its prefader reverb behind. The dry track rises out of the reverb, the vocal becomes too dry, the singer is left too exposed, and the larger-than-life magic combination of dry vocal plus heavenly reverb is lost. The quality of the mix is diminished.

The solution is the postfader effects send. If the source of the signal going to the reverb is *after* the fader (see Figure 2.11), then fader rides will also change levels of the aux send to the reverb. Turn up the vocal, and the vocal's send to the reverb rises with it. The all-important relative balance between dry and processed sound will be maintained.

Effects are generally sent postfader for this reason. The engineer is really making two different decisions: determining the amount of reverb desired for this vocal, and, separately, the level of the vocal appropriate for this mix. Flexibility in solving these two separate issues is maintained through the use of the postfader echo send.

The occasional exception will present itself, but generally cue mixes use prefader sends, while effects use postfader sends.

2.3.3 INSERT

While parallel processing of signals requires the thorough discussion of aux sends above, serial processing has a much more straightforward solution. All that is needed is a way to place the desired signal processor directly into the signal flow itself. One approach is to crawl around behind the gear, unplugging and replugging as needed to hook it all up. Of course, it is preferable to have a patch bay available. In either case, the engineer just plugs in the appropriate gear at the appropriate location. Want an equalizer on the snare recorded on track three? Then, track three out to equalizer in; equalizer out to monitor module three in. This was shown in Figure 2.10. Placing the effects in between the multitrack output and the console mix path input is a pretty typical signal flow approach for effects processors.

Adding additional processing requires only that they are daisy-chained together. Want to compress and EQ the snare on track three? Simple enough: Track three out to compressor in; compressor out to equalizer in; equalizer out to monitor module three in. Elaborate signal-processing chains are assembled in exactly this manner.

If the engineer is hooking up a compressor, and wants to use the console's built-in equalizer, it gets a little trickier: Track three out to compressor in; compressor out to monitor module three in. That's all fine and good if the engineer does not mind having the compressor compress before the equalizer equalizes. There is nothing universally wrong with that order for things, but sometimes one wants equalization before compression. Enter the insert, shown in Figure 2.12. The insert is a patch access point within the signal path of the console that lets the engineer, well, insert outboard processing right into the console.

It has an output destined for the input of the effects device, generally labeled *insert send*. The other half of the insert is an input where the processed signal comes back, called *insert return*. Using this pair of access points, outboard signal processing can be placed within the flow of the console, typically after any onboard equalizer. Using this insert or patching the processing in before it reaches the console are both frequently used serial-processing techniques.

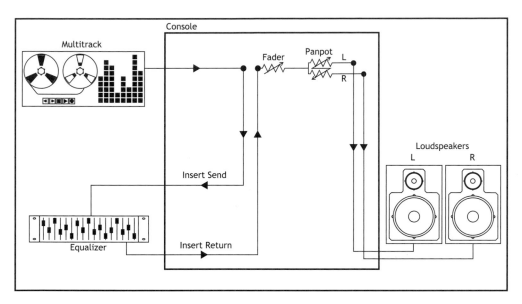

▲ *Figure 2.12 Channel insert.*

2.4 FX Decision Making

Most music recordings are accumulated through the multitrack production process. The final musical arrangement that fills the loudspeakers when the music consumer plays the recording was built up piece by piece; many of the individual musical elements (a guitar, a vocal, a tambourine) were recorded one at a time, during different recording sessions. It is up to the composer, producer, musician, and engineer to combine any number of individual recorded tracks into an arrangement of words, melodies, harmonies, rhythms, spaces, textures, and sounds that make the desired musical statement. This one-at-a-time adding of each piece of the music occurs in overdub sessions. While listening to the music recorded so far, a new performance is recorded onto a separate audio track. Later, during the mixdown session, the various individual tracks are combined into the requisite number of discrete signals appropriate for the release format. The audio compact disc is a stereo format requiring two channels of audio (left and right). The DVD-video and DVD-audio formats currently permit up to six channels for surround-sound monitoring (typically left, center, right, left surround, right surround, and low frequency effects [LFE]). Studios with analog multitrack recorders often have 24 or maybe 48 tracks. Studios with digital multitrack recorders can handle track counts in excess of 96. In the end some 24 to 96 (or more) source tracks are mixed down with added effects to a stereo pair or surround 6-track creation.

2.4.1 FX PROCRASTINATION

Often, it is not until mixdown that the actual musical role of each individual track becomes clear. During an overdub, the artist records a piano performance today, predicting it will sound good with the guitars and violins to be recorded in future sessions. At the time of each overdub, decisions are made about the expected amplitude of each track (e.g., lead vocal versus snare drum versus bass guitar), the appropriate location within the loudspeakers left to right and front to back, and the desired sound quality in terms of added effects. Eventually, during the mixdown session, these decisions are finalized as the audio tracks and signal processing are combined into the final, released recording.

This step-by-step, track-by-track approach to the making of recorded music empowers recording artists to create rich, complicated, musically-compelling works of art that could not occur without a recording studio and the multitrack production approach to help create them. This music is not a single live performance that occurs all at once, in a single space. Popular recorded music is a special kind of synthesis built through multitrack production. The creative process expands in time to include not just the composition phase and the performance experience, but also this multitrack production process. This same step-by-step process desires, if not requires, that many, if not most, decisions about signal processing not be made at the time the track is recorded. Defining the subtle attributes of the various effects is best deferred until the mixdown session, when the greater context of the other recorded tracks can inform the decisions to be made on each individual track.

The recording engineer very often wishes, therefore, to record tracks without effects. The type and amount of effect needed on each track is anticipated, and possibly thoroughly documented. Temporary effects are set up during the overdub sessions so that they may be monitored but not recorded. The control room speakers play the tracks plus temporary effects. The artist's headphones play back the tracks with these first-draft effects, but the signal being recorded onto the multitrack possibly has little to no effects at all. Deferring effects decision making until mixdown can be a wise strategy.

2.4.2 FX ASSERTION

Hedging one's bets, signal-processing wise, and deferring all critical decisions regarding effects to the mixdown session, is not the only valid approach. Many engineers shun the "fix it in the mix" approach and push to track with any and all effects they feel they need. They record, track by track, overdub by overdub, to the multitrack recorder with all the effects required to make each track sound great in the final mix.

It takes a deep store of multitrack experience to be able to anticipate to a fine scale the exact type, degree, and quality of effects that each element of the multitrack arrangement will want in order to finally come together into an artistically valid whole by the end of the project. It takes confidence, experience, and maybe a bit of aggressiveness to venture into this production approach.

Processing to the multitrack might also come from necessity. If the studio has only a few effects devices or limited digital signal processing capability, then it is tempting indeed to make good use of them as often as possible, possibly at each and every overdub session, tracking the audio with the effect to the multitrack. The snare gets processed during that session. The guitar gets the same processors (likely used in different ways) at a different session. The vocal will also take advantage of the same signal processing devices at a future overdub session.

At mixdown, the engineer is not forced to decide how to spend the few effects available. Instead, they have already made multiple uses of these few machines. The faders are pushed up during mixdown, revealing not just the audio tracks, but the carefully tailored, radically altered, and variously processed audio tracks. If the engineer was able to intuit the appropriate effects for the individual tracks, the mix sounds every bit as sophisticated as one created at a studio that owns many more signal processors.

In the end, certain families of effects are better suited to the assertive approach, while others are best deferred until mixdown. Many types of serial effects, such as equalization and compression, are recorded to multitrack during the basics and overdub sessions in which they are first created. Engineers adjust effects parameters as part of the session, in their search for the right tone. These effects and their detailed settings become track-defining decisions very much on par with microphone selection and placement.

Other effects, generally parallel effects like reverb and delay, leave such a large footprint in the mix that they can't be fully developed until all overdubs are complete. The engineer needs to finesse these effects into the crowded mix, in and around the other tracks. This is best done at mixdown, when the full context of the effects and other multitrack sounds is known.

invite them to sit in on their little gig at The Blue Note, with Quincy Jones in attendance. Under high-pressure situations, conscious thought is rather difficult, so one needs to rely on instincts, instincts born and developed in the woodshed.

Recording engineers must also "shed." One of the best ways for a new engineer to fine tune their hearing and attach great knowledge to specific pieces of equipment is to book themselves in the studio, without client work. They lock themselves in the recording studio for some four hours minimum, with no agenda other than to master a piece of equipment, experiment with a microphone technique, explore the sonic possibilities of a single instrument by setting up the entire microphone collection on that instrument and recording each to a different track, practice mix automation moves, etc.

Engineers will get better with time, as more and more session work deepens their knowledge of how to use each piece of equipment. Engineers early in their audio careers can shorten the session-based learning curve by doing some practice sessions.

Restraint

To take one's recording craft to the next level, an engineer needs to learn, grow, innovate, and take chances out of the woodshed and in actual, professional, high-profile recording sessions.

Top-shelf engineers working with multiplatinum bands do not possess a magic ability to, at any moment, create any sound. They do it well now because they have done it before.

The gorgeous sounds, the funky tones, the ethereal textures, and the immersive spaces were all likely born in other sessions in which a band and an engineer invested time and creative energy in search of something. The iconic studio moments documented in the touchstone recordings that all fans of audio collect often came from hours of deliberate exploration and tweaking.

In order to develop the reputation with artists that one really knows how to play the studio as a musical instrument, an engineer needs to pick their spots. Not every band is willing to partake in these hours-long audio experiments. Not every project has the budget. Not every session presents the appropriate moment.

About once every 10 projects, the engineer might strike up a relationship with the band that is so tight, they essentially become part of the band, part of their family. Not just good friends — *best* friends. And they will invite the engineer to contribute, to play, to jam. They will make room for the engineer and the recording studio to make their mark.

On this rare project, an ambitious engineer has the chance to succeed with sonic results sure to get noticed.

But even these projects don't give the engineer studio-noodling carte blanc. Think of the life of a typical album, if there is such a thing.

Preproduction. An engineer can only hope they have time, and that the band invites them.

Basics sessions are generally fast-paced and a little harried. Drums, bass, guitar, keys, vox, etc. Basics takes every ounce of ability and concentration an engineer has. Everybody. Every microphone. Every headphone. Every gobo. Every assistant. With so much happening at once, engineers generally must retreat to a zone of total confidence. If a recordist sets up 24 microphones for a basics session, at least 20 of them are in tried-and-true situations. There is very little room to experiment.

Overdubs. This is the engineer's chance. Some overdubs are calm. Some overdubs move along slowly. One, not all, of the slower, calmer overdub sessions might present the engineer with the chance to dig deep into their creative soul and try out one of the more innovative recording techniques they have been dying to attempt for the last several months.

Practice. Sometimes the band has to practice during a recording session. If the band is in the studio rehearsing because they have brought in a guest percussionist who does not really know the tune yet, this might be the moment to try out some different microphone preamplifiers while they rehearse. If the band is exploring a new idea that has just come to them that looks promising, the clever engineer becomes invisible and starts setting up wacky microphones in wackier places. If the bass player broke a string and needs five or ten minutes to get the instrument back in order, the innovative engineer patches up that compressor they just read about and gives it a test drive.

Mixdown. Some songs need only honest, faithful sonic stewardship. Other mixes hit a roadblock. Here the band wants, needs, and even expects the engineer to come up with the combination of edits, effects, and mix automation moves that lift the song over the roadblock. Great engineers know when to go for it, and when to stay out of the way; when they do go for it, if they have spent enough time in the shed, the band is very likely to like what they hear.

Engineers should not attempt this on every overdub and every mix, every night. These technical explorations for new sounds can be the bane of a good performance. If the vocalist lacks studio experience, if the guitarist lacks chops, if the accordion player lacks self-confidence, then the engineer's first role is to keep the session running smoothly, the performance sounding good, and the musicians feeling comfortable.

Successful studio innovations and an engineer's advanced understanding of all things audio come from just a few well-chosen moments during those special sessions where the planets have aligned, not a wandering moment of distracted tweaking and fiddling. Learn to spot the sessions where this could happen. Build it into the recording budget — more time for overdubs than they should need. Give the session enough slack to make those engineering moments of innovation possible. What seems improvised to the band should come from premeditated moments of inspiration.

Perception

3

"What a fool believes he sees
No wise man has the power to reason away
What seems to be
Is always better than nothing."
— "WHAT A FOOL BELIEVES," THE DOOBIE BROTHERS, *MINUTE BY MINUTE*
(WARNER BROTHERS RECORDS, 1978)

Sitting in a recording studio all day (and most of the night), engineers might easily begin to believe that their music, their art, is the squiggly waveforms drawn on the computer screen before them.

It is important to realize that those signals do not reach humans directly. Humans listen to audio, they don't plug directly into the computer. Those waveforms on the screen represent signals that can be persuaded to make loudspeakers move, which creates sound waves in the air that travel through the room, around each person's head, and into his or her ears. The ears ultimately convert that acoustic wave into some stream of neurological pulses that, at last, are realized within the brain as sound. The music is not in the equipment of the recording studio. Music ultimately lives within the mind and heart of the listener, having begun in the minds and hearts of the composers and performers.

The audio engineer has little control over the processing of sound by the human hearing system. Nevertheless, it is a major part of the production chain. This chapter summarizes the critical properties of the human perception of sound, allowing that engineer to make more informed use of the studio equipment that feeds it.

3.1 Preconscious Listening

Listening as musicians and music fans, we constantly savor and analyze melody, harmony, rhythm, words, etc. Listening as audio engineers and producers, we critically evaluate the qualities of the audio signal, assessing its technical merit and enjoying the production techniques that likely led to the creation of that particular sound.

Remove these conscious aspects of sound perception and cognition, and we find our minds still do a terrific amount of analysis on incoming sounds.

As you read this, perhaps you hear a bird chirping, a hard disk spinning, some people chatting, or a phone squawking. Without trying, a healthy hearing system will determine the direction each of these sounds is coming from, estimate its distance, and identify the source of the sound itself (Was it animal or machine? Was it mother or sister? Is it my mobile phone or yours?). All of those simple observations are the result of a terrific amount of signal analysis executed by our hearing system without active work from us. It is work done at the preconscious level — when we are first aware of the sound, this additional information arrives with it.

3.2 Audible Sound

Figure 3.1 summarizes the range of audible sounds for average, healthy human hearing. Audio engineers often quote this range of hearing along the frequency axis, rounded off to the convenient, and somewhat optimistic, limits from 20 Hz to 20,000 Hz.

The bottom curve, labeled "threshold in quiet," identifies the sound pressure level needed at a given frequency to be just audible. Amplitudes below are too quiet to be heard. Amplitudes above this level are, in a quiet environment, reliably detected. This threshold in quiet does not remotely resemble a straight line. Human sensitivity to sound is strongly governed by the frequency content of that sound. The lowest reach on this curve (near about 3,500 Hz) identifies the range of frequencies to which the human hearing system is most sensitive. The sound pressure level at which a 3,500-Hz sine wave becomes audible is demonstrably lower than the sound pressure level needed to detect a low-frequency (e.g., 50 Hz) sine wave or a high-frequency (e.g., 15,000 Hz) sine wave.

As the threshold in quiet makes clear, the sensitivity of human hearing is a highly variable function of frequency. The equal loudness contours of Figure 3.2 refine this point. Each equal loudness curve traces the sound pressure level needed at a given frequency to match the perceived loudness of any other audible frequency on the curve.

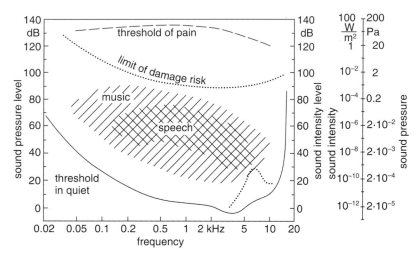

▲ *Figure 3.1 Hearing area, i.e. area between threshold in quiet and threshold of pain. Also indicated are the areas encompassed by music and speech, and the limit of damage risk. The ordinate scale is not only expressed in sound pressure level but also in sound intensity and sound pressure. The dotted part of threshold in quiet stems from subjects who frequently listen to very loud music.*

The bottom curve of Figure 3.2 essentially repeats the data of the threshold in quiet shown in Figure 3.1. The third curve up from the bottom is labeled 20. At the middle frequency of 1 kHz, this curve has a sound pressure level of 20 dBSPL. Follow this curve lower and higher in frequency to see the sound pressure level needed at any other frequency to match the loudness of a 1-kHz sine wave at 20 dBSPL. All points on this curve will have the same perceived loudness as a 1-kHz sine wave at 20 dBSPL. Accordingly, sounds falling on this curve are all said to have a loudness of 20 phons, equivalent to the loudness of the 20-dBSPL pure sine wave at 1,000 Hz. Air pressure can be measured and expressed objectively in decibels (dBSPL). Perceived loudness, a trickier concept requiring the intellectual assessment of a sound by a human, is measured with subjective units of *phons*.

Figure 3.2 shows the results of equal loudness studies across sound pressure levels from the threshold of hearing in quiet up to those equivalent in loudness to a 1-kHz sine wave at 100 dBSPL, or 100 phons.

Even as the sound pressure is changed, some trends in human hearing remain consistently true. First, human hearing is *never* flat in frequency response. Sounds across frequency ranges that are perceived as having the same loudness (a subjective conclusion) are the result of a variety of sound pressure levels (an objective measurement). Uniform loudness comes from

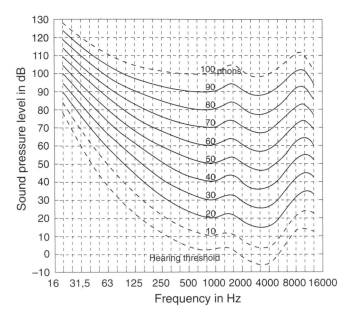

▲ *Figure 3.2 Equal-loudness contours for loudness levels from 10 to 100 phons for sounds presented binaurally from the frontal direction. The absolute threshold curve (the MAF) is also shown. The curves for loudness levels of 10 and 100 phons are dashed, as they are based on interpolation and extrapolation, repectively.*

a range of sound pressure levels across the frequency axis. Similarly, uniform sound pressure level across the frequency axis would lead to a broad range of perceived loudnesses.

Second, there is a consistent trend that human hearing is less sensitive to low and high frequencies and most sensitive to middle frequencies. This trend is shown to exist at a range of amplitudes.

Third, human hearing consistently remains most sensitive to that upper-middle frequency range near about 3,500 Hz, across a range of amplitudes — from quiet, just-audible sound pressure levels (0 dBSPL) up to painfully loud, and possibly unhealthy sound pressure levels (100 dBSPL).

These consistent trends not withstanding, the equal loudness contours do reveal some amplitude-dependent changes in the perception of loudness across frequency. These more subtle properties of human hearing might be the more important points of the equal loudness curves. The sweep upward in amplitude of these equal loudness curves as one follows them from a middle frequency down to a low frequency shows the amount of additional sound pressure level needed at a low frequency to match the loudness of that middle frequency.

Compare the curve passing through 80 dBSPL at 1 kHz (the 80-phon curve) to the curve passing through 20 dBSPL at 1 kHz (the 20-phon curve). The 80-phon curve is flatter than the 20-phon curve. To be sure, the 80-phon curve is far from flat. But the 80-phon curve demonstrates that at higher sound pressure levels, the human hearing system gets better at hearing low frequencies than at lower sound pressure levels. As the overall amplitude of the audio is raised, one does not require as much compensation in sound pressure level to match the low frequency loudness to the middle frequency loudness. The sound pressure difference between a middle and a low frequency at the same perceived loudness decreases as amplitude increases.

Humans are always less sensitive to low frequencies than to middle frequencies. As amplitude increases, however, the sensitivity to low frequencies approaches the sensitivity to middle frequencies.

Every successful audio engineer must be aware of this property of human hearing. Consider a mixdown session, simplified for discussion: a jazz duet consisting of acoustic bass and soprano saxophone. While questionable music might result, compelling audio logic will be revealed.

A fundamental action of the audio engineer is to carefully set the fader levels for every instrument in the mix, two instruments in this case (see Chapter 8). The engineer decides the relative level of the upright bass versus the soprano sax. While both instruments offer complex tone, it would be fair to oversimplify a bit and say that the majority of the acoustic bass signal lives in the lower-frequency range while the soprano saxophone exists predominantly in the middle and upper parts of the spectrum. The engineer finds a balance between bass and sax that is satisfying, musically. The foundation of the mix and the orientation of the harmony might be provided by the bass, while the saxophone offers melodic detail and harmonic complexity filling out the top of the spectrum. When the level of each is just right, the musicality of the mix is maximized; music listeners can easily enjoy the full intent of the composer and musicians.

Imagine the mix engineer listens at very low volume in the control room, at levels near the 20-phon curve. The fader controlling the acoustic bass and the fader controlling soprano sax are coaxed into a pleasing position.

Turn the control room level up about 60 dB, and the same mix is now playing closer to the 80-phon curve. A volume knob raises the level of the entire signal, across all frequencies, by the same amount. So the control room level adjustment from around 20 to around 80 phons has the perceptual result of turning up the bass performance more than the saxophone performance. As discussed above, at the higher level, one is more sensitive to the bass than when listening at a lower level.

Turning up the control room volume knob, the relative fader positions are unchanged. The overall level is increased globally. At the low listening level, maybe the bass fader was 30 dB above the sax fader to achieve a pleasing mix. At the higher listening level, this 30-dB difference has a different perceptual meaning. The 30 decibels of extra bass level now has the effect, perceptually, of making the bass much louder than the sax, shifting the musical balance toward the bass, and possibly ruining the mix.

The human hearing system has a variable sensitivity to bass. When the engineer turns up the level with a volume knob, they raise the level across all frequencies uniformly. Perceptually, however, turning up the level turns up the bass part of the spectrum more than the middle part of the spectrum, because the equal loudness curves flatten out at higher amplitudes.

Not only does this frustrate the mixdown process, it makes every step of music production challenging. Recording a piano, for example, an engineer selects and places microphones in and around the piano and employs any additional signal processing desired. A fundamental goal of this piano recording session might be to capture the marvelous harmonic complexity of the instrument. The low-frequency thunder, the mid-frequency texture, and the high-frequency shimmer all add up to a net piano tone that the artist, producer, and engineer might love. As any recording engineer knows, the sound as heard standing next to the piano is not easily recreated when played back through loudspeakers. The great engineers know how to capture and preserve that tone through microphone selection and placement so that they create a loudspeaker illusion very much reminiscent of the original, live piano.

The general flattening of the hearing frequency response with increasing level as demonstrated in the equal loudness contours portends a major problem. Simply raising or lowering the monitoring volume can change the careful spectral balance, low to high, of the piano tone. Turning it up makes the piano not just louder, but also fuller in the low end. Turning it down leads to a sound perceptually softer and thinner, lacking in low end. The control room level is an incredibly important factor whenever recordists make judgments about the spectral content of a signal. Perceptually, volume acts as an equalizer (see Equalization in Chapter 5).

The hearing system does not tolerate increases in sound pressure level without limit. Figure 3.1 shows that there is risk of permanent hearing damage when our exposure to sound pressure levels reaches about 90 dBSPL (see Decibel in Chapter 1) and above. Many factors influence the risk for hearing damage, including sound pressure level, frequency content, duration of exposure, stress level of the listener, etc. The equipment in even the most basic recording studio has the ability to damage one's hearing, permanently. The reader is encouraged to use common sense, learn the real risks from resources dedicated to this topic, and try to create audio art in an environment that is healthy.

Typical speech and music, if there is such a thing, is shown in the cross-hatched section within the audio window of Figure 3.1. Note that music seeks to use a wider frequency range and a more extreme amplitude range than speech. The history of audio shows steady progress in our equipment's ability to produce wider and wider bandwidth, while also achieving higher and higher sound pressure levels. Musical styles react to this in earnest, with many forms of music focusing intensely on that last frontier of equipment capability: the low-frequency, high-amplitude portion of the audio window.

It is a fundamental desire of audio production to create sounds that are within the amplitude and frequency ranges that are audible, without causing hearing damage. Figure 3.1 shows the sonic range of options available to the audio community. An additional dimension to audibility not shown in Figure 3.1 is the duration of the signal. It is measurably more difficult for us to hear very short sounds versus sounds of longer duration. Figures 3.1 and 3.2 are based on signals of relatively long duration. Figure 3.3 shows what happens when signals become quite short.

The dashed lines show the threshold in quiet for three test frequencies: 200 Hz, 1,000 Hz, and 4,000 Hz. The 4-kHz line is consistently lowest because humans are more sensitive to this upper-middle frequency than to the two lower frequencies tested. This is consistent with all of the equal loudness contours and the earlier plots of threshold in quiet, which all slope to a point of maximum sensitivity in this frequency range. Human hearing is less sensitive to low frequencies, and the 200-Hz plot of Figure 3.3 confirms this.

The effect of duration on the ability to hear a sound is shown along the horizontal axis of Figure 3.3. The shorter the signal, the higher the amplitude needed to detect it. As the signal lasts longer, it can be heard at a lower

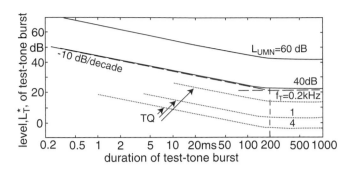

▲ *Figure 3.3 Level of just-audible test-tone bursts, L*$_T$, as a function of duration of the burst in quiet condition (TQ, dotted curves, for three frequencies of test tones) and masked by uniform masking noise of given level (solid curves). Note that level, L*$_T$, is the level of a continuous tone out of which the test-tone burst is extracted. Broken thin lines mark asymptotes.*

level. This trend, that longer sounds are more easily detected than shorter sounds, continues up to a duration of about 200 ms. Beyond 200 ms, the extra duration does not seem to raise the audibility of the sound. At 200 ms or longer, our threshold of hearing remains consistently as reported in Figures 3.1 and 3.2. The solid lines of Figure 3.3 show the same effect for detecting a signal in the presence of a distracting other sound, discussed next.

3.2.1 MASKING

Sounds that are audible in quiet may not be audible when other sounds are occurring. It is intuitive that, in the presence of other distracting signals, some desirable signals may become more difficult to hear. When one is listening to music in the car, the noise of the automotive environment interferes with one's ability to hear all of the glorious detail in the music. Mixing electric guitars with vocals, the signals are found to compete, making the lyrics more difficult to understand, and the guitars less fun to enjoy. When one signal competes with another, reducing the hearing system's ability to fully hear the desired signal, *masking* has occurred.

Spectral Masking

Fundamental to an understanding of masking is the phenomenon like that shown in Figure 3.4. Three similar situations are shown simultaneously; consider the left-most bump in the curve first.

The dashed curve along the bottom is the familiar threshold in quiet. A narrow band of distracting noise is played, centered at 250 Hz. The threshold

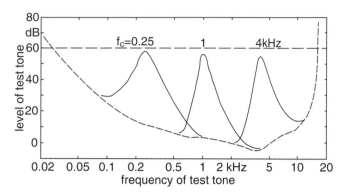

▲ *Figure 3.4 Level of test tone just masked by critical-band wide noise with level of 60 dB, and center frequencies of 0.25, 1 and 4 kHz. The broken curve is again threshold in quiet.*

of hearing, no longer a threshold in quiet, shifts upward in a spectral region around 250 Hz. The departure from the original curve shows the effect of masking. The typical ability to detect a signal is diminished by the presence of the masking signal. Signals at or near 250 Hz need to be a bit louder before they can be heard.

This does not suggest there is any hearing damage, and is not a reflection of unhealthy or inferior hearing ability. Healthy human hearing exhibits the trends shown in Figure 3.4. A listener's ability to detect a faint signal is reduced in the presence of a competing signal.

The upward shift in level needed in order to detect a signal in the presence of a distracting masking signal is at a maximum at the center frequency of the narrowband masking noise. Note also, however, that the masking noise affects our hearing at frequencies both below and above the masking frequency. The masker narrowly confined to 250 Hz casts a shadow that expands both higher and lower in frequency.

Figure 3.4 demonstrates masking in three different frequency regions, revealing similar effects not just at 250 Hz, but also at 1,000 Hz and 4,000 Hz. A distracting signal makes it more difficult to hear other signals at or near the frequency region of the masker.

As the masking signal gets louder, the effect grows a bit more complicated (Figure 3.5). The shape of the curve describing the localized decrease in the ability to hear a sound grows steadily biased toward higher frequencies as the amplitude of the masker increases. Termed "the upward spread of masking," this phenomenon indicates that masking is not symmetric in the

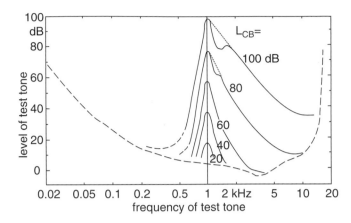

▲ *Figure 3.5 Level of test tone just masked by critical-band wide noise with center frequencies of 1 kHz and different levels as a function of the frequency of the test tone.*

frequency domain. A distracting signal makes it more difficult to hear in the frequency region around the masking signal. Moreover, depending on the amplitude of the masking signal, the masking footprint can work its way upward in frequency, reducing the sensitivity above the masking frequency more than below it.

Temporal Masking

Masking is not limited only to the moments when the masking signal is occurring. Masking can happen both before and after the masking signal. That is, the temporary reduction in hearing acuity due to the presence of a distracting sound occurs for some time after the masking signal is stopped. Perhaps more startling, the masking effect is observed even just before the onset of the masker (shown schematically in Figure 3.6). The downward sloping portion of the curve, labeled "postmasking," on the right side of Figure 3.6 shows hearing sensitivity returning to the performance expected in quiet, but doing so over a short window in time after the masker has stopped. Postmasking (also called forward masking), when the signal of interest occurs after the masking distraction, can have a noticeable impact for some 100 to 200 ms after the masker has ceased. The upward sloping curve, labeled "premasking," on the left side of Figure 3.6 demonstrates the reduction in sensitivity (versus our threshold in quiet) to a signal immediately before the masking signal begins. Premasking (also called backward masking), when the signal of interest happens before the distracting masker, operates at a smaller time scale, meaningful only for the 20 to 50 ms just before the masker begins. Figure 3.6 summarizes a

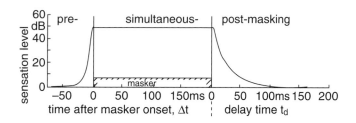

▲ *Figure 3.6 Schematic drawing to illustrate and characterize the regions within which premasking, simultaneous masking, and postmasking occur. Note that postmasking uses a different time origin than premasking and simultaneous masking.*

significant property of human hearing: The masking effect is not limited to sounds occurring simultaneously with the masker. Not only is it more difficult to hear some signals during a masking event, but also it is more difficult to hear those signals immediately before or after the masking signal occurs.

3.2.2 MASKING REDUCTION

The multitrack recordings produced in the studio throw listeners right into a masking conundrum. A simple rock tune consisting of drums, bass, guitar, and vocals presents severe masking challenges. Kick drum, snare drum, three tom toms, hi hat, crash cymbal, ride cymbal, bass guitar, rhythm guitar, lead guitar, and lead vocal — a simple rock arrangement presents an audio engineer with 12 individual elements fighting to be heard. Guitars mask snare drums. Cymbals mask vocals. Bass guitar masks kick drum. An engineer must cleverly minimize unwanted masking.

In multitrack production, the portion of the frequency axis and the range of amplitude used are determined first by the source signal (Is it a piano or a penny whistle, a guitar or glockenspiel?). It is further influenced by the type and quality of musical performance, microphone selection and placement, room acoustics, and signal processing applied.

Among recording studio effects that work the frequency axis directly, the most apt might be equalization (see Chapter 5). Note, however, that all processes can have some influence — directly or indirectly — on the spectral content of the signal.

On the amplitude axis, engineers reach for mute buttons, faders, compressors, expanders, gates, tremolo, and distortion devices for a direct effect.

Identifying the key drivers of audibility is fundamental to creating successful recordings. Audio engineers must balance the duration, amplitude, and spectral content of each and every element of the multitrack production to ensure that the work can be fully enjoyed later on loudspeakers in cars, laptops, headphones, living rooms, department stores, and elevators all over the world.

There is no universal cure for masking, but there are a few levers an engineer can pull.

Spectral Management

One signal can mask another when they occupy similar frequency ranges. Great arrangers, orchestrators, composers, and producers know this well. The very instrumentation chosen, and the playable range of each instrument, are driven, in part, by the spectral competition each chosen instrument presents to others.

When the pianist and guitarist play the same part, in the same range, using the same chord voicings, the instruments might blur into one vague texture. Done well, this creates a single, hybrid sound — a piano/guitar meta-instrument. This can be exactly the intent. The way the two spectrally competing instruments mask each other helps them fuse into a greater whole. Here masking is deliberate and desirable.

More frequently, such masking is unwanted. Each musician would like to hear their own part. Fans of the music want to enjoy each musical element individually. Separation is achieved by shifting the spectral content of the piano and the guitar to more distinct areas, reducing masking. Get the guitar out of the way of the piano by shifting it up an octave. Or ask the piano player to move the left-hand part down an octave, the right-hand part up an octave, or both. Have them play different parts using different chord voicings. This quickly becomes an issue for the band — an issue of arranging and songwriting. Counterpoint is required study for formally-trained musicians, and it applies to pop music as much as it does to classical music. Good counterpoint is good spectral management. Multiple parts are all heard, and they all make their own contribution to the music, in part because they do not mask each other. Choosing which instruments to play, and deciding what part each plays, are the most important drivers of masking. It is important to make those decisions during rehearsal and preproduction sessions.

Of course, the recording engineer influences the extent of the remaining spectral masking. Using equalization (see Chapter 5), engineers carve into the harmonic content of each signal so that competing signals shift toward different frequency ranges.

Level Control

The upward spread of masking serves as a reminder to all engineers: Do not allow the amplitude of any single signal to be greater than it absolutely needs to be. The louder a signal is in the mix, the bigger the masking trouble spot will be. As signals get louder still, they start to compete with all other signals more and more, masking especially those frequencies equal to and greater than themselves.

Worst case: a really loud electric bass guitar. Broad in spectral content, when it is loud, it masks much of the available spectral range. Audio engineers only have that finite frequency space between 20 and 20,000 Hz to work with. Low-frequency dominant, the masking of the bass guitar can spread upward to compete with signals living in a much higher range, including guitars and vocals! The bass must be kept under control.

Panning

All of the masking discussed above diminishes when the competing signals come from different locations. Use pan pots and/or short delays to separate competing instruments left to right and front to back. The reduction in masking that this achieves is immediately noticeable, and can reveal great layers of complexity in a multitrack production. Colocated signals fight each other for the listener's attention. Signals panned to perceptually different locations can be independently enjoyed more easily.

Stretching

Signals that are shorter than 200 ms can be particularly difficult to hear. Most percussion falls into this category. Individual consonants of a vocal may be difficult to hear. Using some combination of compression (see Chapter 6), delay (see Chapter 9), and reverb (see Chapter 11), for example, engineers frequently try to stretch short sounds out so that they last a little longer. So treated, the sounds rise up out of the cloud of masking, becoming easier to hear at a lower level.

Effect Signatures

When a complex sound occupying a range of frequencies is treated with a distinct effect across all frequencies, it may become easier to hear. Tremolo or vibrato (see Chapter 7), for example, helps connect spectrally disparate elements of that instrument's sound into a single whole. The hearing system then gets the benefit of listening to the broadband signal by perceptually grabbing hold of those spectral parts that are not masked. Frequency ranges treated with the same global effect can be perceptually fused to those parts that can be heard. As a result, the overall sound, that entire instrument, will become easier to hear.

When an electric guitar and an electric piano fight to be heard, treat them to different effects: distortion on the guitar (see Chapter 4) and tremolo on the keyboard (see Chapter 7). The spectrally similar signals that had masked each other and made the mix murky perceptually separate into two different, independently enjoyable tracks. Many effects, from reverb to an auto-panning, flanging, wah-wah multieffect can help unmask a signal in this way.

Interaction

When an engineer interacts with the device that affects that signal — when they move the slider, turn the knob, or push the button — it is easier to hear the result. If someone else is making those adjustments, and a listener does not know what that person is changing or when, then the exact same signals can return to a masked state where listeners just can not hear the effect.

There is a difference between what is audible on an absolute basis, and what is audible when one makes small signal processing changes to the signal. There is good news and bad news here.

The good news is that recording engineers can actually listen with high resolution while making the fine adjustments to signal processing settings that are so often necessary in music production. This helps engineers get their jobs done, as that is so much a part of the recording engineer's daily life. Cueing in on the *changes made*, audio engineers are able to detect subtleties that would otherwise be inaudible.

The bad news is that this has no usefulness for consumers of recorded music. People listening to carefully-crafted recordings do not have the same benefit of interacting with the studio equipment that the engineer had when creating the recording. Because they are not interacting with the devices that change the sound, they are less likely to hear subtle changes in the sound. A sound engineer's hard work might go completely unnoticed. Sound for picture (with visual cues) and sound for games (with interactivity), on the other hand, do offer some opportunity to leverage this property and push past some of the masking.

3.3 Wishful Listening

This discussion of what is audible and what is inaudible would not be complete without mentioning what is imaginable. That is, just because good hard scientific research into the physiology and psychology of hearing says humans likely can not hear something does not mean they do not *think* they heard something. A sort of wishful listening can influence the work in the recording studio.

Engineers might be making adjustments to the settings on an effects device, trying to improve, for example, a snare drum sound. They turn a knob. They think it sounds better. The assistant engineer nods with approval. They turn the same knob a bit further. Ooops! Too far. Too much. The producer agrees. They back off a bit to the prior setting and, yes, the snare drum is perfect.

This experience can then be followed by the (hopefully discreet) discovery that the device they were adjusting was not even hooked up. Or, worse yet, the device they were adjusting was affecting the vocal, not the snare!

The desire to make things sound better is so strong that one does not actually need it to sound better in order for it to sound better.

A frustrating part of creating multitrack recordings is that, though science documents when and where certain signals in certain situations will not be audible, there is no clear sign for the hard working engineer that any specific sound or effect is not audible. The hearing system searches earnestly, without fear, for all levels of detail in the recordings being worked on. Human hearing rarely states definitively, "I don't know. I don't hear it." It almost always makes a guess. Subtle moving targets are hard to hear. When one listens for small changes, it can be difficult indeed to tell actual changes from imagined changes.

Section 2
Amplitude FX

Lobster Telephone, 1936. Salvador Dali (1904–1989). Photo credit: Tate Gallery, London/Art Resource, NY © Salvador Dali, Gala-Salvador Dali Foundation/Artists Rights Society (ARS), New York.

Distortion

4

"You can tell me the world is round, and I'll prove to
you its square
You can keep your feet on the ground, but I'll be walking
on air
You're pretty good at waiting, while I go running around
Well that's just the way it is, you know."
— "HOLE IN MY POCKET," SHERYL CROW, *C'MON, C'MON* (A&M RECORDS, 2002)

Those waveforms appearing on the computer screen of any digital audio workstation sketch amplitude versus time. This chapter leaves the time axis well alone and focuses on the sometimes accidental and sometimes deliberate manipulation of the amplitude axis. Amplitude axis distortion represents a family of effects common in many styles of multitrack production. As many decades of rock and roll have shown, distortion is often sought out, deliberately created, and blatantly emphasized.

4.1 Distortion of Amplitude

Consider an audio signal processor with a single input and a single output (Figure 4.1). While the device might be an effects device designed to distort, it might also be a power amplifier, a mixing console, or some other signal-processing device. When the shape of the output waveform deviates in any way from the shape of the input, distortion has occurred. Comparing the detailed shape of anything but very simple waveforms seems difficult at first. Sine, square, sawtooth, triangle waves (see Chapter 1), and other simple periodic waveforms might be visually inspected to confirm a change in shape. Complex musical waves present a challenge.

The comparison of output to input is made easier by plotting the output amplitude versus the input amplitude. Figure 4.2 shows a device that simply passes the signal, without any amplitude change whatsoever. The output

▲ *Figure 4.1 A device that distorts.*

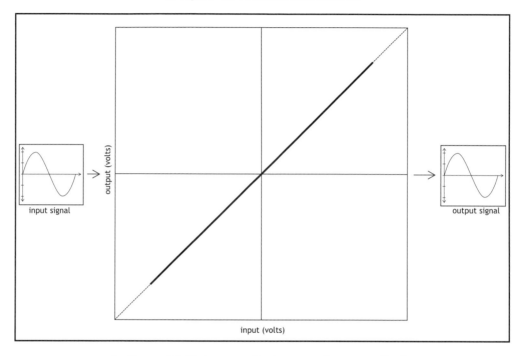

▲ *Figure 4.2 Output amplitude versus input amplitude.*

amplitude is identical to the input amplitude. Because every input value is matched by the exact same output value, this undistorted signal follows a perfect line on this plot, at exactly 45°.

An amplifier that increases the gain without otherwise changing the shape of the waveform simply remains on a straight line, but deviates from the 45° angle (Figure 4.3). The amplitude is consistently scaled up or down in value.

Distortionless unity gain occupies the 45° line. Devices that increase level without distortion are described by lines steeper than 45°. Devices that attenuate without distortion are described by lines shallower than 45°. By preserving the same waveform *shape*, the output versus input plot remains a straight line. For this reason, a device that does not introduce distortion into the signal is said to be *linear*.

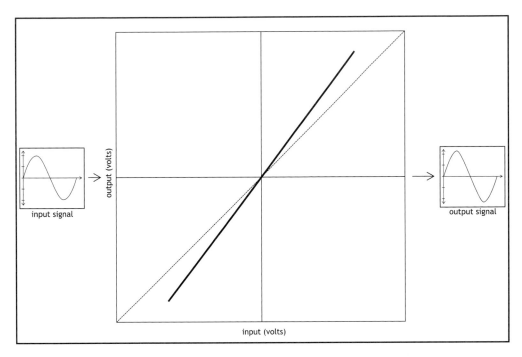

▲ *Figure 4.3 Output amplitude versus input amplitude for linear gain.*

4.1.1 HARMONIC DISTORTION

Recording engineers discover quickly in their careers that a signal cannot be raised in level without bound. All devices have finite capabilities to create amplitude. Each and every device in the recording studio reaches a voltage limit.

Consider a simple domestic light switch — a musical instrument only in the hands of a rare talent. Turn the switch on, and the full voltage supplied by the power company is applied to the light bulb. Flipping the switch harder or faster will not change this. Dimmers can effectively lower the average voltage fed to the light bulb. Barring the addition of other electronic components, the voltage may never be higher than that supplied to the input of the switch. That is the maximum voltage available.

Power amplifiers present a similar situation. The power amp is fed a line-level audio signal. The voltage of that signal is magnified by the power amp and fed to the loudspeakers. That power amp is plugged into the wall. It receives the main's power and converts it into internal power supply voltages. Those internal voltage rails represent the finite limit of voltage capability for the audio output. No matter what sort of voltage swing is supplied on the audio input to the power amplifier, at some level the output can increase no further.

All audio devices, including mixing consoles, guitar amplifiers, and equalizers, possess an output voltage limit. It must be understood what happens when that limit is reached.

Hard Clipping

Consider a device that abruptly hits its output voltage limit (Figure 4.4). Below the point of distortion, there is no warning that the device will run out of output voltage capability. The threshold of distortion is crisp and absolute. The audio below this limit is completely undistorted. As long as the input remains below the threshold of distortion, the performance of the device is perfectly linear.

Send an input whose level seeks to drive the output above this limit, however, and the peak of the wave is flattened. A sine wave input leaves the device looking as if its peaks had been clipped off. The term for this kind of distortion, clipping, comes from a visual inspection of the output

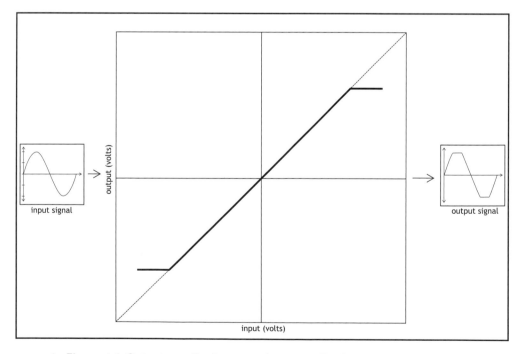

▲ *Figure 4.4 Output amplitude versus input amplitude creating hard clipping.*

waveform. As the transition from linear to nonlinear is abrupt, it is classified hard clipping.

Sine wave inputs below the threshold of distortion remain sine waves on output. Sine wave inputs of ever-increasing amplitude develop more severe clipping. Taken to an extreme (Figure 4.5), the output based on a sine wave input starts to resemble a square wave.

Severely clipping a sine wave until it resembles a square wave has the effect of adding harmonic content to the sine wave signal. That is, send a 100-Hz pure tone through a hard-clipping device and the squareish wave output is a signal that retains some energy at 100 Hz while shifting the rest of that energy up to 300 Hz, 500 Hz, and so on. The act of clipping a waveform essentially synthesizes harmonics, a phenomenon known as *harmonic distortion*. Hard clipping begins to introduce harmonics toward creating the square wave, which was shown in Chapter 1 to follow the equation:

$$Y(t) = \frac{4A_{peak}}{\pi} \sum_{n=1}^{\infty} \frac{1}{2n-1} \sin[2\pi(2n-1)ft] \qquad (4.1)$$

A specific set of odd harmonics makes a square wave square. As the equation makes clear, a true square wave has the impractical requirement that an *infinite* number of harmonics are needed. When a device is driven into hard clipping, beware of the possibility of side effects due to the generation of a broad range of harmonics, possibly reaching into extremely high frequencies, even beyond what is nominally audible.

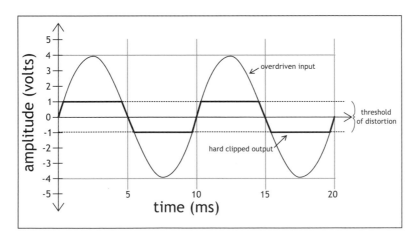

▲ *Figure 4.5 Hard clipping a sine wave in to a square wave shape.*

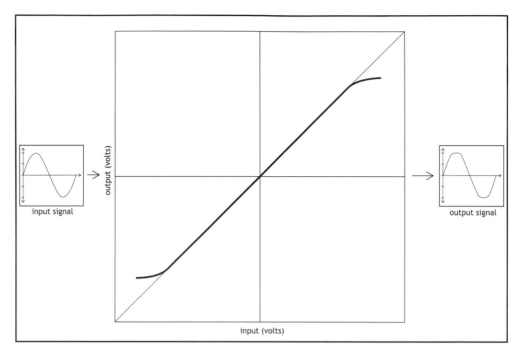

▲ *Figure 4.6 Output amplitude versus input amplitude creating soft clipping.*

Soft Clipping

Clipping is not always so abrupt. Imagine a device that approaches its amplitude limits more gradually (Figure 4.6). There may not be a well-defined threshold of distortion that separates the distorted from the undistorted. As inputs are raised in level, the degree of output shortcoming gets worse too.

This type of distortion also alters the peaks of the waveforms, but in a kinder, gentler way than hard clipping. Call it soft clipping. As the output versus input plot makes clear, soft clipping is nonlinear. It is a kind of amplitude distortion. The peaks are squeezed down, not lopped off. Distortion of a different shape is, not surprisingly, built of a different recipe of harmonics.

By way of analogy, physicists are fond of placing a mass on a spring in a frictionless world. The spring remains linear if not overstretched (i.e., no distortion). If stretched too far, the opposing force of the spring starts to rebel and overreact (i.e., soft clipping). If the spring is stretched so far that the mass bumps into the floor, motion is instantly stopped (i.e., hard clipping). Pressure waves in air can reach similar limits. Voltage oscillations in audio equipment suffer the same indignities.

Asymmetric Distortion

Distortion containing even-order harmonics leads to asymmetry in the output versus input plot, as in Figure 4.7. A sine wave input is modified as shown, with a different shape below zero versus above zero. The simple triode tube gain stage favored in the preamp section of an electric guitar amp often offers this type of distortion.

It is a common misconception that soft clipping comes from even harmonics and hard clipping from odd harmonics. Both hard and soft clipping are in fact made entirely of odd harmonic distortion. Even harmonics will lead to asymmetric traits like that shown in Figure 4.7.

4.1.2 INTERMODULATION DISTORTION

The discussion above, evaluating the spectral changes to a sinusoidal input into a device reaching its amplitude limits is not the whole story. Nonlinear

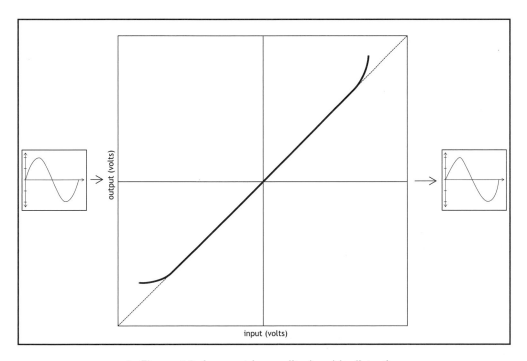

▲ *Figure 4.7 Assymetric amplitude adds distortion.*

devices have a more profound effect on complex signals. That is, when an input contains energy at more than one frequency, the harmonics generated are not just limited to the odd and even multiples of each frequency component. These harmonics are created, to be sure. But additional frequency components, not harmonically related to the input spectrum, are also created. Called *intermodulation distortion*, energy at frequencies related to the sum and difference of the spectral components appears at the output of the distortion device.

An input containing two frequencies, f_1 and f_2, would develop distortion components at frequencies equivalent to $f_1 + f_2$. The differences between frequency components on input lead to distortion artifacts at the arithmetic sum of these frequencies. Similarly, distortion will be found at the differences, such as $f_1 - f_2$ and $f_2 - f_1$.

Depending on the degree of nonlinearity of the distorting device, intermodulation components between the harmonics may also develop. That is, harmonic distortion of frequency f_1 leads to multiples of f_1, such as $2f_1$, $3f_1$, and so on. Sum and difference distortion products built on these harmonics may become significant. This leads to distortion that is still more spectrally complex, containing terms such as $2f_1 - f_2$, $3f_1 - f_2$, $2f_2 - f_1$, $3f_2 - f_1$, etc.

When spectrally-rich signals are subjected to clipping of any kind, harmonic and intermodulation distortion must be anticipated.

4.2 Sources of Distortion

Components of analog circuits such as transistors, transformers, and tubes each exhibit a characteristic type of distortion. Assemble them into various circuit designs and a new type of nonlinearity results. In the digital domain, algorithms may be written to reshape the signal in almost any way desired as headroom is exhausted. The types of nonlinearities are as diverse as the number of people who write the code and design the circuits. Audio systems utilizing these circuits, plus magnets, coils, ribbons, and optical components possess an output versus input behavior that grows more complex still.

Think of hard clipping and soft clipping as two icons of harmonic distortion, but recognize that analog and digital devices can develop unique combinations and variations of distortion that are a complex result of the overall design and layout of the components and algorithms employed.

Some audio devices are purpose-built to distort (guitar amps, guitar amp simulators, and distortion pedals), offering a unique flavor that comes from the combination of harmonics associated with a device's output versus input shortcomings.

Other types of audio equipment possess some kind of distortion more as a side effect. Transducers, compressors, tube gain stages, and analog magnetic tape are among the more notable examples. The devices have other design goals. A microphone must convert acoustic pressure changes into electrical voltage changes, a loudspeaker must convert electrical voltage changes back into air pressure changes, a compressor automatically attenuates audio when the signal is above the specified threshold, etc. In addition to this nominal function, many of these devices also introduce harmonic and intermodulation distortions of varying types which may be taken advantage of by the clever recordist to further refine the sound quality of the audio signal at hand.

Musicians think of timbre as that attribute of sound that separates a piano from a guitar when they play the same note. Engineers must sort out what sonically separates a Steinway from a Bösendorfer among pianos and a Les Paul from a Telecaster among guitars, even as these instruments play the same note. Trained and careful listeners can reliably distinguish very fine properties of a sound. The unique distortion traits of each audio device in the studio endow that device with a sort of sound fingerprint related to distortion. Distorting devices, like fuzz boxes and tube amps, offer their own characteristic qualities of distortion. Some mixture of hard clipping, soft clipping, asymmetric, and other forms of distortion adds up to a distinct and possibly quite desirable sound quality. Any component, pushed beyond its linear limits, offers harmonic and nonharmonic alterations to the signal passing through it that may impart a sonic flavor an engineer seeks.

4.3 Motivations for Distortion

This effect is not just for engineers. Distortion of the right kind is naturally sought out by musicians and music fans alike. Spared the pleasure of reading this text, guitarists, singers, and drummers still employ distortion as a method of musical expression. Fans of many, if not most, styles of popular music react positively to distortion almost instinctively. When a device is overloaded, something exciting must be happening. Someone is misbehaving. Rules are being bent or broken. Distortion in rock and roll is as natural as salsa in Mexican food. It is the caffeine of music. Or perhaps it is the garlic. Used musically, most people can not resist. Accordingly, audio engineers reach for distortion as a staple effect.

Beyond the visceral, distortion has practical benefits. Distortion can influence an engineer's use of the fader. Augmenting a sound with additional, spectrally-related energy likely makes that sound easier to hear (see Chapter 3). In this way, distortion becomes a creative alternative to the more obvious approach of simply pushing up the fader to make it louder. Any single track of a multitrack project fighting to be heard in a crowded mix can achieve distinct audibility through the harmonic lift that comes from at least a little well-chosen distortion.

Distortion offers the engineer an alternative to the more straightforward spectral effect of equalization (see Chapter 5). The fact that amplitude limitations create spectral complexity means that an engineer can reshape timbre in surprising, complicated, and glorious ways through distortion. Equalization, as discussed in Chapter 5, emphasizes and deemphasizes spectral regions present in the signal being processed. Distortion fabricates entirely new spectral components, fodder for further manipulation in the multitrack production.

4.4 Distortion Dos and Don'ts

Two ways to create a square wave are made clear. First, it may be synthesized through the addition of a large number of strategically chosen sine waves, each at just the right frequency with just the right amplitude, diligently following Equation 4.1. Alternatively, a square wave may be created through the hard clipping of a sine wave. The history describing the creation of the square wave does not matter. A square wave is a square wave. In both cases, a waveform looking and sounding like that shown in Figure 4.8 results. Each and every square wave of the same fundamental frequency has the same spectral recipe, whether they were deliberately put there in an additive synthesis effort to build a square wave or they were the inevitable result of hard clipping of a sine wave.

Consider a complex wave with pitch, such as an electric guitar track. Far from a pure tone, electric guitar contains energy at a range of frequencies. Make this the input to a device that distorts. Each and every harmonic component of that timbrally-rich input triggers its own set of clipping-related harmonics. Sum and difference components between energy components at all frequencies also appear. The spectral content spreads

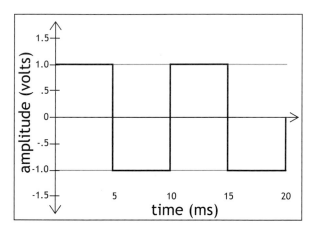

▲ *Figure 4.8 A 100-Hz square wave.*

upward in frequency as harmonics are generated, and spreads further outward up and down in frequency as intermodulation components gain amplitude. Distortion can make a complex signal significantly more complex.

The engineer interested in the creative use of distortion should keep a few trends in mind.

The harmonics generated by the distortion nearest to the frequency content of the original signal can be difficult to hear due to masking (see Chapter 3). The second harmonic may be especially difficult to hear due to the additional affect of the upward spread of masking into this frequency range.

The musical meaning of the harmonic distortion components is important. If the recipe of distortion calls for energy at even harmonics (two times the frequency, four times the frequency, etc.), they form musically-consonant intervals (multiples of the octave) with the source waveform. Odd harmonics (three times the frequency, five times the frequency, etc.) form more dissonant intervals. Table 4.1 shows the approximate musical meaning of harmonic distortion. Note the relative consonant harmony associated with the even harmonics. Three of the first four even harmonics have octave-based relationships with the fundamental. The second, fourth, and eighth harmonics fall exactly on a multiple of the input frequency. The intervening sixth harmonic is also quite consonant in the harmony of the fundamental, falling exactly two octaves plus one perfect fifth above it. The odd harmonics not only appear at more dissonant intervals, but also possess pitches that

Partial	Harmonic	Frequency (Hz)	Musical Interval	Error (cents)*
Table 4.1 Musical Meaning of Harmonic Distortion				
0	1	110		
1	*2*	*220*	*1 octave exactly*	*0.00*
2	3	330	1 octave + P5	1.96
3	*4*	*440*	*2 octaves exactly*	*0.00*
4	5	550	2 octaves + MAJ 3	−13.69
5	6	660	2 octaves + P5	1.96
6	7	770	2 octaves + min 7	−31.17
7	*8*	*880*	*3 octaves exactly*	*0.00*
8	9	990	3 octaves + MAJ 2	3.91
9	10	1,100	3 octaves + MAJ 3	−13.69
10	11	1,210	3 octaves + dim 5	−48.68
11	12	1,320	3 octaves + P5	1.96
12	13	1,430	3 octaves + min 6	40.53
13	14	1,540	3 octaves + min 7	−31.17
14	15	1,650	3 octaves + MAJ 7	−11.73
15	*16*	*1,760*	*4 octaves exactly*	*0.00*

*The frequency of the harmonic is actually the listed number of cents above or below the musical interval indicated.

are detuned from these intervals by many tens of cents versus equal tempered tuning. The distortion associated with even harmonics can be more pleasing to the ear, just adding musically-sensible, high-frequency emphasis. The distortion associated with odd harmonics can sound harsher, creating a stronger effect that stands out more.

The harmonic energy associated with distortion can cause masking of other signals. That is, a vocal that had been perfectly intelligible and a snare drum that formerly cut through the mix can both become more difficult to hear when distortion is added to the electric guitar. Spectral masking is very much in play (see Chapter 3). Distortion, like equalization, must be strategically used. Guitar processing affects not only guitar, but also those instruments that occupy similar spectral regions, after distortion.

More aggressive clipping generally increases the amplitude of distortion components. Distortion can become easier to hear as the energy associated with the clipping is increased. Three phenomena are at work:

1. Each individual harmonic increases in amplitude.
2. Harmonics of ever-higher frequencies are introduced.
3. Intermodulation components grow in amplitude with the harmonic components.

When harmonic distortion is used as an effect, the engineer must be aware that the spectral result is the generation of harmonics above and below all the spectral components of the source waveform, not just the fundamental. Distinguishing one element of a multitrack production with a thoughtfully chosen type of distortion may cause unwanted masking of other signals fighting to be heard in those spectral regions now occupied by distortion harmonics. The thoughtful engineer listens for these possible side effects and makes only strategic, not opportunistic, use of distortion.

4.5 Selected Discography

Artist: Green Day
Song: "Boulevard of Broken Dreams"
Album: *American Idiot*
Label: Reprise Records
Year: 2004
Notes: Tremolo enhanced distorted guitar (see Chapter 7) is a hook here. The introduction reveals a quarter note burst of distortion over a sixteenth note amplitude modulating pulse. Most of the tone of the guitar is removed, leaving the harmonics associated with the distortion in plain view. This leaves plenty of room spectrally for the vocal, the snare, the bass, the acoustic guitar, and the piano. Then when the full guitars come in later, it is huge!

Artist: The Rolling Stones
Song: "Sympathy for the Devil"
Album: *Beggars Banquet*
Label: ABKCO
Year: 1968
Notes: An iconic guitar solo sound whose unique guitar tone is built on aggressive distortion. An otherwise thin, piercing tone commands respect through this sonic barbed wire.

Artist: U2
Song: "Zoo Station"
Album: *Achtung Baby*
Label: Island Records
Year: 1991
Notes: While distortion can be applied to any track, the lead vocal is the rare target here, sitting front and center, with the first words sung on the album. Do not miss those tasty fragments of distorted drums in the intro too.

Artist: Stevie Ray Vaughan
Song: "Pride and Joy"
Album: *Texas Flood*
Label: Epic Records
Year: 1983
Notes: Not the only song and guitarist, but this distorted tone is a touchstone for every guitarist who tries to play the blues.

Artist: The Wallflowers
Song: "One Headlight"
Album: *Bringing Down the Horse*
Label: Interscope
Year: 1996
Notes: Lead vocal gets emotional lift through aggressive distortion in the chorus. Chorus harmonies get an even stronger dose; hard clipping is a significant part of this flavor of distortion.

Equalization

5

"Your heart is the big box of paints."
— "WRAPPED IN GREY," XTC, *NONSUCH* (CAROLINE RECORDS, 1992)

The equalizer. It is a peculiar name for an audio signal processor. While it may sound like the title of an action movie or a utopian political movement, it is a genuine part of the audio lexicon. In the early days of audio (beginning in the late 1800s), the output signal was often clearly inferior to the input signal. Transmission lines, transducers, and storage media of the time took their toll on the quality of the signal. Equalizers were invented to try to compensate for this, to make the output more closely *equal* to the input.

Participants in audio today have it much easier. The quality of every piece of the audio recording, storage, processing, transmission, and reproduction chain has improved significantly. While the audio industry still makes frequent use of equalizers, it is rarely motivated by a need to correct for the inherent defect of an audio device. The use of an equalization processor is driven more by creative desires, and less by technical shortcomings. Equalization (EQ) is less about making things equal and more about creating things beautiful, unusual, or functional. Yet the original name has stuck. A more descriptive term would be *spectral modifier* or *frequency specific amplitude adjuster*. Such unwieldy names have not been welcome in the recording studio, and audio tradition seems to insist we refer back to those early years of signal processing. So, like it or not, we *equalize*.

This chapter presents a summary of the topologies and user interfaces likely to be encountered on equalizers in the recording studio. Key properties are discussed and adjustable parameters are defined. The chapter then inventories the most common technical and creative motivations for and approaches to equalization.

5.1 Spectral Modification

The audio goal of an equalizer is to adjust the relative levels of the various frequency ranges present in any signal passing through it. If the signal is dull lacking high-frequency sparkle, the equalizer is the tool used to amplify what little high-frequency energy is there. If the sound is painfully bright, harshly assaulting the ears with too much high-frequency sizzle, the equalizer offers the solution by attenuating the level of the exaggerated high-frequency portion of the audio signal. Engineers use equalizers to adjust the relative amplitude of a signal within specific and controllable frequency ranges. The console master fader adjusts the amplitude of the entire audio signal. Think of an equalizer as a frequency specific fader; it increases or decreases the amplitude of a signal at certain frequencies only.

The frequency response of a device describes its consistency in creating output signals across the entire audio frequency range (see Chapter 1). Figure 5.1 shows an example of a device with an output that emphasizes the low frequencies (if you are the "glass half full" type) or attenuates high frequencies (for you "glass half empty" types).

A pair of sine wave oscillators or a synthesizer with a sine wave patch is all that is needed to see and hear this effect. Consider a middle frequency, 1 kHz, to be a reference. Play two sine waves of equivalent amplitude simultaneously: the 1-kHz reference and a low-frequency sine wave around 100 Hz. Route both sine waves at equal level into a device with a frequency response like that shown in Figure 5.1. When one observes the output of

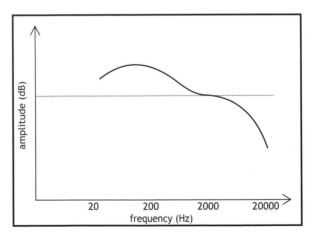

▲ *Figure 5.1 Frequency Response.*

the device, the low frequency tone will be higher in amplitude than the middle-frequency reference. The two frequencies went into the device at the same amplitude, but the low-frequency sine wave is hotter (greater in level) than the 1-kHz middle-frequency at the output of the device.

Now play the same 1-kHz middle-frequency reference and a *higher* frequency sine wave of equivalent amplitude, maybe 10 kHz (if you can stand to listen to it). Using the device with the frequency response of Figure 5.1, one will find that for the same input level, the higher frequency becomes reduced in amplitude relative to the 1-kHz reference. The device with a frequency response like that of Figure 5.1 favors low frequencies over high.

As any audio engineer who has ever had to listen to sine waves for very long (as in the experiment above or when aligning analog magnetic tape recorders) knows, pure tones can create a rather unpleasant, distinctly nonmusical listening experience. It is worth taking another look at the meaning of the frequency response plot of Figure 5.1. Consider an input that is not just a simple pair of sinusoids, but is instead an entire mix — a gorgeous, full bandwidth mix. The mix is a careful blend of instruments and effects that fills the audio spectrum exactly to the mix engineer's liking, with a beautiful, detailed midrange; an airy, open high end; and a rich, controlled low end. Sent through the device of Figure 5.1, that spectral balance is altered. The mix, which had been pleasing in its use of spectrum, becomes too heavy in the low frequencies and loses carefully prepared energy up high.

For devices like microphone cables and mixing consoles, it is commonly desirable for the device to treat the amplitude of all signals the same way at all frequencies. A flat-frequency response (the dashed line in Figure 5.1) is the goal. These devices should not change the frequency character of the signal being recorded or the mix being monitored. A key exception to the desire for a flat-frequency response is the equalizer. In fact, the equalizer is designed specifically to have a nonflat frequency response.

The goal of the equalizer is to alter the frequency response. If a vocal track, or an entire mix, needs a little more low end, an engineer simply runs it through an equalizer with a frequency response like that of Figure 5.1. To understand equalization, one need only understand this: This process changes the frequency content of a signal by deliberately routing it through a device whose frequency response is assuredly nonflat. The trick, discussed in more detail below, is to alter the frequency response in ways that are tasteful, musical, and appropriate to the sound. One warning: It is easy to get it wrong. Dialing up just the right EQ "curve" for a given situation will require experience, good ears, a good monitoring environment, and good judgment.

5.2 Parameters for Spectral Modification

In order to make the frequency response like that of Figure 5.1 adjustable from flat (the dashed line) to the specific contour shown (the solid curve), some set of controls is needed. This may be accomplished in a number of ways.

Consider the most flexible type of all: the parametric equalizer. It is named a parametric EQ because it offers the user three useful *parameters* for manipulating the spectral shape of any audio signal. Understanding the three parameters here makes understanding all other types of equalizers a breeze. The other equalizers will have one or two of these three parameters available to the user. When engineers learn how to use a parametric equalizer, they are learning how to use all types of equalizers.

5.2.1 FREQUENCY SELECT

Perhaps the most obvious parameter needed is the one that determines the frequency range to be altered. The center frequency of the spectral region being affected is dialed-up on a knob labeled *frequency*. In search of bass, a sound engineer might decide that the signal needs additional low-frequency content in the area around 100 Hz. Or, is it closer to 80 Hz? These decisions are made at the *frequency* select control.

5.2.2 CUT/BOOST

The recordist next decides how much to alter the amplitude of the frequency selected. The addition (or subtraction) of bass happens via adjustment of the second parameter: *gain* or *cut/boost*. This control determines the amount of decrease or increase in amplitude at the center frequency chosen with the frequency selection parameter above. To reduce the shrillness of a brassy horn track, select a high frequency (perhaps around 8 kHz) and *cut* a small amount (maybe about 3 dB). To add a lot of bass, *boost* 9 to 12 decibels at the low frequency that sounds best, somewhere between 40 and 120 Hz perhaps. These two parameters alone — *frequency select* and *cut/boost* — give the engineer a terrific amount of spectral flexibility.

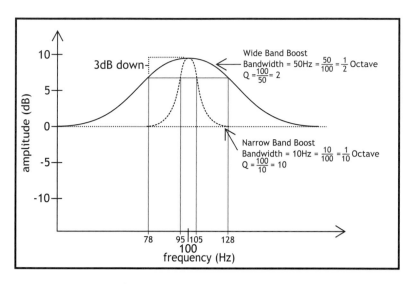

▲ *Figure 5.2 Bandwidth and Q.*

5.2.3 Q

The parametric equalizer's third and final parameter is perhaps more subtle. Consider a boost of 9 dB at 100 Hz. This could be the sonic alteration needed to make the guitar sound powerful. For such a precise decision, it fails to fully define how the frequency response is altered. Figure 5.2 demonstrates two possible results for the same frequency and cut/boost settings. When a center frequency for this increase in level is selected, it affects not just that single frequency, but also the neighboring frequencies as well. The degree to which it also boosts other frequencies nearby is defined by the third parameter, *Q*. The Q describes the width of the cut or boost region. A better understanding of Q comes from first better defining a related concept: *bandwidth*.

Bandwidth is the frequency region on either side of the center frequency that is within 3 decibels of the center frequency's cut or boost. Starting at the center frequency and working outward both above and below along the frequency axis, one can find the point where the signal is 3 decibels down from the amplitude at the center frequency. The *width* of a cut or boost at a specific frequency describes the frequency range bounded by these 3 decibel down points, measured in units of hertz.

In the example of a 9-dB boost at 100 Hz, the width is based on the frequencies that are boosted by 6 dB (9 dB − 3 dB = 6 dB) or more. Figure 5.2 reveals that the wide 9-dB boost at 100 Hz has 3-dB down points at 78 Hz and 128 Hz. The width then is 50 Hz (the spectral difference between 78 Hz and 128 Hz). The narrow boost is 3-dB down at 95 Hz and 105 Hz, giving a smaller width of just 10 Hz.

Expressing values in actual hertz is rarely very useful in the studio. Humans do not process music that way. When an arranger creates a horn chart, they do not decide to add a flute part 440 Hz above the tenor sax. Instead, they describe it in musical intervals, saying that the flute should be perhaps one *octave* above the tenor sax. Music pitches are described in terms of musical ratios or intervals, the most famous of which is the octave. The octave represents nothing more than a mathematical doubling of frequency, whatever the frequency may be. Borrowing from musicians, engineers stay true to this way of describing spectral properties on the equalizer.

One way to do this is to express the width in octaves. A 50-Hz width around a 100-Hz center-frequency boost represents a bandwidth that is half an octave wide; the bandwidth is half the value of the center frequency.

In the course of a session, it is common for the engineer to use the frequency select parameter to sweep up or down in frequency, an ears-based search for exactly the right place to boost. The equalizer does not lock in to a spectral width that is, for example, always 50-Hz wide as the frequency control is swept. Instead the equalizer circuit maintains the same ratio of width to center frequency as the engineer works, providing a bandwidth in the EQ contour that retains a more consistent subjective impression. Rather than staying a fixed width in number of hertz, it retains a fixed perceptual width musically, in number of octaves. If one sweeps the center frequency down from 100 Hz to 50 Hz, the bandwidth of the bump in frequency response needs to retain the same approximate perceptual value; therefore, its width in hertz changes to remain exactly half the value of the new center frequency. Had it remained at a mathematically rigid bandwidth of 50 Hz as the frequency was swept down from 100 Hz to 50 Hz, it would have *sounded* like a wider, less-precise equalization adjustment. With a band width of half an octave, sweeping the center frequency down from 100 Hz to 50 Hz would be accompanied by an actual width that decreases from 50 Hz to 25 Hz, preserving the half octave ratio. This narrowing of bandwidth as measured in hertz ensures that the perceived *character* of the equalization does not change much as the recording engineer zeroes in on the desired center frequency.

That is the idea of bandwidth, and that is almost the end of the math in this chapter. But there is one more layer to consider. While expressing the bandwidth of an equalizer boost or cut in octaves makes good sense, the tradition is to flip the ratio over mathematically: *center frequency* divided by *width* instead of *width* divided by *center frequency*. The spectral "width" described this way is the parameter *Q*. It is the reciprocal of *bandwidth* expressed in octaves. The wide boost discussed above and shown in Figure 5.2 is 50-Hz wide at a center frequency of 100 Hz. The Q, therefore, is 2 (center frequency of 100 Hz divided by the width of 50 Hz). The narrow boost has a Q of 10 (100 Hz divided by the narrow 10-Hz width). Engineers frequently say "low Q" and "high Q" to describe wide and narrow equalization settings, respectively.

5.2.4 MULTIBAND EQ

Frequency select, *cut/boost*, and *Q* are the three basic, adjustable parameters needed to achieve almost any kind of alteration to a frequency response, from broad and subtle enhancements to aggressive and pronounced notches. This trio of controls makes up a single parametric band of equalization. A parametric equalizer may then offer three or four of these bands, with overlapping frequency ranges. The engineer coordinates the action of these multiple bands of EQ to achieve a single, beneficial affect on the signal.

A four-band parametric EQ has *12* controls on it (3 controls × 4 bands = 12 controls in all). It offers the three parameters four different times so that the recording engineer can select four different spectral targets and shape each of them with their own amount of boost or cut, and each with a unique bandwidth. Where they overlap, the equalization changes accumulate. Boosts overlapping with boosts lead to still more of a boost at those frequencies. Cuts overlapping with cuts behave similarly. When the boost of one parametric band overlaps with the cut of an adjacent band, they work against each other, and the net effect is simply the algebraic sum the boost minus the cut. The larger magnitude wins.

The result, if one's ears can follow it all, is the ability to affect a tremendous amount of change on the spectral content of a signal. Figure 5.3 shows a possible result of four-band parametric equalization. The terrific amount of sonic shaping power that four bands of parametric equalization offer makes it a popular piece of gear in any studio. However, other useful EQ options exist.

5.2.5 TAKE AWAY THE Q

Some equalizers set the bandwidth internally, providing the user access only to the frequency select and cut/boost parameters. Because of this

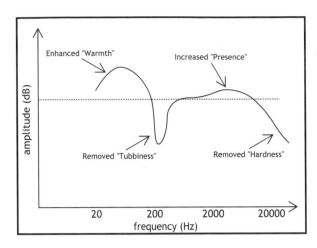

▲ *Figure 5.3 The creative power of a four band parametric equalizer.*

reduction from three user-adjustable parameters to two, this type of EQ is sometimes called a *semi-parametric equalizer.* Emphasizing the adjustability in the frequency domain, this type of equalizer is also sometimes called a "sweepable" EQ. The audio engineer cannot directly adjust the bandwidth. This configuration in which each band offers only two adjustable parameters (frequency and cut/boost) is common as it is easy for the recordist to use (removing controls that might be distracting is liberating for the creative engineer), easier/cheaper/better sounding for the manufacturer to design than a full parametric, and still very useful in the production of music.

A semi-parametric equalizer's lack of a Q control does not mean the concept of a broad or narrow boost or cut is gone. There is still a region of neighboring frequencies affected by the selection of frequency and gain on a semi-parametric equalizer. The spectral width of any EQ adjustment is set by the design of the equalizer, not by an adjustable control available to the user. Through careful listening, experienced engineers develop a sense of the width of each make and model of semi-parametric EQs they use. This knowledge influences their thinking when they consider applying EQ to a signal at any given point in a recording session. Much is made about the sound quality of many types of equalizers. Some sound better than others, it is said. To be sure, the quality of the design, components, and manufacturing processes can have a noticeable effect on the sound quality of the device. But in the case of semi-parametric equalizers, a critical distinguishing characteristic is the internally-determined Q. Some music production situations leave the engineer in need of a wide boost or cut — they reach for the low-Q equalizer. Other production challenges are solved with a sharp notch to more surgically remove a troublesome resonant frequency. In this case, a high-Q processor is needed. The better-sounding EQ is the one that is right for the task at hand.

A semi-parametric equalizer's lack of a Q control does not mean this parameter is fixed. Some equalizer topologies deliberately allow the Q to increase (narrower bandwidth) as the cut or boost becomes more extreme. The idea, and it is a good one, is that a small gain setting, say +/−3 dB, reflects an engineer's need for subtle reshaping of spectrum. A low Q accompanies such a gain setting. A larger gain setting, perhaps +/−12 dB, implies more specific frequency ranges must be highlighted or removed. A high Q is offered. This useful design carefully allows Q to increase as the absolute value of gain increases to extreme cut/boost settings. In the hands of a thoughtful engineer, semi-parametric equalizers offer great spectral shaping flexibility.

5.2.6 TAKE AWAY THE FREQUENCY

There is room for further simplification of the user interface on an equalizer. Sometimes we only have control over the amount of cut/boost, and can adjust neither the frequency nor the Q of the equalization shape. Generally called *program EQ*, this is the sort of equalizer found on, for instance, mom's stereo (labeled "treble" and "bass"). This type of EQ is also found on many consoles, vintage and new. It appears most often in a two- or three-band form: three knobs labeled high, mid, and low that are fixed in frequency and Q and offer the recording engineer only the choice of how much cutting or boosting they are going to apply. In the case of consoles, remember that there may be the same equalizer repeated over and over on every channel of the console. If it costs an extra 20 dollars to make the equalizer sweepable, that translates into a bump in price of more than $600 on a 32-channel mixer. If it costs 50 quid to make them fully parametric, and it is a 64-channel console . . . well, you do the math. The good news is that well-designed program equalization can sound absolutely gorgeous. It often offers frequencies that are close enough to the ideal spectral location to get the job done on many tracks. Often the engineer does not even miss the frequency select parameter.

5.2.7 TAKE AWAY THE KNOBS

A variation on the equalizer above is the graphic equalizer. Like program EQ, this device offers the engineer only the cut/boost decision, fixing Q and frequency. On a graphic EQ, the several frequency bands are presented, not as knobs, but as sliders, like faders on a console. The visual result of such a hardware design is that the fader positions provide a good visual description of the frequency response modification that is being applied, hence the name "graphic." Handy also is the fact that the faders can be made quite compact. It is not unusual to have from 10- to 30-band graphic equalizers that fit into one or two rack spaces.

When the graphic equalizer lives in software, adjustments can be as simple as clicking on a frequency response plot and dragging the line into the shape that is desired.

Graphic EQ is an extremely intuitive and comfortable way to work. Being able to see an outline of the current frequency response modification will make it easier and quicker to achieve the spectral goal at hand. Turning knobs on a four-band parametric equalizer is more of an acquired skill.

There are times in the course of a project when one must reshape the harmonic content with great care, using a parametric EQ. In other instances, there is no time for such careful tweaking and a graphic EQ is the perfect, efficient solution. Plan to master both.

5.2.8 SOME KNOBS ARE SWITCHES

In all types of EQ, one will find models where the knobs move smoothly and continuously across their available range of settings, while other models have knobs "click" to discrete values. Frequency select may be continuously sweepable from, say 125 Hz to 250 Hz. Alternatively, the knob may instead click directly from 125 Hz to 250 Hz. If, in the latter case, the engineer wanted the equalization contour to be centered on exactly 180 Hz, they are out of luck.

How could this be? What seems to be a reduction in engineering flexibility may offer an improvement in equalizer sound quality. Clicking a knob on the faceplate may physically select different electronic components inside the device. The equalizer is literally swapping components in the circuit path for different frequency selections! It is not just adjusting some variable piece of the circuit, it is physically changing the circuit. In choosing which type of equalizer to acquire and use, engineers have to trade-off sound quality versus price and processing flexibility versus ease of use. Some companies have such high standards for sound quality that they take away a little bit of user flexibility (continuous controls) to get a better sound. Conversely, if one finds an equalizer that is fully parametric and sweepable across four bands yet costs less than a large pizza, it would be wise to investigate how the manufacturer made the EQ so infinitely adjustable and how much sound quality was sacrificed in the name of this flexibility. Do not value an equalizer based on the number of controls it has. A simple program EQ with cut and boost controls might contain only extremely high-quality components inside.

5.2.9 SHELVING EQ

The most elaborate, feature-rich equalizer envisioned so far, the one with the most knobs on the faceplate, is the multiband parametric equalizer. With four bands of parametric EQ, the device provides the user 12 knobs. There is room for still more processing flexibility. Each parametric band offers a region for spectral emphasis when boosting or deemphasis when attenuating. This shape is called a *peak/dip contour* because of the visual change it makes in the frequency response. Roughly shaped like a bell curve, it offers a bump up or down in the frequency response. Two alternatives EQ contours exist.

The *shelving equalizer* offers the peak/dip response on one side of the selected center frequency, and a flat cut or boost region on the other (Figure 5.4). A broad equalization desire might be to add brightness to the sound in general. A high-frequency shelving EQ bumped up 6 dB at 8 kHz will raise the output from about 8 kHz and above. It is not limited to a center frequency and its associated bandwidth. The resulting alteration in the frequency response is flat (like a shelf) beyond the selected frequency. As Figure 5.4 shows, the concept of shelving EQ applies to low frequencies as well as high, and cuts as well as boosts. In all cases, there is a flat region beyond (above or below) the selected center frequency, which is boosted or attenuated.

The transition back to the unprocessed region is simply half of the bell curve of a peak/dip contour. Therefore, the idea of a 3-dB down point and the concept of bandwidth and Q apply to shelving EQ as well. A shelf boosting by 9 dB at 10 kHz and above is in fact 3-dB down (+6 dB) at 10 kHz, achieving the full boost (+9 dB) a little bit above 10 kHz. Below 10 kHz, this shelving EQ is thought not to be processing the signal. This is true well below 10 kHz, but just below 10 kHz, the shelf EQ starts to lift the signal amplitude up. A high-Q shelf abruptly transitions from the region of unaltered frequency response to the region of the 9-dB boost. A low Q spreads that transition out across a broader spectral region. Both approaches have production value, in the right circumstances, so an engineer must pay careful attention and listen intensely.

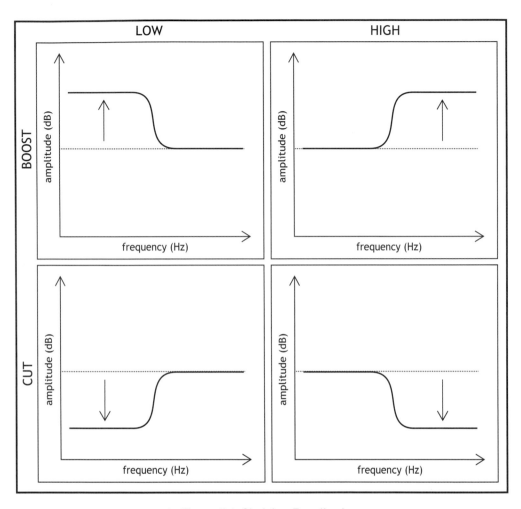

▲ *Figure 5.4 Shelving Equalization.*

5.2.10 FILTERS

An important third option exists for reshaping the frequency content of a signal: the filter. Engineers speak generally about filtering a signal whenever they change its frequency response in any way. Under this loose definition, all of the equalizers we have discussed so far are audio filters. But to be more precise, a true filter must have one of the two shapes shown in Figure 5.5. A high-pass filter (Figure 5.5a) allows high frequencies through with no change in amplitude, but attenuates lows. A low-pass filter does the opposite. A low-pass filter (Figure 5.5b) allows low frequencies to pass through the device without a change in amplitude, but high frequencies are attenuated.

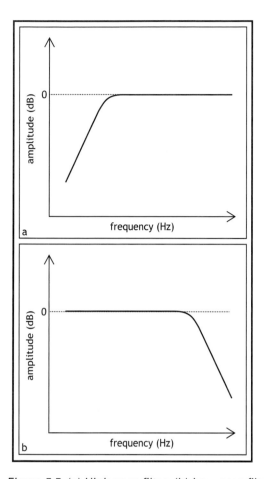

▲ *Figure 5.5 (a) High pass filter, (b) Low pass filter.*

Because the sonic result can be rather similar to shelving filters cutting out extreme high or low frequencies, there is the potential for some confusion between them. Filters distinguish themselves from shelving equalizers in two key ways. First, filters are cut-only devices. They never boost at any frequency. Shelf EQ can cut or boost. Second, and this is important, filters offer a never-ending amount of attenuation beyond the selected frequency. They do not flatten out like the shelf. They just keep cutting and cutting. If there is some unwanted low-frequency air conditioner rumble on a track that you never, ever want to hear, a filter can offer significant attenuation. A shelf equalizer will have a limit to the amount of attenuation it can achieve, perhaps only 12 or 16 dB down. The weakness of using a shelving equalizer in this case is easily revealed on every quiet passage whenever that track is being played, as it might still be possible to hear the air conditioner rumbling on faintly in the background.

Filters and shelving equalizers introduce some complexity to the faceplate of an equalizer. The four-band parametric (12 knobs) gets upgraded with a low-pass and high-pass filter at each end of the spectrum, as well as a switch that toggles the lowest parametric band from peak/dip bell curve to a low-frequency shelf and the top parametric band from a peak/dip bell curve to a high-frequency shelf. Such an equalizer contains a rich amount of capability with which the engineer can freely alter the spectral content of any signal in the studio. These knobs and switches enable the recordist to bend and shape the frequency response of the equalizer into almost any contour imaginable.

5.3 Applications

The strong, creative drive to push the limits of a sound must be balanced by musical and technical knowledge of the sound and the equipment being used. Experience, opinions, and strategies will help the engineer with all of their EQ processing decisions.

Discussing EQ starts to sound like a weather report. With temperatures, it gets confusing when the *low* is in the *upper* forties while the *high* is in the *lower* seventies. It can be similarly confusing with frequencies. What is a lower mid? Is there such a thing as an upper low?

Such confusion is only enhanced by the mysticism of the audio business. Conversations like this abound in studios around the world:

"I hear they used a Spastastic X-3000 Equalizer on that kazoo track."
"No waaaay."
"Yup, and that album went triple platinum."
"No waaaay."
"And I saw on a website that they gave it 4-dB boost in the upper lows."
"Wait a second. Lemme grab a pencil. . . . What was that killer EQ move again?"

Serious students of audio are not fooled by the confusion and the hype. Accurate information and deep experience will empower audio creativity.

5.3.1 TECHNIQUE

Equalizers are typically incorporated into the audio signal flow using the insert of the console or digital audio workstation. Alternatively, EQ can be patched between the multitrack return and the mixer line input. The use of auxiliary sends or buses is unusual for this type of effect (see Chapter 2).

The most common approach for finding a useful EQ setting is to *boost*, *search*, and then *set* the equalizer. *Boost* by a clearly audible amount, maybe 9 dB to12 dB or more. *Search* by sweeping the frequency select knob until the desired spectral region is audibly highlighted. Finally, *set* the EQ to the level and bandwidth needed. Either cut the frequency if it is unwanted, or back off to just the right amount of boost and bandwidth if it is a positive contributor to the mix. It is that simple.

Over time, through experience and ear training, engineers find they can still be productive in this process while boosting by less than 12 dB, eventually skipping the boost and search steps entirely and instead reaching immediately for the frequency range they wish to manipulate. Growing engineers should occasionally challenge themselves to first listen to the problematic tracks. Then, before touching any equipment, they imagine — in their mind's ear — the frequency ranges in need of alteration. They reach for the equalizer and start immediately at those frequencies. The goal is to skip the boost and search parts.

This premeditated approach to equalization is the method of the savvy, experienced engineer. The less-experienced engineer should try this from time to time, during low-pressure recording sessions. In fact, audio listening skills will be developed more quickly when engineers discipline themselves to make an informed guess of the target frequency without boosting and searching first. Learning from mistakes made this way will help the engineer develop a deep feeling for the qualities of different spectral regions.

However, until one has had several years behind a console, know that there is no harm in taking advantage of those first two steps. It is fast, relatively easy, and has the extra benefit that the producer and any band members or label executives who might also be in the control room will be able to follow along sonically and be supportive of the EQ decisions. Even the famous, expensive engineers resort to the boost, search, and set approach on occasion. Remember, the music-buying public does not care how it is done, just that it sounds great.

When does a recording engineer boost, search, and set? What is s/he listening for? Why and when do they equalize? EQ is simple in concept but difficult in application. Some important words of comfort: All engineers have a lot to learn about EQ. All engineers — apprentices, hobbyists, veterans, and award winners are still exploring the sonic variety and musical capability of equalization. Three things slow one's progress in mastering this topic. The first challenge is the sheer range of possibilities that EQ offers. Using boosts and cuts, high Q and low Q, peak/dip and shelf, EQ represents a vast range of options from 20 Hz to 20,000 Hz.

Second, it is difficult to learn about EQ. Critical listening skills are developed over a lifetime and require careful concentration, good equipment, and a good monitoring environment. No one learned the difference between 1 kHz and 1.2 kHz overnight.

Third, interfering with this already challenging learning process is the temptation to imitate others or repeat equalization moves that worked on the last song. "Magic" settings that make every mix sound great simply do not exist. If a rookie got the chance to write down the equalizer settings used on, say, Jimi Hendrix's guitar track on *The Wind Cries Mary*, it might be tempting to apply it to some other guitar track thinking that this unique equalizer setting goes a long way toward improving the tone. But the fact is, the tone of Jimi's guitar is a result of countless factors: the playing, the tuning, the type of strings, the kind of guitar, the amp, the amp settings, the placement of the amp within the room, the room, the microphones used, the microphone placement chosen, etc. The equalizer alone does not create the tone. In fact, in the scheme of things, EQ plays a relatively minor role in the development of the tone. Those many other factors have a much bigger influence. So repeating EQ moves that worked in the past simply will not ensure a good sound today.

The way to gain fluency with this infinitely variable, difficult-to-hear thing called EQ is to develop a *process* that helps the engineer strategize on when and how to equalize a sound. Armed with this organized approach, one can pursue a more complete understanding of EQ. The audio needs and desires that motivate an engineer to reach for equalization fall into four categories: the fix, the feature, the fit, and the special effect.

5.3.2 FIX

A major motivation for engaging an equalizer is to correct things spectrally, getting rid of problems that lie within specific frequency ranges. For example, outboard equalizers, console channel inputs, microphone pre-amplifiers, and even microphones themselves often have low-frequency roll-off filters. Their prevalence suggests a common need for a filter. These devices remove low-frequency energy less for creative "this'll sound awesome" reasons, and more to fix the common problems of rumble, hum, buzz, pops, and proximity effect. We will take these in order.

Rumble

In many recording situations, the microphone picks up a very low-frequency (40 Hz and *below*) rumble. This low-end energy likely comes from the building's heating, ventilation, and air conditioning system, or the vibration of the building due to traffic on nearby highways and train tracks. (Note to self: Do not build the studio next door to major freeways or train stations.)

Rumble can be really low-frequency stuff, so low that singers and most musical instruments are incapable of creating it. Rumble is so deep that many loudspeakers, even expensive professional-tier models, can barely reproduce it, if at all. Because the loudspeakers will not reproduce it, rumble can be difficult to notice from the comfort of the control room. Rumble must be diagnosed in the recording room, using one's ears.

As it is so difficult to hear and nearly impossible to reproduce, it is appropriate to wonder whether or not removing rumble is worth the trouble. If the rumble is easily overlooked, can the problem be ignored? There are two motivations for keeping deep rumble out of our recordings: dynamic range and production standards.

The dynamic range issue is straightforward. Although music listeners can not easily hear rumble, the audio system uses up precious dynamic range in its effort to process the signal. In fact, an inaudibly low-frequency signal might drive an audio device into distortion. The engineer will not hear the cause, but they will hear the effect. The harmonics of the distortion (see Chapter 4) may be well within the more audible mid and high frequencies. The audio engineer's strategic use of the dynamic range of all of their audio devices is corrupted by the presence of unwanted, low-frequency rumble. If one is to record correctly a vocal, snare drum, guitar, or whatever signal a session presents, the engineer will want to raise the level of the signal well above the noise floor of the system, but not so much that the device runs out of headroom and begins to clip. Rumble will use up some of that valuable headroom, reaching overload at lower signal levels. To prevent rumble from causing overload at any stage — within the microphone, the preamp, the console, any signal processors, the storage device, the storage medium, the power amplifiers, or the monitors — it should be attenuated or removed if possible.

Regarding production standards and rumble, is it really acceptable to allow rumble into your recorded production knowing that maybe 80–90 percent of listeners will not notice it because their systems and listening environments will not reveal it? Most audio engineers have a natural desire to look out for that 10–20 percent of listeners who will hear it. If the reader sets the production standards high enough to satisfy the pickiest listeners and to far exceed the expectations of the rest, then s/he shares the audio standards of most successful engineers. Rumble is unwelcome because it undermines those high standards.

To prevent problematic rumble, recording engineers listen carefully to the recording space, even as the band is being situated and the microphones are being selected and placed. During the sometimes-chaotic time of getting ready to record, one must assess the noise floor of the recording space. If a deep sound is heard and/or possibly felt (in the chest cavity or through the feet), try to identify the source and, if possible, stop it. It is not uncommon in some facilities and especially in some on-location recording sessions (e.g., in symphony halls or houses of worship) to turn off the heating or air conditioning and/or plan to use one of the bass roll-off features of your recording chain. Because very little music happens at such low frequencies, it is often appropriate to insert this high-pass (i.e., low cut) filter to remove most of the problematic, very low-frequency energy entirely. The intent is for this filter to have little to no effect on the musical signal sitting higher up the frequency spectrum. Listen carefully for this. Filters can have a noticeable effect on the signal well above the cutoff frequency. Compare the signal with and without the high pass filter engaged. Try not to be seduced by the obvious benefit of rumble removal and pay careful attention to the audio signal, even though it may be well above the filter frequency. If the signal is not diminished in quality, the rumble removal is a success. Unwanted attenuation of non-rumble portions of the signal and phase distortions that blur the transients of the signal, removing realism and detail, are a potential side effect of these filters.

Hum and Buzz

That's rumble. A slightly different problem is hum. Hum is typically the interference from our power lines and power supplies that is based on 60-Hz AC power (or 50 Hz, depending on your locale). The alternating current in the power provided by the utility company often leaks into our audio through damaged, poorly designed or failing power supplies and audio interconnects. It can also be induced into our audio through proximity to electromagnetic radiation of other power lines, transformers, electric motors, light dimmers, and such. As more harmonics above 60 Hz appear — 120 Hz, 180 Hz, and 240 Hz — the hum blossoms into a full-grown buzz. Buzz finds its way into almost every old guitar amp, helped out a fair amount by florescent lighting and single-coil guitar pickups.

The best solution is to stop the hum or buzz by identifying and removing the cause. This can be done through things as simple as turning off all lights on dimmers, turning off all florescent lights, or asking the guitarist to move or turn slightly. Sometimes plugging the equipment into a different outlet helps. On the other hand, chasing hum and buzz problems can lead to more significant, difficult solutions that can not be implemented during a recording session: rewire the patch bay, redesign the power distribution and grounding for the entire recording studio, modify the equipment, etc. Such solutions require deep knowledge of audio circuitry and power and grounding. These solutions can be dangerous and expensive if done incorrectly. Clearly, a specialist is needed. There is no single cure.

When the source of the hum can not be found and the cause of the hum directly stopped, a high-pass filter helps. To remove hum, engage a roll-off starting at a frequency just above 60 Hz (50 Hz) or perhaps an octave above, 120 Hz (100 Hz). This is high enough in frequency that it can audibly affect the musical quality of the sound. Exercise care and listen carefully when filtering out hum. Many instruments (e.g., some vocals, most saxophones, a lot of hand percussion, to name a few) are not changed much sonically by a high-quality filter. But low-frequency based instruments (e.g., kick drum, bass guitar) are not going to tolerate this kind of spectral alteration. The lowest note on a six-string guitar has a fundamental of about 80 Hz, frustratingly close to the fundamental hum frequencies.

Fortunately, the hum might be less noticeable on these instruments anyway as their music can mask a low-level hum. Buzz, with its larger spectral footprint due to upper harmonics built on 60 Hz, is more challenging. The higher frequency content of buzz makes removing it even more musically destructive. Drive carefully.

Pops

Other low-frequency problems fixed by a high-pass filter are the woofer-straining pops of a breath of air hitting the microphone whenever the singer or spoken word artist articulates a plosive sound. One must pay attention to words containing a "P" or a "B." A similar problem can occur whenever your recording work takes you outdoors (gathering news, doing live sound in outdoor venues, or collecting natural sounds in the field). Any breeze across the microphone creates unwanted, distracting, dynamic-range-consuming, low-frequency noise. The typical defense is a pop filter and/or windscreen. These devices attempt to keep the breath and the breeze off of the capsule allowing the desired audio signal through. In fact, pop filters and windscreens are acoustic high-pass filters. If these acoustic devices can not keep the wind off the microphone sufficiently, then filter the low frequencies out with a low-frequency roll-off switch, preferably at the microphone itself.

Proximity

When the instrument being recorded is positioned very near a directional microphone, proximity effect appears. Sometimes this is sonically beneficial. Radio DJ's love it — it makes them sound larger than life. Sometimes the low-frequency boost is just too much.

Place a directional microphone witin about a foot of an acoustic guitar. The resulting sound will have a pulsing, low-frequency thud associated with each full strum of the guitar. This thump generally sounds unpleasant and can mask or interfere with the otherwise great sound of the rest of the captured tone of the instrument. If possible, reduce or remove the proximity effect by backing the microphone away from the instrument or switching to an omnidirectional microphone (omnidirectional microphones based on pure pressure transduction do not suffer from proximity effect). Alternatively, reach for a filter. Roll-off some low end with a high-pass filter and tame the unwanted proximity effect.

The list of "fixes" motivating the use of filters and equalizers goes on. Ever faced a snare drum with an annoying ring? In some tunes, the ring sounds great. In others, such a snare sound is decidedly unpleasant. If the ringing tone is unwanted, it is of course best to fix it at the drum. Before recording the music, dampen or retune the drum. All too often, however, the mixing engineer is not the same person as the tracking engineer. Mixers can inherit problems not fixed at the earlier recording sessions. In this situation, find the frequency range (boost, search, etc.) most responsible for the ring and try attenuating it (as much as −12 dB or more) at a narrow bandwidth.

What qualifies as a narrow Q? Greater than 4. Many equalizers do not offer a Q any sharper. Rare EQs will offer a Q of 10 or greater. These are powerful tools for this kind of EQ approach. A notch at the troublesome frequency that is wide enough to remove the unwanted ring, but narrow enough not to diminish the rest of the instrument's tone, is ideal. Often, turning down that ring reveals an exciting snare sound across the rest of the frequency range.

Ever track a guitar with old strings? Dull and lifeless, it is unlikely this is fixable, but do not rule it out until you have tried a bit of a boost somewhere up between 6 kHz and 12 kHz.

Sometimes a gorgeous spectral element of a sound is hidden by some other, much less appealing frequency component. A good example of this can be found in drums. Big-budget commercial releases often have wonderfully powerful, punchy drum sounds. Yet when the less-experienced engineer goes searching for the right frequency to boost on their lower-budget projects, it never quite sounds right. The low-frequency stuff that makes a drum sound punchy often lives just a few hertz lower than some rather muddy energy. Boosting the lows invariably boosts some of the mud.

The solution is to keep the low end power but remove the mud. Search at narrow bandwidth for the ugliest, muddiest component of the drum sound (likely somewhere between about 180 and maybe as high as 400 Hz) and *cut* it. As this problematic frequency is attenuated, listen to the low end. Often this approach reveals plenty of low-end power and punchiness that just was not audible before the well-placed cut was applied. This application highlights low-frequency elements of the sound by applying a strategic EQ cut at a frequency range just above it.

Removal of a narrow band somewhere within this often undesirable 200- to 400-Hz "mud-range" must be done carefully. Cutting too wide an area or attenuating too deeply can rob the instrument of fullness. This effect must be used sparingly as well. Applying it to all tracks leaves the entire production thin and powerless. This spectral region must have some energy. Save it for the vocals, guitars, strings, and/or keys. But pull it out of some of the drums (e.g., kick drum and tom-toms).

5.3.3 FEATURE

A natural application of equalization is to enhance a particular part of a sound in order to bring out the positive spectral components of the sound.

Readers are encouraged to build their own templates but here are some ideas and starting points:

- *Voice.* It might be fair to think of it as sustained vowels and transient consonants. The vowels happen at lower mid frequencies (200–1,000 Hz) and the consonants happen at the upper mid frequencies (2 kHz on up). Want a richer tone to the voice? Manipulate the vowel range. Having trouble understanding the words? Enhance a bit of the consonant range. Watch out for overly sizzling "S" sounds, but do not be afraid to emphasize some of the human expressiveness of the singer taking a big breath right before a screaming chorus.
- *Snare.* It is a burst of noise (see Chapter 13). This one is tough to EQ as it reacts to almost any spectral change. A starting point might be to divide the sound into two parts. One is the low-frequency energy coming from the drum itself. Second is the mid- to high-frequency energy up to 10 kHz and beyond due to the rattling snares underneath. Narrow the possibilities by looking for power in the drum-based lows and exciting, raucous emotion in the noisy snares. Vocals and guitars and pianos and reeds are going to want their own various upper-midrange areas to themselves. The snare, with much to offer in the mids, welcomes a complementary set of gentle cuts to make room for these instruments, balanced by a gentle boost in any midrange area that remains so that it can express itself.
- *Kick drum.* Like the snare, consider reducing this instrument to two components. There is the click of the beater hitting the drum followed by the low-frequency pulse of the ringing drum. The attack lives broadly up in the 3-kHz range. The tone is down around 80 Hz and below. These are two good targets for tailoring a kick sound.
- *Acoustic guitar.* Try separating it into its musical tone and its mechanical sounds. Listen carefully to the tone as you seek frequencies to highlight. Frustratingly, this covers quite a range from lows (100 Hz and below) to highs (10 kHz and above). In parallel, consider its more peculiar noises that may need emphasis or suppression: finger squeaks, fret buzz, pick noise, and the percussive sound of the box of the instrument itself that resonates with every aggressive strum. Look for these frequency landmarks in every acoustic guitar you record and mix. EQ is a powerful way to gain control of the various elements of this challenging instrument. Manipulate the relative loudness of these elements as part of shaping the tone and character of the acoustic guitar.

For the instruments an engineer plays and records often, it is wise to spend some time examining their sounds with an equalizer. Look for defining characteristics of instruments and their frequency ranges. Also look for the less-desirable noises some instruments make and file those away on a "watch-out" list. These mental summaries of the spectral qualities of some key instruments will save the recording engineer time in the heat of a session when the client wants more punch in the snare (aim low) and more breathiness in the vocal (aim high).

5.3.4 FIT

A key reason to equalize tracks in multitrack production is to help fit all these different tracks together. One of the simplest ways to bring clarity to a component of a crowded mix is to get everything else out of the way, spectrally. EQ is the tool that lets the engineer directly challenge spectral masking (see Chapter 3). Wanting listeners to be able to hear the acoustic guitar while the string pad is sustaining, an engineer might find a satisfyingly present midrange boost for the guitar and perform a complementary cut in the mids of the pad. This EQ cut on the string pad keeps the sound from competing with or drowning out the acoustic guitar. The engineer pieces together a spectral jigsaw puzzle. The trick is to find a spectral range that highlights the good qualities of the guitar without doing significant damage to the tone of the synth patch. It will take some trial and error to get it just right, but this approach allows clever recording engineers to layer several details into a mix.

This EQ strategy often motivates engineers to simply narrow the frequency range of some tracks. Consider a crowded mix, with 48 or more tracks of audio. With so much going on, it is difficult to create a mix in which the various tracks are not competing with each other and undermining each other, spectrally. There is room for some aggressive equalization. Maybe a piano track is part of a particularly full part of the musical arrangement, when guitars, strings, synths, and background vocals have all joined in. When this sort of crescendo happens in a pop production, the piano might be subjected to severe EQ. A low-frequency shelf might be engaged to cut by 12 dB or more everything from about 200 Hz and below. Solo the piano track and it likely has a low-quality, cheap, and thin sound. Listen to the piano in the context of the mix, however, when the drums, bass, guitars, strings, synths, vocals, etc. are all playing along, and the piano sounds, well, like a piano. With so much else happening, the listener's experience seems to fill in or assume the low-frequency energy is there.

So fitting tracks together with complementary boosts and cuts can happen at even a crude level. It is not always a complicated jigsaw puzzle of spectral

line into a digital delay that repeats at the rate of a quarter note triplet: "Gouda . . . Gouda . . . Gouda." For maximum effect, it is traditional to equalize the signal as it is fed back to the delay for each repetition. The first "GOUDA!" is simply a delay. It then goes through a low-pass filter for removal of some high-frequency energy and is fed back through the delay. It is delayed again: "Gouda!" It goes again through the same low-pass filter for still more high-frequency attenuation and back through the same delay: "gouda." The result is (with a triplet feel): "GOUDA! . . . Gouda! . . . gouda." With some help from the high-frequency roll-off, the echoes seem to grow more distant, adding depth to the mix.

Obviously, this EQ approach applies to signals other than echoes, and it even works on nondairy products. In composing the stereo or surround image of the mix, one not only pans things into their horizontal position, but the thoughtful engineer pushes them back, away from the listener by adding a touch more reverb (obvious) and removing a bit of high end (not so obvious). This EQ move is the sort of subtle detail that helps make the stereo/surround image that much more compelling.

Speaking of stereo, a boring old monophonic track can be made more interesting and more stereolike through the use of equalization. What is a *stereo* signal after all? It is difficult to answer such an interesting question without writing a book, or least an entire chapter, dedicated to the topic. But the one-sentence answer is: A stereo sound is the result of sending different, but related, signals to each loudspeaker. Placing two microphones on a piano and sending one microphone output left and the other right is a clear example of stereo. The sounds coming out of the loudspeakers are similar in that they are each recordings of the same performance on the same piano happening at the same time. But there are subtle (and sometimes radical) differences between the sounds at each microphone due to their particular location, orientation, and type of microphone. The result is an audio image of a piano that is more interesting, and hopefully more musical, than the monophonic, single microphone approach would have been.

If an engineer is presented with just a single microphone recording of a piano and wishes to create a wider, more realistic, or just plain weird piano sound in the mix, one tool at his disposal is equalization. Split the single track onto two different channels on the mixer or digital audio workstation and EQ them differently. If the signal on the left is made brighter than the same signal sent right, then the perceived image will shift slightly to the left, brighter side (remember distance removes high frequencies). Consider EQ differences between left and right that are more elaborate and involve several different sets of cuts and boosts so that neither side is exactly brighter than the other — just *different*. Then the image will widen without shifting one way or the other. The piano becomes more unusual (remember, this section of the chapter is called "Special Effects," so anything goes); its image is more liquid, less precise. Add some delays, reverbs, and other processing (see Chapters 9, 10, and 11) and a single microphone, monophonic image takes on a rich, stereophonic life.

5.3.6 AUDIO EAR TRAINING

Of all recording studio signal processors, no effect will benefit more from audio ear training than equalization. Any engineer's ability to use EQ, knowing when and when not to use it, and which specific frequency regions to manipulate in order to achieve a desired effect, requires intellectual mastery of the innate sound qualities of the entire audible spectrum, from about 20 Hz to 20,000 Hz. Audio schools and tonmeister programs almost universally include courses on critical listening, and a detailed aural study of the frequency domain is a fundamental step toward mastery of audio engineering.

If academic studies are not part of your plan for developing recording skills, you can take advantage of the self-guided audio ear-training tools available. You can also develop work habits that fast forward your spectral education.

The talented audio engineers who created the touchstone recordings at the birth of high-fidelity pop music in the 1950s and 1960s captured and created fantastic sounds without going to school or studying audio ear training. How did they do it? More importantly, how can we do it too? Dedication and repetition.

These engineers had careers in the field. It was not a part-time job. It was not an easy job. Long hours, day after day, was the work ethic. Through sheer brute force and round-the-clock immersion, engineers developed an innate ability to identify all regions of the audio spectrum reliably.

It helps also to work in the same studio, with the same gear. If every session happens on the same mixing console, with the same equalizers on every channel, one stops looking at the EQ knobs and just dials in the desired settings. When using the same console EQ every day, across all kinds of tracks, from acoustic guitar to zither, engineers will start to "feel" the frequency ranges. Peak back at the knobs at the end of each session and register the actual center frequencies used. Over time one can intellectually connect what was heard to be correct sonically to the spectral numbers that must be internalized.

The challenging and sometimes subtle art of equalization need not be surrounded in mystery. Whenever one encounters a track with a spectral problem to be removed or a frequency feature to be emphasized, grab it with EQ. If a mix is getting crowded with too many instruments fighting for too little frequency real estate, carve out different spectral regions for the competing instruments using EQ. Sometimes an engineer just wants to radically alter a sound and make it more interesting. EQ is the technical, musical, and creative tool that accomplishes these things in the studio.

Compression and Limiting

6

"A connecting principle,
Linked to the invisible
Almost imperceptible
Something inexpressible.
Science insusceptible
Logic so inflexible
Causally connectible
Yet nothing is invincible."

— "SYNCHRONICITY I," THE POLICE, *SYNCHRONICITY* (A&M RECORDS, 1983)

The typical recording studio has several — sometimes upwards of 100 — faders. The simple fader, nothing more than an adjustable gain stage, is essential to music production.

There are times when a fader does not satisfy an engineer's need for gain control. Music signals are rarely consistent in level. Every crack of the snare, syllable of the vocal, and strum of the guitar produces a signal that surges up and recedes down in amplitude. Figure 6.1 shows the changing amplitude during about one bar of music. Signals like this one must fit within the amplitude constraints of the entire audio chain (see the discussion of "Dynamic Range" in Chapter 1) without damage: the microphone, microphone preamp, console, outboard gear, multitrack recorder, two-track master recorder, power amp, and loudspeakers. The highest peak must get through these devices without clipping, while the detail of the lowest, nearly silent bits of music must pass through without being swamped by noise. When the recording engineer aims for 0 VU on the meters, they are targeting an amplitude that just avoids distortion when the music peaks, yet is not lost in noise when the signal relaxes.

With agile fingers and intense concentration, one could constantly adjust the faders, in reaction to every snare hit, vocal phrase, and guitar flourish. Better yet, let a machine do it. To help narrow the sometimes extremely

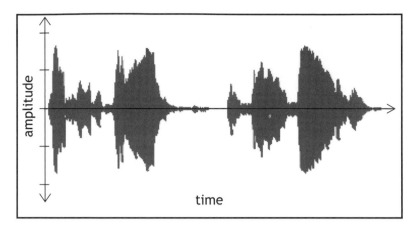

▲ *Figure 6.1 The range of amplitude in two bars of music.*

wide dynamic range of typical audio signals so that they better fit within the amplitude limits imposed by the studio, engineers reach for a compressor — a sort of automatic, semi-intelligent fader. Less intuitively, perhaps, the compressor is also the basis for a wealth of other effects having very little to do with the control of dynamics.

This chapter reviews the common parameters found on compressors/ limiters, summarizes the core technologies used, and studies in depth the various production techniques — both practical and creative — built on this device.

6.1 Parameters

What is the compressor's task? Quite simply, when a signal gets too loud, it turns it down. What counts as too loud? How much should it turn it down? How quickly? For how long? Some compressor controls are needed.

6.1.1 THRESHOLD

A line is drawn separating too loud from not too loud using the *threshold* control. The threshold setting on the compressor sets the amplitude above which compression is to occur, and below which compression is to stop. As long as the signal remains below this threshold, the compressor is not triggered into action. When the signal exceeds the threshold however, the compressor begins to attenuate the signal, like a fader automatically pulled down. Once the compressor is attenuating a signal, the signal must fall back below this threshold before the compressor will stop attenuating and return to unity gain.

6.1.2 RATIO

When the audio is above this threshold, the compressor begins to attenuate. The amount of compression is primarily determined by the *ratio* setting. Mathematically, the ratio compares the amount of the input signal above the threshold to the amount of the attenuated output above the threshold. For example, a 4 : 1 ratio describes a situation in which the input level above the threshold is to be four times higher than the output above the threshold: 4 dB above threshold in becomes 1 dB above threshold out, and 8 dB above threshold in becomes 2 dB above threshold out. A ratio of X : 1 sets the compressor so that the input must exceed the threshold by X dB in order for the output to achieve a level just 1 dB above the threshold.

It is important to note that the ratio only applies to the portion of the signal above the threshold. When the input is below the threshold, the compressor is not applying this ratio to the signal. It is in the business of not compressing signals below the threshold. Only that part of the input that exceeds the user-defined threshold is multiplied by the compression ratio.

It is worth making the distinction now between a compressor and a limiter. The websites, catalogs, and showrooms selling compressors often tout them as compressors/limiters. This is basic marketing; one device seems to have two functions. Yes, dear reader, that is two for the price of one.

Most of this chapter is dedicated to discussing the production potential of the compressor/limiter. Readers of this chapter will find it easy to tap into more than a dozen different effects, all created by this one effects device. The sonic output is very much a function of all of the settings dialed in by the engineer. The compressor/limiter in the hands of a talented engineer is much, much more than two effects for the price of one.

The defining parameter separating compression from limiting is the ratio setting. A ratio below 10 : 1 indicates compression. A ratio above 10 : 1 makes the effect limiting. It is simply a matter of degree. For very tight control of the amplitude of a signal, *limit* the dynamic range with a high ratio. For less abrupt modification of the amplitude of a signal, gently *compress* the dynamic range with a low ratio.

As the principle technology for compression and limiting is the same, most compressors are also limiters, and most limiters are also compressors. Therefore, most faceplates and brochures declare the device both: a compressor/limiter.

Switch on a light, and there is a time element to how long it takes the light to achieve full illumination. Any good commuter has noticed the difference between light-emitting diodes and incandescent bulbs in the stoplights and taillights around them. LEDs snap on and off all but instantly. Incandescent bulbs turn on more slowly, and linger briefly with a little glow even after they are turned off. Luminescent panels exhibit a memory effect where, if they have been bright for a period of time, they remain illuminated even after the illuminating voltage has been removed.

The particular type of photovoltaic cell similarly contributes to the compressor's attack and release characteristics. A compressor's attack and release characteristics are ultimately determined by the complex interaction of the light source and light receiver chosen by the equipment designers.

6.2.3 VCA COMPRESSORS

Solid-state compressors might leverage a field effect transistor (FET) or other transistor-based voltage-controlled amplifier (VCA) to change the gain of the compressor. The Urei (now Universal Audio) 1176LN compressor is a particularly famous use of an FET gain stage. The dbx compressors (old and new) take advantage of high-quality audio VCAs for compression. Using transistors, the detector circuit is empowered to drive the gain via a control voltage.

6.2.4 DIGITAL COMPRESSORS

Compressors in the digital domain are liberated from the constraints of real-time analog circuit components. The qualities of digital compression are as varied as the companies that design them. When audio becomes a string of numbers, digital compressors become a bank of calculations.

One might be tempted to conclude — numbers being numbers — that all digital compressors sound alike. This simply is not the case. Humans write the code that governs the behavior of a digital compressor. There is no right answer, no single solution to compression.

A digital compressor represents one algorithm for achieving compression through calculations, an algorithm that is the result of countless decisions, trade-offs, and moments of inspiration by the creative software engineers who write the code.

Software might be written with the expressed goal of trying to emulate old analog compressors. With so many desirable analog compressors, it makes sense to attempt to capture some qualities of tube, optical, or solid-state compressors digitally. Another valid approach is to create entirely new algorithms that essentially reinvent the compressor-leveraging opportunities not achievable through analog topologies.

6.3 Nominal Application: Dynamic Range Reduction

The compressor is used to create a range of effects, each pursuing different goals artistically, through different means technically. The most intuitive use of compression is to reduce, sometimes slightly other times radically, the audio dynamic range of a signal.

6.3.1 PREVENT OVERLOAD

When singers get confident and excited, they may start to sing the choruses very loud — louder than during all the other takes in rehearsal. Uh oh. If this great performance overloads the analog-to-digital converters or the analog tape machine, the track is unusable. Without some amplitude protection, a killer take is easily lost to distortion. Be ready for this with some gentle (around 4 : 1 or less) compression across the vocal. In this way, the equipment will have no trouble accommodating the adrenaline-induced increase in amplitude that comes from musicians when they are "in the zone."

During the course of a song, some snare hits are harder than others. The slamming that goes on during the chorus might be substantially louder than the delicate, ghost-note-filled snare work of the bridge. A limiter is employed to attenuate the extreme peaks and prevent unwanted distortion. A limiter is nothing more than a compressor taken out to rather extreme settings: The threshold is high so that it only affects the peaks, leaving the rest of the music untouched; the ratio is high, greater than 10 : 1, so that any signal that breaks above the threshold is severely attenuated; and the attack is very fast so that nothing gets through, even briefly, without limiting.

Figure 6.3 gives an example of peak limiting, the sort of processing used to prevent distortion and protect equipment. Fitting a signal on tape without overloading or broadcasting a signal without overmodulating requires that the signal never exceed some specified amplitude. Limiters are inserted to ensure these amplitude limits are honored. In live sound, exceeding the

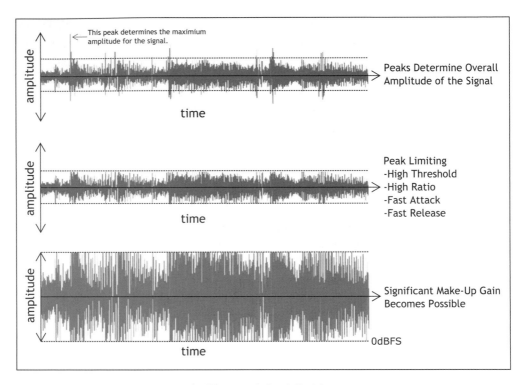

▲ *Figure 6.3 Peak limiting.*

amplitude capability of the sound reinforcement system can cause feedback, damage loudspeakers, and turn happy crowds into hostile ones. Limiters offer the solution again. They guard the equipment and listeners downstream by confining the signal amplitude to safer levels.

Analog-to-digital converters, radio transmitters, and tape recorders can be particularly unforgiving of being overdriven. Compressors and limiters are a regular process to add to the signal chain, giving the engineer a protective buffer before troublesome distortion ruins the music.

6.3.2 OVERCOME NOISE

Frequently music recordings must consciously fight the noise floor of the storage medium (e.g., analog tape) or the noise floor of the listening environment (e.g., an automobile). Compression offers much needed help, narrowing the higher amplitude portions of a signal to a level that is a little nearer the lower amplitude parts. Specifically, with the high-amplitude parts attenuated some, the engineer can turn the overall level of the signal up some without subsequent risk of overload distortion. This makes the softest parts of the performance a little louder, lifting them up out of the noise floor.

Some engineering finesse is required, as this application of compression generally does not intend to change the *musical dynamics* of the signal. The dynamics intended by the composer and the performer should not be undermined. Overcoming noise with compression requires the *audio dynamics* to be narrowed without harm to the musicality of the piece.

6.3.3 INCREASE PERCEIVED LOUDNESS

Compression is inserted into perfectly fine tracks in order to make them louder. A handy side effect of compressing — reducing the overall dynamic range of the signal — is that now it can be turned up. That is, one can make the track louder as a whole if the more extreme swings in amplitude have been compressed. Figure 6.3 demonstrates this sort of compression. It is counterintuitive at first, but this gain reduction device is used to make a track louder.

With the peaks attenuated, the overall signal can then be raised in level (see the lower part of Figure 6.3). The result, when desired, is an increase in the overall loudness of the signal. The average amplitude is higher, the area under the amplitude curve is increased, and the perceived loudness is raised.

There is no doubt that this effect is overused in contemporary pop-music productions at the beginning of the twenty-first century. The reader is encouraged to use mature musical judgment to find the appropriate amount of peak limiting, creating productions that bear repeated listening by passionate fans without fatigue, yet sound exciting and competitive in the market.

Maximizing loudness is often taken to radical extremes where mixes are absolutely crushed (i.e., *really* compressed; see also squashed, smooshed, squished, thwacked, etc.) by compression so that the apparent loudness of the song exceeds the loudness of all the other songs on the radio dial or shuffled playlist. Selling music recordings is a competitive business. Loudness does seem to help music sales figures, at least in the short term. And so it goes. Often the music suffers in this commitment to loudness and hope for sales. Artist, producer, and engineer must make this trade-off carefully: long-term musical value versus the short-term thrill of loudness. One must not forget that the consumer has the ability to adjust the level of each and every song they hear by operating the volume knob on their playback system. If they want the song louder, than can make it so. Even in small measures, a little bit of gentle compression buys the tune a little bit of loudness.

6.3.4 IMPROVE INTELLIGIBILITY AND ARTICULATION

There is real artistry and technique in singing and speaking for recording and broadcast. That singers need vocal lessons and practice is accepted. Voiceover artists do too. Part of what is studied and rehearsed is diction. It is essential in almost all styles of music and forms of speaking that each and every word be plainly understood. The intelligibility of words, spoken or sung, depends in part on the ability to accurately hear consonants. Vocal artists are trained to control their vowels and make distinct each and every consonant. Pronouncing a word, the letter B must not be confused with the letter D; N's and M's must be differentiated.

Imagine looking at a meter on the mixing console while a singer sings "bog" and "dog." The meters would reveal no visible difference between these two words. Imagine looking at the waveform of each word on the screen of your digital audio workstation. They appear, to the naked eye, incredibly similar. There is precious little difference between these two words, differences that aren't easily measured or confirmed by any device available in the typical recording studio — except our hearing. The incredibly subtle difference is extracted from the signal our ears pick up through the work done by our hearing physiology and the analysis performed by our brain.

Add in hiss, hum, static, and other noises, and distinguishing "bog" from "dog" becomes more difficult. Add drums, guitars, and didgeridoo to the mix, and the intelligibility is further threatened. Add delay, reverb, and other effects, and hearing detail in the vocal may become more challenging still.

Compression can help. Adjusting the gain of the vocal, syllable by syllable and word by word, the engineer is able to make sure that no important part of a vocal performance gets swallowed by competing sounds. Set the threshold low enough that it is below the average amplitude of the vocal. Reach for a fast attack time and medium to fast release time, and raise the ratio as needed, possibly as high as 10 : 1. The effect is not usually transparent. The vocal is clearly being processed. Done well, though, it can be worth it. Intelligibility rises.

Compression is used often, particularly on the dialog portion of film and television soundtracks, to tame the loudness inconsistencies of the vocal performance so that no portion of the track is difficult to hear. Every syllable of every word is tamed in amplitude by the compressor so that the intellectual meaning of every syllable of every word is communicated clearly.

The articulation of each and every note by a melodic instrument may also carry significant musical value, and yet be drowned out by other elements of a crowded mix. The intelligibility of a horn line or a guitar part can be every bit as important as the intelligibility of the lyrics. Many styles of music, from jazz to progressive rock, make their artistic living based on communicating the exact phrases — the timing and the pitch — of sometimes several instruments at once. Careful control of dynamics through compression may be, at times, a required part of the production.

As with so many effects, tread carefully. This compressor move alters the musical dynamics of the track processed. If the sax player wanted some notes to be softer, with a more impressionistic and expressively vague form of articulation, this kind of compressor effect will be unwanted. Apply only as needed, where artistically appropriate.

6.3.5 SMOOTH PERFORMANCE

When the guitarist gets nervous, she leans to and fro on the guitar stool, leaving the engineer with the challenge of a static microphone aimed at a moving target. A compelling performer, nervous in the studio, still deserves to be recorded. Without the constant gain riding of a compressor, listeners will hear the guitarist moving on and off the microphone. A little gentle compression might just coax a usable recording out of an inexperienced studio performer.

Beware of the vintage guitar. When the bass player pulls out that wonderful, old, collectible, valuable, sweet-sounding, could sure use a little cleaning up, are those the original strings, could not stay in tune for eight bars if you paid it, gorgeous beast of an instrument, we can be sure that — even in the hands of a master — some strings are consistently a little louder than other strings. The instrument is not balanced, a sad fact that might be revealed by abrupt level changes in the bass line being played. Of course, one solution is compression. Careful, note-by-note, precision amplitude adjustments must be made to the signal or the very foundation of the song becomes shaky.

The equal loudness contours (see Chapter 3) highlight the human sensitivity to bass levels. When full-bandwidth signals increase in level, the perceptual result at low frequencies is exaggerated — the lows sound even louder. The perceptually volatile low-frequency portion of our music productions requires particular focus on levels. On bass, audio engineers almost always need the careful level adjustments of gentle compression.

When a drummer's foot gets (understandably) tired during what may be several takes of the same song, the kick drum performance can become a little ragged. Some individual kicks are noticeably louder than others. Use compression to make the performance more consistent.

It is important to note that smoothing any performance with compression is a signal processing "fix" that would never be needed in an ideal world. That is, it is always best to fix these issues at the time of performance, if at all possible. Teach the singer microphone techniques in which they back off the microphone for louder parts of the performance, and lean in slightly on the quieter portions. Find a way to get the guitarist to remain in a relatively steady position by the microphones. After all, the quality, not just the level of the sound, is changed when the acoustic guitar moves toward and away from the microphone.

Find a bass player who is so sensitive to their instrument and performance that they adjust note by note, string by string, for level differences in their playing technique. Great players can create a performance with musically-expressive dynamics across the entire range of their instrument.

If the drummer is getting tired, coordinate with the producer to have strategically-timed breaks so that a consistent performance is recorded to the multitrack and need not be chased with smoothing compression. There is no doubt that a kick drum hit hard has a different spectral content than a kick drum hit softly. Having a compressor match levels of various kicks will not lead to a believable timbre.

6.4 Advanced Applications

When the answering machine was invented, its intended purpose was to answer the phone and take messages while we were out. But the day after the first one was sold, the answering machine took on a new, more important role: call screening. The most common message on these devices is something like, "It's me. Pick up. Pick up!" The use of a device in ways not originally intended occurs all too often, and the compressor offers a case in point. While dynamic range reduction and peak limiting are effective, intended uses for the device, recording engineers use them for other, less obvious, more creative reasons as well.

6.4.1 ALTERING AMPLITUDE ENVELOPE

Compression can be used to change the amplitude envelope of the sound. The envelope describes the "shape" of the sound, how gradually or abruptly the sound begins and ends, and what happens in between. Envelope is a step back from the fine-level, cycle-by-cycle detail of the waveform, looking instead at the more general signal amplitude fluctuations as in Figure 6.3. Drums, for example, have a sharp attack and nearly instant decay. That is, the envelope resembles a spike or impulse. Synth pads, on the other hand, might ooze in and out of the mix, a gentle envelope on both the attack and decay side. Piano offers a combination of the two. Its unique envelope begins with a distinct, sharp attack and rings through a gently changing, slowly decaying sustain. All instruments offer their own unique envelope.

Consider the sonic differences among several instruments even as they play the same pitch, say A440: piano, trumpet, voice, guitar, violin, and didgeridoo. There are obvious differences in the spectral content of these instruments; they have a different tone, even as they play the same note. But at least as important, each of these instruments begins and ends the note with its own characteristic envelope — its amplitude signature.

The compressor is the tool used to modify the envelope of a sound. A low threshold, medium attack, high-ratio setting can sharpen the attack. The sound begins at an amplitude above threshold (set low). An instant later (medium attack), the compressor leaps into action and yanks the amplitude of the signal down (high ratio). Such compression audibly alters the shape of the beginning of the sound, giving it a more pronounced attack.

This approach can be applied to almost any track. A good starting point for this sort of work is a snare drum sound. It is demonstrated in Figure 6.4. Be sure the attack is not too fast or the compressor is likely to remove the sharpness of the snare entirely. Set the ratio to at least 4 : 1, and gradually pull the threshold down. This type of compression has the effect of morphing a spike onto the front of the snare sound. Musical judgment is required to make sure the click of the sharper attack fits with the remaining

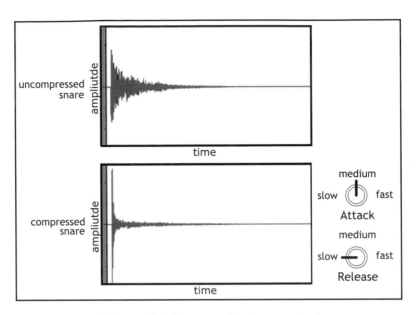

▲ *Figure 6.4 Compress to sharpen attack.*

ring of the snare. Adjusting the attack time while trading off a low threshold with a high ratio offers the engineer precise control over the shape of the more aggressive attack. This compressor effect works well on any percussive sound from congas to pianos. It can also sharpen the amplitude onset of less percussive tracks, from the saxophone to the electric bass. Done well, this creates a more exciting sound that finds it place in a crowded mix more easily.

Another unusual effect can be created using the compressor's release parameter. A fast release pulls up the amplitude of the sound even as it decays. This is shown in the snare example of Figure 6.5. Notice the raised amplitude and increased duration in the decay portion of the waveform. Dial in a fast enough release time, and the compressor can raise the volume of the sound (i.e., uncompress) almost as quickly as it decays. This increases the audible duration of the snare sound, making it easier to perceive each drum hit in and among the distorted guitars and ear-tingling reverberation (see "Masking Reduction" in Chapter 3). Altering the decay of instruments can be taken to radical extremes. Applied to piano, guitar, and cymbals, the instruments can be coaxed into a nearly infinite sustain, converting them into bell or chimelike instruments in character, while still retaining the unmistakable sound of the original instrument. File under "Special Effects." It can be just the sort of unnatural effect a pop tune needs to get noticed.

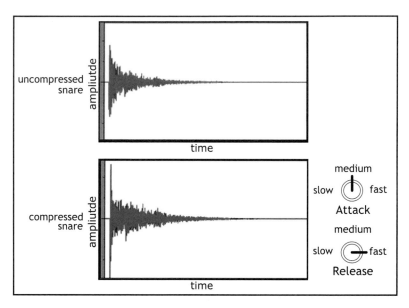

▲ *Figure 6.5 Compress to lengthen sustain (attack remains sharp).*

6.4.2 AMBIENCE AND ARTIFACTS

A coordinated adjustment of compressor parameters (threshold, ratio, attack, and release) enables the audio engineer to manipulate, however indirectly, the shape of the amplitude envelope toward still other goals.

Consider some extreme (i.e., high ratio, low threshold) compression with a very fast release time. If the compressor pulls down the peaks of the waveform and then quickly releases the signal as it decays, listeners may be able to hear parts of the sound that were previously inaudible. Through fast-release compression, the engineer effectively turns up the later parts of the sound, revealing more of the decay of a snare, the expressive breaths between the words of a vocal, the ambience of the room in between drum hits, the delicate detail at the end of a sax note, and so on. Clues about space, emotion, and performance intensity can be brought forward through compression for a powerful musical result.

6.4.3 DE-ESSING

Pop music standards push productions to have bright, airy, in-your-face, exciting vocal tracks. This vocal sound must rise above a wall of distorted guitars, tortured cymbals, shimmering reverb, and sizzling synth patches. Needless to say, it is common to push vocals with a high dose of high-frequency hype (through EQ, see Chapter 5). Engineers can get away with this aggressive equalization move except for those instances when the vocal was already bright to begin with: during hard consonants like S and F (and, depending on the vocalist and the language, even Z, X, T, D, K, Ch, Sh, Th, and others). These sounds are naturally rich in high-frequency content. Run them through the equalizer that adds still more high-frequency energy and the vocal zaps the listeners' ears with pain on every S. It is unacceptable when everyone in the room blinks each time the singer hits an S.

It is a tricky problem. Tracks on the radio have wonderfully bright and detailed vocals. When an audio engineer tries this for the first time in their studio, it comes with a price. While much of the track may be improved, the S's are too loud, so loud it hurts. Clever compression will solve this problem.

In all of the compressor applications discussed so far, the settings of threshold, ratio, attack, and release have been based on the signal being compressed. What if the compressor adjusted the gain of one signal while "looking at" another?

Specifically, consider compression of the lead vocal. But instead of compressing it based on the vocal track itself, use a different signal to govern the compression. Called a *side chain*, a modified vocal signal is fed into this alternative input direct to the level detector (Figure 6.6). The vocal itself is what gets compressed, but the behavior of the compressor — when, how much, how fast, and how long to compress — is governed by the side chain signal. To get rid of S's, we feed a signal into the side chain that has enhanced S's. That is, the side chain input is the vocal track equalized so as to bring out the S's, and deemphasize the rest.

This S-emphasized version of the vocal never makes it into the mix. Listeners do not hear this signal. The compressor does. When the singer sings an S, it goes into the compressor's level detector loud and clear, breaking the threshold and triggering the compressor into action. The side chain signal is the vocal with a high-frequency boost, maybe a +12 dB high-frequency shelf somewhere around 4–8 kHz, wherever the particularly painful consonant lives for that singer. The side chain vocal can be cut at other frequency ranges; the lows are not needed. The compressor is set with a mid to high ratio, fast attack, and fast release. The threshold is adjusted so that the compressor operates on the loud S's only, and ignores the side chain vocal the rest of the time.

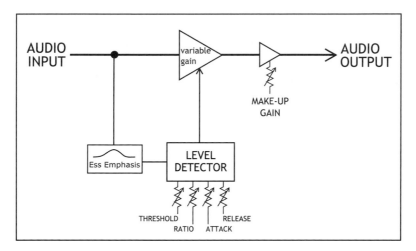

▲ *Figure 6.6 Signal flow of de-esser.*

The result is a de-essed (as in, no more letter S) vocal. The compressor instantaneously turns down the great sounding vocal every time it sings an S. In between the S's, the compressor does not effect the level of the vocal. This vocal can be made edgy and bright without fear: use EQ to boost the highs, knowing the S's will be reliably pulled down in level at just the right instant and by just the right amount using a de-esser.

6.4.4 DISTORTION

It is generally most accurate to think of compression acting on the amplitude envelope of the signal, not the fine, cycle-by-cycle detail of the waveform. Picture the waveform when the digital audio workstation is zoomed out looking at a few bars of music, not zoomed in looking at a few individual cycles of an oscillation. All of the compression effects discussed above happen at the envelope level. But sometimes compression is made to attack and release so quickly that it chases individual peaks and valleys of a waveform (Figure 6.7).

With a threshold set below the peak amplitude of a sine wave, and an attack and release setting faster than a fraction of the period of the sine wave, the compressor will attenuate during each peak (positive or negative) and uncompress in between the peaks. The output from the compressor is clearly no longer a sine wave. If the ratio is high enough (as was used in the creation of Figure 6.7), the sine wave starts to turn into a square wave possessing the same fundamental frequency as the uncompressed sine

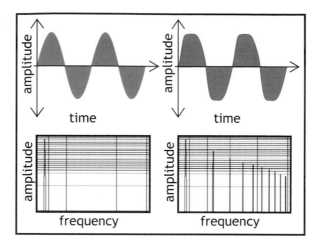

▲ *Figure 6.7 Distortion through compression.*

wave, but containing all the additional harmonics known to exist in a square wave (see "Complex Waves" in Chapter 1). In this application, compression is used to generate related harmonics, hopefully creating a pleasing kind of harmonic distortion.

Engineers apply a strong dose of compression — to an individual track or the entire mix — for the well-loved effect of distortion. There is something visceral and stimulating about the sound of distortion that makes the music more exciting. The distortion typically dialed in on most electric guitar amps adds an unmistakable, instinctively adrenalizing effect. Compression, with settings that deliberately modify the detailed shape of the waveform, creates a kind of distortion; it seems to communicate an intense, on-the-edge, pushing-the-limits sort of feeling.

A profoundly effective example of this is Tom Lord-Alge's mix of "One Headlight" by the Wallflowers. At each chorus, there is a compelling amount of energy. It feels right. Listen analytically, not emotionally, and one finds that there is no significant change in the arrangement. The drummer does not start swatting every cymbal in sight. No wall of additional distorted guitars comes flying in. Jakob Dylan's voice is certainly raised, but it is well short of a scream. Mostly the whole mix just gets squashed (compression has many nicknames), big time. A critical listen suggests that parts of the song actually get a little quieter at each chorus, with the compression across the lead vocal and the entire stereo mix pushing hard. But musically, the chorus soars. That is the sort of compression that sells records.

6.5 Advanced Studies: Attack and Release

Attack and release parameters are frequently misunderstood. Mastering the sonic implications of these parameters can help the engineer tap into the full creative potential of compression.

6.5.1 DEFINITIONS

The concept behind attack time is straightforward: It should describe the length time it takes for the compressor to compress — the time it takes the gain change to reach the necessary level so that the user-defined ratio has been reached. An attempt to plot the gain of the compressor as it goes from its fully uncompressed state to fully compressed reveals the problem (Figure 6.8). The compressor might march directly from uncompressed to compressed (Figure 6.8a). It might begin compression slowly and accelerate into full compression (Figure 6.8b). It might begin attenuation quite quickly and approach the ultimate compressed state asymptotically (Figure 6.8c). It might wait some specified period of time and then snap instantly into compression (Figure 6.8d). Or, it might follow any path, however random, from uncompressed to compressed (Figure 6.8e). Attack time defies a single number description. Release time presents the same issue. The subtle variation in the shape of the gain change can have an audible result. In fact, this is part of what distinguishes one make and model of compressor from another.

While no single number fully describes attack and release parameters, many compressors offer a knob or switch for each, labeled "attack" and "release." Each manufacturer reduces it to a single, adjustable parameter so that audio engineers may interact with this setting. Any audio student's understanding of attack and release will depend in part on this set value. As discussed below, it depends on many other variables. The frustrating conclusion is that the attack and release parameter values are not useful on their own. They must be considered in the context of the other compressor settings, the type of gain change element, and the type of signal being compressed. An attack time setting of 4 on one compressor is not equivalent to an attack time setting of 4 on a different make or model.

Drivers can rely on the speedometer to accurately quantify the speed of the automobile, no matter who made the car, or what year it was manufactured. No such standard exists for attack and release values. They are best understood as indicating a relative range, from fast to slow, for the attack

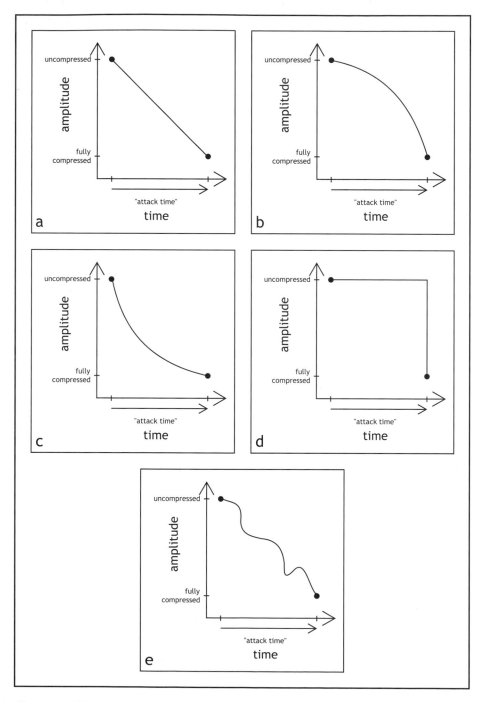

▲ *Figure 6.8 Time-based definition of attack. A single number description is incomplete.*

and release rates. On an absolute scale, they are essentially meaningless to the practicing sound engineer.

6.5.2 VISUALIZING ATTACK AND RELEASE

In order to better understand the all-important attack and release settings, consider the contrived demonstration signal in Figure 6.9. This signal is a burst of pink noise that begins abruptly, sustains briefly, and then decays. It is relatively stable in amplitude during the sustain portion, and the decay is chosen to appear linear. Compression with various attack and release time settings leads to new, and potentially useful, amplitude envelope shapes.

In all of these examples, the threshold is set low so that the sustained portion of this demonstration signal is well above it. The ratio is generally high (6 : 1 or more) so that the compression effect is clearly demonstrated. For the figures that follow, the top plot will always be the original test signal, uncompressed. The middle figure is the signal after compression. The bottom curve shows the gain of the compressor in reaction to the signal, per the attack and release settings discussed.

As a starting point, consider compression with slow attack and slow release settings (Figure 6.10). Because of the relatively slow attack time setting, it takes a visibly obvious amount of time for the compressor to reach its desired amount of attenuation. The resulting compressed signal now has a modified amplitude envelope, one with something of an initial transient in front of the overall signal.

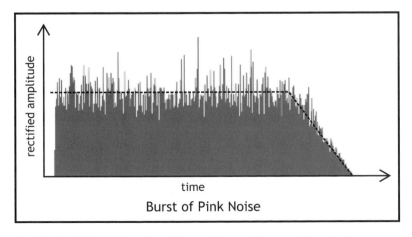

▲ *Figure 6.9 Test signal for understanding attack and release.*

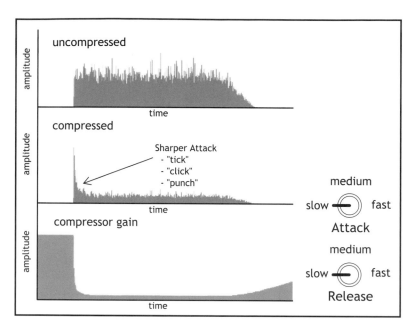

▲ *Figure 6.10 Slow attack and slow release.*

The approach of Figure 6.10 is what was used to sharpen the snare envelope shown in Figure 6.4. The compressor has been used to mold a sharp attack onto the signal, which, depending on the exact setting of attack time, might be perceived as a "tick," "click," or "thump" at the beginning of the sound. Sounds lacking a noticeable attack can be sharpened so that they cut through a crowded mix. Slow the attack time just enough, and that ever-elusive "punch" that rock and roll so often favors — especially on drums and bass — is found. From a high-frequency-oriented "click" to a low-frequency-oriented "punch," attack time is the key parameter engineers use to alter the attack portion of the audio waveform.

Speed the release up to a fast setting (Figure 6.11), and notice that the previously observed modification to the initial part of the waveform is gone. Welcome to the clumsy user interface of the compressor. By modifying the *release* setting (from slow to fast), the audio engineer changes the way the signal responds to the *attack* setting. The fast release instructs the compressor to uncompress as fast as it can. Therefore, it constantly keeps attempting to restore gain every instant the signal falls back below the threshold. Keep in mind this noise signal is a randomly changing waveform with a distribution of energy across the audible spectrum. This waveform swings back and forth through zero amplitude with a frequency ranging from 20–20,000 cycles per second. The fast release setting drives the

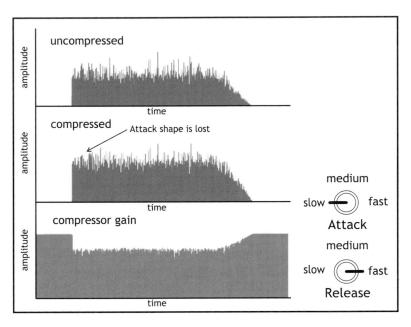

▲ *Figure 6.11 Slow attack and fast release.*

compressor back toward unity gain for that portion of the noise signal that is below the threshold.

The sharpened attack of the waveform relies on a slow release time setting to even exist. The compressor must be slow enough in its release that it stays compressed for the duration of this signal, once the attack time setting has allowed the compressor to reach its specified ratio and achieve full compression.

Consider now compression with fast attack and medium release (Figure 6.12). The attack portion of the amplitude envelope of this test signal after compression is now very much smoothed out. Speeding up the attack time setting can have the effect of removing any rough texture or transient properties of the signal, giving it a smooth, steady, unobtrusive entrance. This approach can be used to dull the attack of a drum, such as the snare drum, for example. Fast attack compression can make an overly angular acoustic guitar that is distracting in the mix sound smooth and more integrated into the mix.

Notice the change in the decay of the test signal, when subjected to this fast attack, medium release compression. The previously straight-line decay has been reshaped. Zooming in on the late portion of the audio waveform

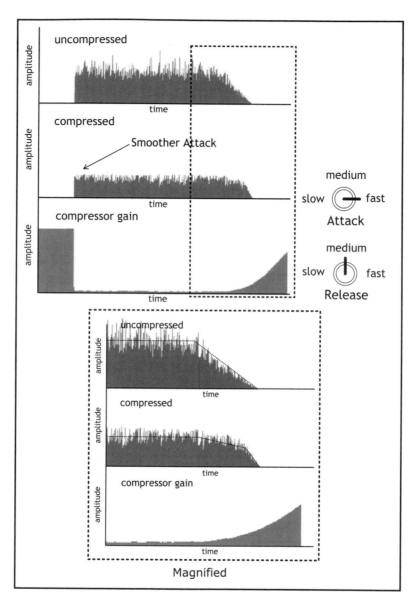

▲ *Figure 6.12 Fast attack and medium release.*

(the lower portion of Figure 6.12) reveals two slopes during the decay. The amplitude envelope remains flat during the sustain of the signal. While decaying, the compressor releases (turns the signal back up, when it can), leading to a shallower slope. Finally, once the signal is fully below the threshold, a still steeper decay slope is observed as the signal reaches its end after being compressed and subjected to some make-up gain. Increase

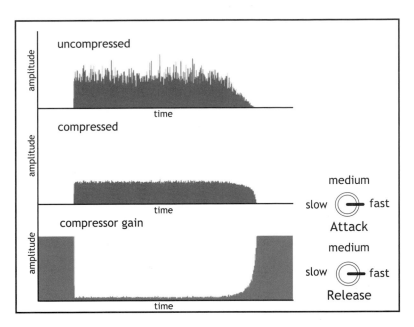

▲ *Figure 6.13 Altering amplitude envelope.*

release to a faster setting and the decay is further modified (Figure 6.13). A linear slope becomes a complex curve.

The release time parameter provides a way (if the attack time is fast enough) to change the decay of the signal. Human hearing is incredibly sensitive to the reshaped amplitude envelope that compression can create. If a flute and a violin play the same note, we still distinguish the flute from the violin in large part because of the amplitude envelope templates we have in mind for these instruments. Modifying the shape of the onset and the Idecay of the envelope allows the engineer to challenge the listener's basic assumptions about timbre. In addition, the release parameter on the compressor enables the recording engineer to extract and make audible some previously difficult to hear low-level portions of the signal — highlighting important details, exaggerating intimacy, and heightening realism.

6.5.3 PARAMETER INTERDEPENDENCE

The parameters of compression (very much *un*like the parameters of a parametric equalizer, Chapter 5) are a rather clumsy set of controls for achieving an automatic reduction in signal amplitude. Engineers must pay careful attention. For example, if the signal is not exceeding the threshold, no adjustment to ratio, attack, or release will have an effect.

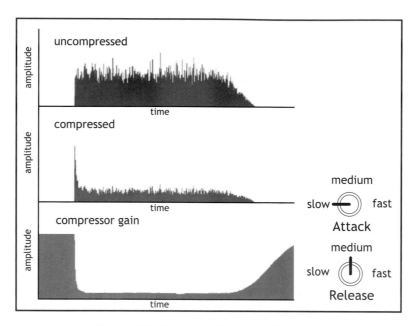

▲ *Figure 6.14 Slow attack and medium release.*

There is a particularly challenging tug of war between the attack time and release time settings. Consider the slow release setting already shown in Figure 6.10. It may be difficult, if not impossible, to hear the result of any small adjustments to release time (compare with Figure 6.14), because the bulk of the release happens after the audio signal has ended. With no signal present, how can one hear a change to the compression effect caused by a modification to the release time? Moreover, if the onset of the next portion of the waveform occurs before the compressor has released sufficiently, any desired reshaping of the onset of the next waveform may not occur or be audible. That is, the effects of any attack time compressor settings may not be revealed because of the current release time settings.

Figure 6.10 demonstrates the sometimes useful sharpening of the onset of a waveform that a slow attack time can create. Figure 6.11 points out that slow attack time does not guaranty this effect will occur. In this case, because the release time is so very fast, the compressor uncompresses fast enough to prevent that sharpening of the waveform. An engineer adjusting compressor attack times in search of the sharp waveform of Figure 6.10 will do so in vain, if the compressor's release time is as fast as that shown in Figure 6.11.

Attack settings may make release adjustments difficult or even impossible to hear. Release settings can obscure or make sonically invisible any adjustments to the attack parameter. Depending on the type of signal present and the particular parameter settings in use, some adjustments will have no effect, or might have an unintended result.

The reader should know that the author has wasted many a late hour in the studio puzzling over compression. It is a difficult effect to hear in the first place, and the sometimes counterintuitive and surprising interaction of compression parameters sneaks up on all engineers.

6.5.4 PROGRAM DEPENDENCE

Just as earnest engineers new to the field start to get their heads around the twisted logic and sonic surprise of compression generally, and attack and release specifically, yet another layer of complexity appears. Most, if not all, compressors have some amount of program-dependent behavior. That is, even as the most effective attack and release values are set up on a compressor, the compressor then has the audacity to flex those attack and release parameters based on the signal being compressed. Program-dependent compression, while making compression a more challenging effect to master, is a feature that generally leads to more musical results.

It is not unusual for compressors to treat transients with faster attack and release times than steady-state signals, even when the audio engineer has not modified the attack or release settings. On some models, changing the ratio modifies the threshold.

Moreover, the attack and release trajectories are sometimes a function of the amount of gain change currently implemented by the compressor. Slow settings as the compressor initiates compression are often desirable. However, when a compressor is in the midst of compressing a signal, already applying say 6–12 dB of attenuation, the attack and release times might start to increase. More intense attenuation begets quicker reaction. In the meantime, the attack and release parameters themselves, as set by the engineer, remain misleadingly fixed on the faceplate of the device.

The attack and release settings set by the user amount to starting points driving the general behavior of the compressor. They rarely represent rigid values strictly followed by the compressor at all times.

6.5.5 PUMPING AND BREATHING

One of the few times compression can be clearly heard, even by untrained listeners, is when it starts *pumping*. Pumping refers to the audible, unnatural level changes associated primarily with the release of a compressor. The audio signal contains material that changes level in unexpected ways. It might be steady-state noise, the sustained wash of a cymbal, or the long decay of any instrument holding a note for several beats or bars. Listeners expect a certain amplitude envelope: The noise should remain steady in level, the cymbals should decay slowly, etc. Instead, the compressor causes the signal to get noticeably louder. This unnatural increase in amplitude occurs as a compressor turns up gain during release.

Of course compressors always turn up the gain during release. Pumping is the generally unwanted, audible artifact of compressor release revealed by the slow or unchanging amplitude of the audio signal being compressed. The fun part is that the release value is not necessarily too slow or too fast. The problem may be because the release parameter on the compressor is too, um, *medium*.

A much faster release time could remove pumping by having the compressor release immediately and unnoticeably after the compression-triggering event (as in Figure 6.11). A snare drum hits. The compressor attenuates. The snare sound ends immediately. The compressor releases instantly. If this all happens quickly on each snare hit, the snare sound itself makes it difficult to impossible to hear the quick releasing action of the compressor (see "Masking" in Chapter 3). The result is a natural sound, despite compression. The release happens so soon after the snare hit and occurs with such a steep release slope that pumping is not easily heard.

More often, a slower release time is a good option for removing unwanted pumping. Slow the release time substantially, so that the level change of the compressor through release takes a couple of seconds or longer, and the gradual increase in gain by the compressor is slower than the gradual reduction in amplitude during the decay of the sound. The result is a natural-sounding decay of the audio signal, even as the compressor slowly turns up the level. The cymbal takes longer to decay, but it still sounds like a naturally-decaying cymbal. Very slow release is another way to achieve compression without pumping.

Beware of pumping whenever a steady-state, near-steady-state, or sustained sound occurs within the signal you wish to compress: cymbals, long piano notes, tape hiss, electronic or acoustic noise floors, synth pads, whole notes, double whole notes, or longer. Signals that have a predictable, slowly changing amplitude envelope can inadvertently reveal compression, forcing the recordist to abandon compression altogether, back-off the degree of the effect, or at least modify the release settings.

Note, on the other hand, that the pumping artifact may, on occasion, be an interesting effect. If each snare hit causes the cymbals to pump, the result can be an interesting, unnatural envelope in which it sounds as if the cymbals have been reversed in time. Used sparingly, such an effect can add great interest to the production.

6.6 Learning Compression — Some Personal Advice

This single device, the compressor, is used for a wide range of very different sounding, difficult-to-hear effects, using counterintuitive parameters that are a function of the type of signal being compressed, the amount of compression occurring, and the type of components or algorithms within the device. Compression is a difficult effect to master.

6.6.1 UNNATURALLY DIFFICULT TO HEAR

There is nothing built in to human hearing that makes it particularly sensitive to the sonic signatures of compression. Humans are generally pretty good at hearing pitch, easily identifying high versus low pitches in music. Humans react naturally to volume. Without practice, any listener with healthy hearing can separate a loud sound from a sea of quieter sounds. Our hearing mechanism can do deeper analysis, for example, identifying the source of a sound and the location and approximate distance of that source, without conscious thought. The hearing system figures out the angle of arrival of sounds by analyzing, among other things, differences in amplitude and differences in time of arrival at the two ears. Our hearing system tells us which way we need to turn our head in order to face the sound source. Human hearing also extracts seemingly subtle (and darn difficult to measure with a meter or oscilloscope) information from a signal, such as the emotional state or personal identity of a sound source. Who is calling out for us? Are they happy, sad, or angry? Identifying pitch, volume, arrival angle, emotion, and identity are all easier than riding a bicycle; humans do it naturally, instinctively, preconsciously, with no deliberate, intellectual effort or training.

This simply is not the case for compression. There is no important event in nature that requires any assessment of whether or not the amplitude of the signal heard has been slightly manipulated by a device. Identifying the audible traits of compression is a fully learned, intellectual process that audio engineers must master. It is more difficult than learning to ride a bicycle. It is possible that most other people, including even musicians and avid music fans, will never notice, nor need to notice, the sonic fingerprint of compression. Recording engineers have this challenge to themselves.

Early in their careers, most engineers make a few compression mistakes. Overcompressing is a common problem. An effect that is difficult to hear must be overdone to become audible. Compression falls easily into that category of effect.

Young engineers will try compressing until it is clearly audible, and then backing off, so as not to overdo it. Not a bad strategy. At the early stages of learning compression it is not unusual to find that sometimes over-compression is not noticeable until the next day. The effect of compression is at times quite subtle and at other times quite obvious. Spending all day mixing one song, with ears wide open, can make it hard to remain objective. A fresh listen to the mix the next day can be an effective education. Try to learn from these mistakes, but cut yourself some slack. It is not easy.

Wait, it gets worse: It is often the goal of the engineer to set up the compressor so that its affect is inaudible. For many types of compression, we wish the effect to be transparent. Engineers carefully adjust threshold, ratio, attack, and release until they do not even notice it working. Do not let anyone tell you otherwise — tweaking a device until it sounds so good that you can not even hear it isn't easy.

6.6.2 HYPE AND HYPERBOLE

You are gonna hear people say, "Hear it, man? It's beautiful. That's definitely the Squishmeister G6vx and it sounds so cool." And you might think, "I don't hear it. What are they talking about? How do they know this? How do they know which compressor it is when I don't even know it's compressed? Should I get out of audio and perhaps become a banker?" Again, some compression is difficult to hear and requires experience. Perhaps they have had the chance to hear this kind of compression before. All you need is time between the speakers immersed in compression of all kinds and you will pick it up. Keep in mind also — and I speak from experience here — that some people in the music business are full of hooey.

6.6.3 IMITATION

Learning to use compressors, and knowing with confidence when to reach for a specific make and model, is a nearly impossible challenge very much like microphone selection.

At first, it makes sense to imitate other engineers whose work we admire. Microphone X on snare drum, microphone Y on female vocals, etc. We take in the vast complexity of the sounds that result over many recording and mixing experiences. Then, when we reach for a different microphone in a familiar application, the sonic contribution of the microphone becomes more obvious.

So too with compressors. We observe successful use of certain compressors for vocals, others for snare drums, and still others for electric bass. As a starting point we can do the same. Once we feel we are getting satisfying sonic results using these specific compressors for these specific applications, we can start to branch out and explore, reaching for different compressors in tried-and-true applications. The sonic fingerprint of the compressor begins to be revealed. Over time, we start to internalize the general sound qualities of every compressor we own, and every make and model of every studio we hire. That is our task as advanced users of compression.

Compression is not a single effect. It is used to prevent overload, overcome noise, increase perceived loudness, improve intelligibility and articulation, smooth a performance, alter amplitude envelope, extract ambience and artifacts, de-ess vocals, and add distortion. In order to correctly learn from the way other engineers use compression, we need to know which effect they sought to achieve. There is no such thing as a vocal compressor or a snare drum compressor. When an engineer reaches for a specific compressor for a specific instrument, they are not demonstrating the universally correct compressor for that instrument. When we learn by watching and listening to others, we need to somehow know, or at least have a good guess, what they were trying to accomplish from a production point of view. It is rarely appropriate for an assistant engineer to interrupt the productive and creative flow of a session and ask this directly. The discreet assistant watches, guesses, experiments, and eventually learns. If you have a good relationship with the seasoned engineer, you might be able to have a discussion on a specific compression strategy in a calm moment after the session. Try to find what they were going for when they patched-up that particular compressor on that particular track at that particular moment. That can be the seed for your eventual success with that kind of effect on that kind of compressor.

6.6.4 MULTIPLE PERSONALITIES

Obtaining a complete understanding of compression grows more complex still. Every compressor with adjustable parameters offers a great range of sound qualities. Compressors do not have a single personality for us to discover and codify. In fact, for every make and model of compressor we wish to skillfully wield, we need to develop an understanding of its unique capability across the adjustable parameters. Memorize each compressor for low, medium, and high ratios, across fast, medium, and slow attack times, using fast, medium, and slow release settings. There is a frustratingly large range of permutations to work through. Experienced engineers have done this already. The reader is encouraged to choose a couple of compressors and develop this kind of deep knowledge for themselves. With experience, you will find this task less daunting. Then as you seek to become proficient with additional compressors, you will assess their behaviors and sonic traits relative to the compressors you already use comfortably.

This is not difficult to do, but it is a slow process. There is no substitute for experience. Any pianist can tell within an eighth note whether the piano is a Steinway, Bösendorfer, Baldwin, Falcone, or Yamaha. Any guitarist can identify a Stratocaster, Telecaster, Les Paul, and Gretsch. Seasoned guitarists can also make good guesses as to the pickups used, the gauge of the strings, and if an alternative tuning is being used. Audio engineers need to develop a similar, instinctive ability to identify specific compressors. The distinctions, indescribable at first, become second nature.

6.7 Achieving Compression Success

Dialing-in the right settings on the right compressor as needed in the course of a production is not easy. Experience is essential, but that is about as fair as requiring experience for a job before you can get a job. At first, you are stuck. Assert yourself with confidence even when you are new to compression. Listen carefully to your productions and compare them to recordings you admire. Assist more experienced engineers whenever you get the chance.

The process of using a compressor begins with vision and a strategy: which type of effect to you wish to achieve? Recall again that compressors are capable of a vast range of results; target a specific effect or two from the long list of possibilities: clipping prevention, noise suppression, loudness maximization, intelligibility and articulation enhancement, performance smoothing, amplitude envelope alteration, ambience and artifact exaggeration, de-essing, and distortion.

The type of compression effect selected, choose the make and model that is most effective at achieving this. It takes experience here. As with microphone selection, you will get better at this over time. It feels like a trivia contest at fist. Through experience on many different compressors, though, this step will become easy and intuitive.

Next, set the parameters to settings appropriate for the type of effect. With an understanding of the overall strategy for the effect, threshold, ratio, attack, and release can be pre-set to values near where they need to be before any audio is run through the compressor. For example, sharpening the attack portion of the amplitude envelope requires a low threshold, high ratio, medium attack, and medium to slow release.

Now, at last, listen carefully to the resulting compression and tweak the parameters until you have fine-tuned the effect to what is needed. Glance at the meters on the compressor to make sure you are not causing unwanted overload and to confirm the intended amount of gain reduction is occurring.

That is all there is to it. If no amount of parameter adjustment leads to the desired effect, reach for a different compressor. As with microphone selection, the initial choice of equipment determines the range of possible results. Getting this wrong can make certain goals unachievable. Especially early in your career, expect compression to be a highly iterative, sometimes frustratingly slow process. Steady progress is sure to make the overall quality of your productions improve.

6.8 Selected Discography

Artist: Sheryl Crow
Song: "Everyday is a Winding Road"
Album: *Sheryl Crow*
Label: A&M Records
Year: 1996
Notes: The vocal ranges from impossibly intimate to triple forte, yet the objective loudness stays essentially the same. We feel the emotional dynamics, but the vice grip control over level guarantees every syllable is understood.

Artist: Death Cab for Cutie
Song: "Marching Bands of Manhattan"

Album: *Plans*
Label: Atlantic Records
Year: 2005
Notes: Listen to solitary piano note signaling the end of the tune. Compressed for unusual decay.

Artist: Los Lobos
Song: "Mas y Mas"
Album: *Colossal Head*
Label: Warner Bros. Records
Year: 1996
Notes: Heavy compression is part of the sound of the production/engineering team of Mitchell Froom and Tchad Blake. While the entire album offers a wealth of examples, this tune is chosen because, well, the tune grooves hard. Compression is heavy on the vocal; no matter how much he screams, the vocal stays locked-in at one level. It is on the guitars; listen to the spring reverb rise up in level between the wailing notes on the rhythm guitar panned right.

Artist: Liz Phair
Song: "Polyester Bride"
Album: *whitechocolatespaceegg*
Label: Matador
Year: 1998
Notes: Snare is compressed for, among other things, exaggerated sustain. First snare hit of the tune is particularly revealing of the effect, before the rest of the mix kicks in.

Artist: The Wallflowers
Song: "One Headlight"
Album: *Bringing Down the Horse*
Label: Interscope
Year: 1996
Notes: Lead vocal is compressed aggressively throughout. Note especially chorus vocals with added distortion, brought about in part through extremely fast compression. Snare also offers a classic demonstration of extracting maximum timbral detail, in part, through fast release compression.

Expansion and Gating

7

"If there's no one beside you
When your soul embarks,
Then I'll follow you into the dark."
— "I WILL FOLLOW YOU INTO THE DARK," DEATH CAB FOR CUTIE, *PLANS*
(ATLANTIC RECORDS, 2005)

A cousin of the compressor/limiter is the expander/gate. Expanders and gates do very much the opposite of what compressors and limiters do, but it is easy to get tangled up in the logic of this. This chapter clarifies the function of these devices and explores their production potential.

7.1 Increasing Dynamic Range

Recalling the significance of audio dynamic range (see Chapter 1), Table 7.1 lays out the logic of these devices. Readers new to this family of signal processors may initially feel frustrated. Keeping track of which device boosts and which device cuts, based on whether or not the signal is above or below the threshold, seems a bit trivial at first. The musical potential of these devices is not captured in the least by Table 7.1. However, knowing these rules is essential to their successful use in multitrack production.

As discussed in depth in Chapter 6, compressors and limiters turn the signal down when the music drifts above the threshold — a simple concept capable of a terrific variety of effects. Compressors and limiters *reduce* dynamic range. Expanders and gates, on the other hand, *increase* dynamic range.

By definition, the upper limit of the audio dynamic range is the peak level obtained before distortion (see "Dynamic Range," Chapter 1). The lower bound of dynamic range is the noise floor. If the peak amplitudes of the audio signal are increased without distortion, or the level of the noise can some how be decreased, then the dynamic range is stretched. One might say it is *expanded*.

Dynamics Effect	Gain ↑ Boost/Attenuate ↓	Threshold ↑ Above/Below ↓	Dynamic Range ↑ Increase/Decrease ↓
Compressor	↓	↑	↓
Limiter	↓	↑	↓
Upward expander	↑	↑	↑
Downward expander	↓	↓	↑
Noise gate	↓	↓	↑

Table 7.1 Modifying Audio Dynamic Range

There exist two types, two philosophies of expanders. *Upward* expanders increase the amplitude when the signal exceeds the threshold in an attempt to raise the level of the signal without distorting. *Downward* expanders decrease the amplitude when it falls below the threshold, in an attempt to lower the level of the noise. Both approaches increase dynamic range.

That is the theory. In practice, it is quite rare to find an upward expander in the racks of gear available at any great studio. Most devices labeled "expander" are in fact downward expanders. Upward expanders are more difficult to control. Think about it. Taking the higher amplitude portions of a complex signal and *increasing* the amplitude further still makes the chances for distortion much more likely. No safety cushion is built into this process. It challenges the very idea of headroom. It is inherently unstable. Clipping distortion (see Chapter 4) is a likely result if the upward expander is not used very carefully.

Downward expansion is a bit better at taking care of itself. Downward expansion, working the other end of the dynamic range limit, attenuates the lower amplitudes. If accidentally overused, the lower amplitudes get too low. Distortion is not the result — silence is. By covering its own overuse, downward expansion is the more stable, easier to use effect. Downward expansion is, therefore, the more common type of expansion.

As is customary in the recording industry, when this book uses the moniker "expander" without indicating upward or downward, it is assumed to be a downward expander.

Gating is to expanding as limiting is to compression. A gate is simply a downward expander set to more extreme settings. Rather than slightly attenuate that part of the signal that falls below the threshold, it aggressively attenuates. In the case of gating, it might in fact fully *mute* the signal when it falls below the threshold.

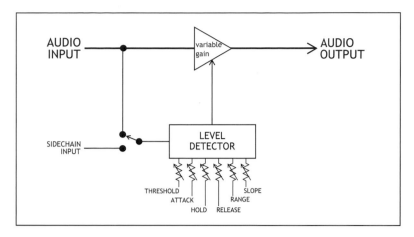

▲ *Figure 7.1 Signal flow through an expander/gate.*

7.2 Parameters

Comparing expanders and gates to compressors and limiters reveals that they have many of the same controls and use the same basic idea, but achieve a very different volume effect (Figure 7.1).

Audio flows through a gain element. This variable gain stage is instructed to boost or attenuate based on instructions from a level detector circuit. The level detector might look at the very audio signal being processed. As with compressors, a side chain input permits the level detector to look at one signal while the gain change element processes a different signal. This provides the flexibility needed for more creative signal processing.

7.2.1 THRESHOLD

The expander must sort out that portion of the waveform that is to be attenuated, and the portion whose amplitude is to remain unchanged. For downward expansion, the threshold control determines amplitude below which attenuation is triggered into action. As long as the signal remains above this threshold, no expansion is initiated, and the gain stage of the expander stays at unity. When the amplitude of the signal sinks lower than the threshold, however, the expander begins to attenuate the signal, very much like a fader being pulled down automatically. Once the expander is attenuating a signal, the threshold identifies the instant when the level has returned to a high enough amplitude that the expander should stop attenuating and return to unity gain.

In the case of upward expansion, the threshold control specifies the amplitude above which the gain is to be further increased. The signal above the threshold is to be amplified. When the signal is below the threshold, no upward expansion is needed. The device returns to unity gain.

7.2.2 SLOPE

When the audio falls below threshold, the downward expander begins to attenuate. The degree of expansion is determined by the slope setting. The *slope* compares the level below the threshold of the input to the level below the threshold of the output (Figure 7.2). For example, a 1 : 2 ratio describes a situation in which the output level below the threshold is two times lower than the original input below the threshold: 1 dB below threshold in becomes 2 dB below threshold out, and 10 dB below threshold in becomes 20 dB below threshold out. A ratio of 1 : X sets the expander so that the output must fall below the threshold by X dB whenever the input is just 1 dB below the threshold.

It is no coincidence that the math above sounds reminiscent of the ratio setting on a compressor. The logic is quite similar. In fact, some expander manufacturers choose to label this parameter ratio instead of slope. The terms are interchangeable. Recording tradition has it that the ratio for an expander is less than one, while the ratio of a compressor is greater than one.

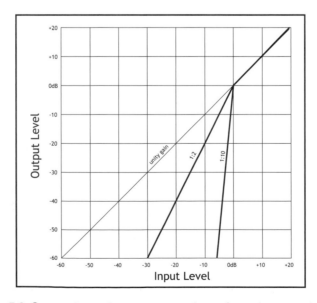

▲ *Figure 7.2 Comparison of output versus input for a downward expander.*

The ratio (or slope) applies only to the portion of the signal below the threshold. When the input is above the threshold, the expander is not applying this slope to the signal. The expander is specifically designed not to change the gain for signals above the threshold. Above the threshold, the hope is that the expander is sonically transparent, that the audio sound is completely unchanged when its amplitude is above the specified amplitude. Only that part of the input that falls below the user-defined threshold is processed according to the expansion slope. The expander applies its attenuation based on the decibel difference between the amplitude of the audio and the threshold.

It is often the case that the slope is not user adjustable. While many compressors offer this control, many expanders and gates do not offer the engineer access to this parameter. It is determined internally. While essential for compression, many expansion and gating processes do not require a user adjustable slope. The other parameters are more important to how this device is used.

7.2.3 ATTACK

How fast the signal is *un*attenuated is a function of the attack setting. Attack describes how quickly the expander can return to unity gain after the amplitude of the signal has passed upward through and above the threshold. As with compression, attack describes the speed of the imagined fader as the signal increases in level. Fast attack times will enable the expander to react very quickly, snapping nearly instantly to unity gain. Slow attack times are lazier, sneaking the level up gradually to the unity gain of the unexpanded signal.

7.2.4 RELEASE, FADE, OR DECAY

The speed of the imagined fader within the machine as it moves down is determined by the *release* setting (identified by some manufacturers as the fade or decay control). When a signal falls below the threshold, for example, during the brief sustain of the snare or at the last breath of trumpet note, the expander must begin attenuating. The release time setting governs the speed of this reduction in level. That is, the release parameter sets the speed that an attenuating expander can turn the gain down from unity toward the level required by the slope setting.

Attack and release parameters for an expander are similar in concept to the attack and release parameters of a compressor. Attack describes the agility of the dynamics device when the amplitude of the audio is increasing — the onset of each note. Release refers to the speed of action while the amplitude of the audio is decreasing — the end of each note.

When expanding or gating, the parameters listed above — all of which appear in very similar form on a compressor — leave the engineer wanting a bit more control. Enter two additional parameters: hold and range.

7.2.5 HOLD

Threshold is not the sole determinant of whether or not an expander should attenuate the signal. The *hold* setting is a minimum length of time that the expander/gate must wait after the signal crosses the threshold before any attenuation is allowed to occur.

Consider a sound with a long decay, such as a single, sustained piano note. If the piano is recorded in a noisy environment, it might be helpful to expand the signal. When the piano is not playing, the expander could attenuate the signal, pulling unwanted noise out of the mix. When the piano does play, however, the expander needs to return to unity gain and let the piano performance through. The noise hopefully is not distracting while the piano is playing — the piano masks the signal (see "Masking," Chapter 3). The threshold setting balances desired signal versus unwanted noise. The engineer carefully adjusts the threshold so that the noise is below the threshold, causing expansion of the noise down to an unobtrusive level. The threshold must also be set so that the piano is above the threshold, allowing the expander to get out of the way when the piano plays.

During the sustained portion of any piano notes, the desired piano sound will eventually fall below the threshold. The piano sound is still audible even as it falls down to a level equivalent to or even below the level of the noise. Humans have no trouble hearing music even when it is swimming in noise. Listening to music in the car and having a one-on-one conversation in a noisy bar are common examples (at least for your author) of enjoying a signal even when it is below the amplitude of road noise and cocktail chatter by others.

The trouble is, an expander will start attenuating according to the setting of the slope parameter, and at the rate specified by the release setting, as soon as the piano amplitude falls below the threshold. Any portion of the piano note with an amplitude below this threshold will be attenuated thusly. The hold parameter lets the engineer outsmart this. Setting a hold time that is perhaps a second or more in duration instructs the expander to wait at least this long before expanding, even if the signal falls below the threshold.

In this way, audio signals of all kinds can remain convincingly natural sounding, full of low-level detail and nuance as they decay naturally, yet still be processed by an expander to conveniently remove unwanted parts of the signal between the notes and phrases of the performance.

7.2.6 RANGE

Sometimes, it is necessary in audio to avoid extreme amplitude limits. When all settings above instruct the expander to attenuate by 60 dB or more, the signal at the device's output is essentially silent. When the expander starts to unexpand later, and the gain later increases, the expanded signal eventually becomes audible again. That transition — from inaudible to audible — might be noticeable and distracting not only to recording engineers but also to casual listeners.

It is not always necessary or desirable to fully mute the level in these sections of parameter-determined gating. Lowering the level 20–30 dB instead of the more extreme 60 dB might be enough to make the undesired sounds unobtrusive. One solution is to set a maximum amount of attenuation. In this way the expander attenuates only as far as the engineer allows. When the parameters of threshold, slope, attack, hold, and release have instructed the expander to attenuate, that attenuation hits a stop at the level specified by the range setting. Specifying a milder amount of expansion using the range control enables the engineer to avoid the awkward transition from silence to just audible. The range control prevents the unwanted artifacts associated with expanding into and out of deep, dark silence.

7.3 Applications

The expander/gate is used in a number of related ways, discussed next. Typically, serial processing is preferred, so the expander is inserted in-line with the signal flow. In order to prevent any accidental attenuation or removal of the musical portion of the audio signal, one rarely records to the multitrack through expanders and gates. This signal processor is often saved for mixdown so that more finesse may be achieved in dialing in the right settings. Of course, in live broadcast and sound reinforcement, the mix occurs live, so the signal processing does too.

7.3.1 EXPANSION

Listen to an audio track when the musician is not playing and plenty of potentially undesirable sounds will be heard: hiss from microphones, preamps, and analog tape may be present; the whoosh and rumble of the heating, ventilation, and air conditioning system; the purr of a refrigerator chilling the beverages in a nearby room; the sounds of traffic, construction, neighbors, or residents present during the recording session; leakage of sound from other instruments in the studio playing during the same take; the noodling, practicing, and pitch-finding that happens between parts; foot-tapping, clothes rustling, and even knuckle cracking. The list goes on.

Working in well-designed, professional recording spaces, working with experienced musicians, and using good recording craft, these noises can be minimized and made essentially inaudible. Working in lesser studios with talent new to the studio in noisy urban environments, these sounds between performances may become a nuisance. An expander minimizes the problem.

For example, recording a horn quartet using close mikes (one each on trumpet, alto sax, tenor sax, and trombone) will no doubt lead to at least a little leakage. The trumpet microphone picks up mostly trumpet, but it can also have a good amount of sound energy from the other horns. Using isolation booths, gobos, and clever musician and microphone placement, the engineer can minimize this.

It is likely the section will want to play together as a group standing side by side or, perhaps, in a circle, so that they can see and hear each other easily. An intimate physical layout generally helps create a tighter ensemble performance. Moreover, making the musicians comfortable is much more important than getting clean tracks. The frustration is that recording them physically closer together also leads to more leakage.

This may not be an acoustic problem when they are all playing at once. Good microphone placement probably enables the engineer to get a great section sound in part *due* to the leakage across the multiple microphones. That is, with care and experience, one hopes to engineer it so that the sound of the trombone in the trumpet microphone helps the overall sound of the section.

The trouble starts when the section stops and someone plays a solo. Then, the sound of the trumpet all alone in the trombone microphone may not sound so good. Use an expander to further separate the volume of the wanted signal (the target instrument of the close microphone) from the volume of the unwanted signal (the trumpet solo leaking into the nearby trombone microphone).

On the trombone microphone signal, patch up an expander and set the threshold so that the trombone playing directly into the microphone is well above the threshold but the trumpet leakage into this microphone falls below the threshold. When the trombone plays, the gate stays open and the signal is not attenuated. When the trombone stops and the trumpet solos, the expander steps in and turns down the unused trombone track: leakage reduced, automatically.

7.3.2 NOISE GATING

More extreme cleaning up of tracks might be warranted, for example, on close-microphone pop/rock drums. Here, a single musician plays multiple instruments (kick, snare, hi-hat, toms, etc.) all at once, in close proximity. It is inevitable that there will be some snare sound in the tom mikes, for example. When they are not being played, attenuate the tom tracks a little (expansion) or a lot (gating) by patching an expander/gate across them during mixdown.

The decision whether to expand or gate is a creative one. The engineer must do whatever sounds best for the tune at hand. Through persistent tweaking of the threshold, attack, hold, and release settings, an engineer is able to make the gate turn the tom track down when the toms are not being played and to leap out of the way to let each and every tom hit come through. The kick drum and snare drum are automatically removed from the tom microphones by this kind of signal processing.

Through gating, a tighter drum sound can result, but one should plan to spend a fair amount of studio time getting the gates to cooperate. It takes experience and good recording technique to get the gates to open and close when they should. As always, pay attention to the music. Some tunes welcome the clarity such fine-tuned gating brings to a drum kit. Some songs, on the other hand, sound better with the leakage front and center. The snare may sound best when heard not just through the close snare microphones, but also through all the other microphones that happen to be in the room at the time. A gate across those other microphones mutes that beneficial leakage. Good engineering has a lot to do with anticipating what a given microphone selection and placement on one instrument will do to the sound of a nearby instrument. So gate, not out of habit, but out of necessity. Only gate out leakage when the music requires, and have the courage not to if it sounds good without the gate.

Taken to an extreme, an engineer can make the quiet parts silent. Some low-level noises (e.g., tape hiss and amp buzz) can be automatically shut off by a noise gate. The threshold is set so that it is lower in level than all the music. This prevents the gate from trying to attenuate the signal while someone is playing. But the threshold must also be carefully set so that it is above the amplitude of the noise you are trying to remove.

The music stops. The signal falls below the threshold. But, instead of hearing the faint hiss or buzz of the track, the noise gate sneaks in and turns it down. Bye bye noise. As soon as the music resumes on that track, the gate gets back out of the way and lets the sound of the instrument back in. Sure the hiss, buzz, or whatever is there while the music is playing, but it is masked by the much louder sound of the instrument. Between the noise gate in the quiet parts and the music cranking the rest of the time, practically speaking, any perceived noise is removed.

As with expansion, the threshold specified when gating is set to an amplitude above these unwanted noises, but below the level of the musical performance. In this way, the gate attenuates the nuisance sounds to a lower level, while leaving the actual performance untouched.

Attack time must be set to taste. Very quick attack times are often desirable so that the gate snaps open immediately in the presence of threshold-exceeding music. However, the very act of the gate opening can create an audible "click" or other sound artifact. This click most likely occurs when gating signals that contain very low frequencies. The abrupt opening of the gate during the slow cycling of a low-frequency signal radically reshapes the waveform. Such a click might be helpful to the music. More often, however, it is perceived as a distracting nuisance. Remove the click by slowing the attack time down. A slower opening of the gate makes for a smoother waveform.

Musical material with a fairly slow onset — strings or vocals — generally benefits from a slow attack time. Sounds with highly transient onset — drums, percussion, or piano — tend to respond well to fast attack times. In fact, the click associated with a gate opening quickly might help intensify the sharpness of the transient in a way that is beneficial to the mix. The engineer dials in an attack time setting that is slow enough to prevent unwanted clicking or thumping, but fast enough not to miss the natural onset of the musical waveform. As always, some finesse, patience, and musical judgment is required.

Release time is influenced by the nature of the decay of the musical signal. Slowly decaying sounds — strings, vocals, and piano — sound most natural with slow release times. Percussive sounds can tolerate or even be improved by fast release gating without artifacts.

It should be re-emphasized that the noises that occur between musical passages are not always a negative. The leakage from other instruments can benefit the overall sound of the mix. Applying a gate across each and every track so that pristine silence exists between every musical note may make a beautiful musical statement, revealing detail and subtlety. On the other hand, it may do damage, musically, creating a sound that is too sterile and lacks a sense of cohesion or musical ensemble. The engineer must use good judgment in determining whether or not noise gating is appropriate to the style of music, the band, or the song being produced.

7.3.3 GATING THROUGH WAVEFORM EDITING

In a digital audio workstation, noise gating shifts to a different paradigm. The engineer willing to spend the time could work through each and every track and edit out any and all unwanted noises. This is *gating through waveform editing*. Find the parts when the singer is not singing and simply cut them out.

Better yet, let the computer do all the work. Once the audio has been recorded into a computer, the digital audio workstation may have an algorithm that simply crunches the numbers and nondestructively deletes all audio below a specified threshold. As a result, a formerly continuous single waveform, with unwanted noises in between musical phrases, is chopped up into separate pieces, each a musical phrase with total silence in between.

The algorithm is made more powerful if it can impose a fade-in and fade-out at each transition, analogous to attack and release time settings on a gate. In this way, the computer effectively removes the undesired low-level portions of any track. The patient audio engineer can then manually go through each event and tailor the fade-in and fade-out as desired. The result is a playlist full of audio events separated by silence.

Analog noise gates are forced to work in real time: The audio goes in. The audio is immediately processed. The audio goes out. The digital audio workstation has the luxury of being able to read the audio files well ahead of playback. This can inform the actions of any signal processor, such as a gate. In this way processes that simply could not work in the analog domain become feasible in a digital audio workstation; new creative capabilities follow.

In addition, the digital audio workstation has the ability to work "off-line." That is, it can implement signal processing such as noise gating across the entire file ahead of time and offer the engineer the chance to accept, reject, or at anytime refine the result.

7.3.4 GATING

Noise gating, as discussed above, focuses intensely on keeping a natural, realistic sound while removing unwanted noise. Sometimes noise isn't the issue. Gates may also be used to manipulate the reverberant sound of the room. This is simply called *gating*. The approach is similar to noise gating, but the engineer seeks consciously to alter the attack and decay envelop of the signal for an ear-tickling effect.

Consider the room tracks that are often recorded onto separate tracks during the close-microphone recording of a drum kit. Even in a great-sounding room, those room tracks can be very difficult to use effectively in a mix. Too often, the room tracks posses a thrilling snare drum sound, but a loose kick drum decay and a messy wash of cymbals. In some projects, there is no appropriate placement in the mix for these room tracks that works. Place them at a level that sounds right for the snare drum, and the kick drum and cymbals will become disappointing. Turn the room tracks down so that the kick and cymbals aren't undermined, and the snare receives no benefit.

Gating is the answer. A gate is inserted into the room tracks. The plan is to have the gate open on each snare hit but remain closed in between. In this way, the gated room tracks can be placed loud enough in the mix to decorate the snare without any loss of clarity in between snare hits. A few complications, easily solved, arise.

In stereo productions, room tracks are often recorded via a matched pair of microphones that seek to capture a stereophonic image of the room. In surround productions, room tracks are four or five tracks. Gating these tracks requires as many gates.

Stereo gating simply uses two identical noise gates, each inserted into one of the tracks that make up the stereo pair. In the example above, it would be hoped that upon each snare hit, the two gates would open and close together. Matching the settings on the two gates can help achieve this.

If the gates open at different times, listeners will localize toward the track that opens earlier. As the snare drum is almost always panned dead center in a stereo mix, gated pairs of tracks that do not open and close in lock step are problematic. A stereo link feature is available on many gates. The *stereo link* ties the gates together so that they both look at a single detector circuit. Each gain element of each noise gate therefore receives identical instructions, causing the two gates to open and close in unison, even as the audio passing through them is different. The snare image remains centered.

7.3.5 KEYED GATING

Getting room tracks to open on each snare hit is easier when the close microphone on the snare feeds the detector circuit of the gate (Figure 7.3). To be sure, the room tracks are the audio flowing through the gain change element. It is the room tracks that are attenuated by the gate.

But decisions about when to open the gate, how quickly to open and close the gate, and by how much to open the gate, are based on the settings of all parameters and the gate's analysis of the close-microphone snare track. This is gating one track while looking at another.

As with the microphones placed out in the room some distance from the drum kit, the microphone placed close to the snare drum will receive plenty of sound leakage from other drums and cymbals. However, unlike the room tracks, the sound of the snare drum will dominate. This gives the engineer better opportunity to place the threshold at a level that is below the amplitude of the snare hits but above the amplitude of the various other elements of the drum kit leaking into the track. The gate can be made to open only on snare hits, and to remain closed no matter what happens on the kick, toms, or cymbals.

Filter the side chain input for even more control (Figure 7.4). A high-pass filter could attenuate the low-end-dominated kick drum leakage. A low-pass filter could attenuate much of the cymbal leakage. This further sharpens the ability of the gate to detect the snare and ignore the rest.

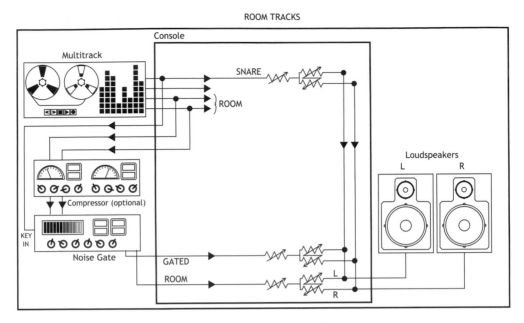

▲ *Figure 7.3 Signal flow for gated room tracks.*

This keyed gating effect can be made more pronounced if the room sound being gated is compressed (see Chapter 6). The compression, inserted before or after the gate, can radically alter the room decay associated with each snare hit, exaggerating it in level and duration. Such compression also makes the problems of a sloppy kick drum sound and a chaotic cymbal wash even worse. Keyed gating becomes essential when the ambience tracks are aggressively compressed.

While the gates might be patched across the insert of the room tracks, it is often desirable to route them as shown in Figures 7.3 and 7.4. Here, the room tracks make their way to the mix, un-gated, for use in any way desired by the recording engineer. In addition, the room tracks are split and sent through an additional path. This additional path is compressed and gated (keyed open by the close microphone on the snare) and introduced to the mix on separate faders. The mix now places many useful room track production variables at the engineer's fingertips. The drum sound in the final mix is made up of the close microphones, the overhead microphones, the leakage from various other microphones, the room microphones, and the gated room microphones. The pros and cons of all of these elements are carefully considered by the mix engineer, who orchestrates the overall drum sound they desire.

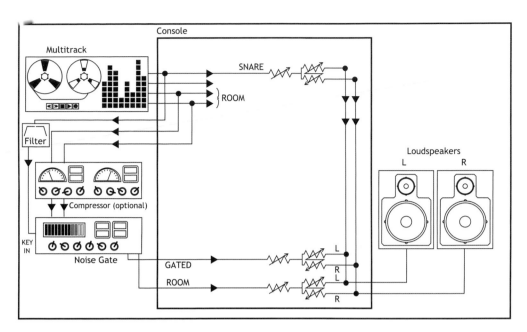

▲ *Figure 7.4 Filtering the key input for gated room tracks.*

What works for room tracks likely works for reverb as well (see Chapter 11). Gate and (optionally) compress the reverb returns in exactly the same way, using a close-microphone key input for controlled opening and closing of the gates, and a sophisticated effect results (Figure 7.5). This sort of sound just does not happen in nature. It is one of many creations that only exists in the music created in the studio and enjoyed through loudspeaker playback.

The inexperienced engineer may view keyed gating as a heavy-handed effect, appropriate only to certain styles of music, or revealing of specific eras of music. It is true that many forms of dance music and electronic music make more aggressive use of this type of effect than others. It is also true that a pronounced gated snare effect is reminiscent of the synth-pop music of the 1980s. To reduce keyed gating to these musical trends and styles misses many opportunities. This effect may be made quite subtle; it exists in more recordings than the casual listener might at first expect.

Using the snare to key open room tracks or reverb returns might be considered a method of enriching the sound of the snare, without altering it in a way that is noticeable to the untrained ear. The snare becomes easier to hear in a crowded mix if each hit of the snare has a bit of stereo width associated with it. Room tracks and reverb returns gated by the close microphone on the snare can sound quite natural if the attack and release

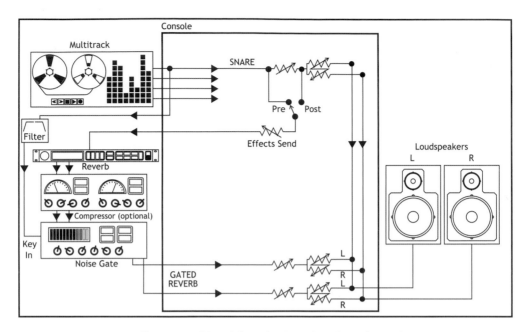

▲ *Figure 7.5 Signal flow for keyed gating of reverb.*

times are slowed down a bit. Push the gated sounds up in the mix until just audible, and then back off a little. Muting and unmuting the returns, the engineer's goal is to have a sound that changes almost indescribably when the gated ambience is on versus off. The snare becomes a little easier to hear without being louder. The stereo image of the snare is broadened; it stays center, but feels wider than the single close-microphone track.

The sound of the snare drum captured by the overheads often sounds better than the sound captured by the close-placed microphone. Split the overheads off to an additional pair of inputs. Gate them with the snare, and layer this into the mix. It is a multitrack production cliché: gated room, reverb, and/or overhead tracks, keyed by the close-microphone snare. Do this to an obvious degree where desired, or explore the benefits of using it even subtly (see "Selected Discography" below).

Naturally, snare is only one obvious example among many. Gate the ambience tracks (recorded in a room or created in a reverb device) associated with the toms, congas, claves, or any impulsive sound for a unique change in quality or a subtle enhancement in audibility. Used on less-impulsive sounds, a vocal can be made to sound more live, a piano can be made to sound unnatural, etc. Keyed gating rewards the engineer willing to explore.

7.3.6 DUCKING

Using a gate with key inputs makes possible another effect. *Ducking* turns down one track by a specified amount in the presence of another signal. Voiceovers provide a good example. Any radio jingle with background music likely wants quality, attention-getting, memorable music. But the music is subservient to the content of the words. The voice over the music must be perfectly audible, always. The music is not permitted to drown out any syllable of any word. Required for advertisements, perfect vocal intelligibility is compulsory for most forms of pop music too. The intellectual content of the lyrics is generally the most important single ingredient in any song. That the voice is the most important element of the jingle is an indisputable fact in the advertising business.

A sensible goal then is for the music to be pleasingly forward, loud, and attention-getting. At the instant when the voice speaks, however, the music must immediately be attenuated. When the voice stops, the music should rise up quickly to grab the listener's attention and keep them interested in the soap, beer, or mobile phone service being offered. The music is ducked under the voice.

A noise gate with a ducking function is the solution. The music bed to be ducked is patched through the expander/gate. The voice of higher priority, in addition to feeding the mix and all other vocal effects, is supplied to the key input of the gate. Whenever the voice speaks, the jingle music is attenuated. Adjust attack time to be very fast, so that the music snaps out of the way as soon as the voice begins. Adjust the release time to as fast as possible without sounding unnatural, so that the jingle does not sag with quiet moments, losing the listener's attention.

The alert reader may notice that a compressor with a side chain input should achieve the nearly the same thing. The ducking process described above is attenuating a signal when the threshold is exceeded — the very goal of a compressor. Background music is attenuated when the voice goes above the threshold. The trouble with using compression for this effect, and the reason some expanders and gates provide this feature, is the presence of that critical parameter, *range*.

Not available on most compressors, range sets a maximum amount of attenuation, useful in many expansion and gating effects. In the case of ducking, it is likely that the music should be turned down by a specific, fixed amount in the presence of the speaking voice. No matter the level of the voice, the music should simply be attenuated by a certain finite amount based on what sounds appropriate to the engineer, perhaps 10–15 dB. Compression would adjust the level of the music constantly based on the level of the voice. The amplitude of the music would modulate constantly in complex reaction to the level of the voiceover. Compression does not hit a hard stop because compressors do not typically possess the range parameter. Therefore, look to noise gates for the ducking feature.

Ducking is not just for jingles. Engineers occasionally find applications in music productions as well. It is rarely if ever appropriate to have the lead vocal duck the entire rest of the mix. Ducking takes a different form in multitrack music. Consider the situation in which an engineer has recorded drums and finds the leakage among the many drum microphones to be generally beneficial to the production. The sound of the kick drum in the overheads helps the power and excitement of the kick drum. The sound of the cymbals and the room reflections in the tom microphones helps keep the overall drum sound convincing, organic, and consistent with the live feeling that the band wants. To be clear, there is no rule that says this leakage into neighboring microphones is bad and must be noise-gated out. In fact, it is the presence of this leakage that can help push away the sound of the studio and artistically reduce the precision of multitrack production. Jazz and folk music often benefit from this looser, more integrated sound. Many styles of pop music do too.

One very loud instrument in the drum kit may make this strategic use of leakage a problem: the snare drum. The drum kit, recorded through several close microphones, may sound great with all the microphones up and open, ungated. Except, every snare hit may be a murky mess. Duck the tom tracks on each snare hit, even as little as 6 db, and the clarity of the snare can be preserved even as leakage elsewhere flatters the rest of the drum sound.

7.3.7 ENVELOPE FOLLOWING

Courtesy of the side chain input, the amplitude envelope of one sound can be made similar to the amplitude envelope of another through a process called *envelope following*. Insert a gate across a guitar track, but key the gate open with a snare track. The result is a guitar tone burst that looks a lot like a snare, yet sounds a lot like a guitar.

This kind of synthesized sound is common in many styles of music, particularly those where the sound of the studio — the sound of the gear — is a positive, such as trance, electronica, and many other forms of dance music (always an earnest adopter of new technologies).

A low-frequency oscillator, carefully tuned, can be keyed open on each kick drum for extra low end thump. Sine waves tuned to 60 Hz or lower will do the job. Distort slightly the sine wave for added harmonic character or reach for a more complex wave (e.g., saw tooth). The resulting kick packs the sort of whollup that, if misused, can damage loudspeakers. Done well, it draws a crowd onto the dance floor.

A noise source (pink noise, white noise, or electric guitar) can be gated on and off with each snare hit, adding significant spectral complexity to the naturally noisy buzz and rattle of any snare drum.

Less obvious pairings can be made. Use a vocal track as the key input to an expander with guitar feedback. Let a high hat open and close on a piano. Engineering requires imagination.

7.3.8 TREMOLO

Rather than attenuate between notes and phrases, one can attenuate *during* the notes, with a more regular rhythm. The result is tremolo, the familiar wobble of slow amplitude modulation. Many guitar amps and keyboard patches offer this feature. It helps draw attention to the instrument and set the vibe for the track. It gives a track a signature effect that can help reduce masking (see "Masking Reduction," Chapter 3).

A throbbing, twangy guitar approach is common, but in the studio tremolo effects are applied freely to any signal. Engineers do not need a tremolo knob on the instrument or the amp to get this effect. Effects units can do this instead.

A multi-effects unit may very well have a patch labeled "tremolo." Done. But tremolo is hidden in another effect, the autopanner (see Chapter 8). The autopanner is found on many multi-effects devices. A single input is fed to a pan pot that can be programmed to move in some desired patterns: left to right, right to left, back and forth repeatedly, etc. The speed, depth, and shape of the panning are often controllable as well (much like the modulation section of a delay; see Chapter 9).

Patch up an autopanner set to pan continuously, but use only one output. The volume will increase as it pans toward the active output. It will decrease as it pans toward the unused autopanner output. Tremolo results.

The Leslie cabinet (see "Obvious Effects" in Pitch Shifting, Chapter 10) is an acoustic signal processor that, among other things, causes a form of tremolo.

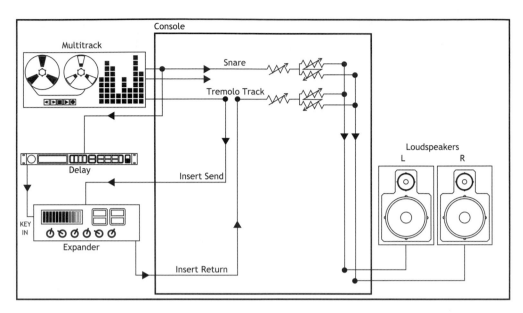

▲ *Figure 7.6 Creating tremolo through keyed expansion.*

The most sophisticated tremolo effect comes through clever use of an expander (Figure 7.6). Insert the expander into the signal to be amplitude modulated. Feed the key input of the expander with a musically relevant, consistent performance. The snare drum back beat is an ideal candidate. Feed a bit of snare to a delay. Set the delay to an eighth note time interval. Raise the delay feedback so that it repeats several times. Feed the delay output to the side chain input of the expander. A little (okay, more than a little on the first few tries) messaging of parameters and the result is tremolo, locked into the tempo of the song. Each snare hit and each delayed repetition of the snare hit enters the level detector of the expander and triggers some attenuation. The expander returns to unity smoothly in between these snare triggers. The amplitude of the signal is continuously and regularly turned down and back up again.

Adjust the delay time for slower or faster tremolo. Select a different side chain track for possibly more interesting, less consistent tremolo (e.g., hi hat).

An expander offers the opportunity to add tremolo to *any* track. Add tremolo to a piano. Obviously, no real piano can do this at a symphony hall. But any piano recording can do this in the control room. Apply tremolo to an extremely long reverb tail, turning a 30-second decay into a pad of reverberant energy that seems to play along like a musical instrument. Clearly, the expander/gate is a source of a broad range of musically valuable effects.

7.4 Selected Discography

Artist: Counting Crows
Song: "Mr. Jones"
Album: *August and Everything After*
Label: Geffen Records
Year: 1993
Notes: Classic tremolo guitar in the second rhythm guitar, left side, entering in the middle of verse one. A single note counter melody pulses in the background of the arrangement, thoughtfully entering just before the word, "guitar."

Artist: Sheryl Crow
Song: "Redemption Days"
Album: *Sheryl Crow*
Label: A&M Records
Year: 1996
Notes: Touchstone tremolo, electric guitar on the left channel.

Artist: Peter Gabriel
Song: "Big Time"
Album: *So*
Label: Geffen Records
Year: 1986
Notes: The snare has gated reverb, which is unmistakable, but not so overdone that it has not aged well. The clap sound is more extreme. It's a song about hype, after all.

Artist: Green Day
Song: "Boulevard of Broken Dreams"
Album: *American Idiot*
Label: Reprise Records
Year: 2004
Notes: That distorted rhythm guitar probably was not tracked with tremolo. It is almost certainly manufactured at mixdown through use of a delay, with feedback, input to the sidechain.

Artist: Imogen Heap
Song: "Hide and Seek"
Album: *Speak for Yourself*
Label: Megaphonic Records
Year: 2005
Notes: Multiband envelope following, sometimes called vocoding, is a key part of the sound of this song. The vocal drives the envelope of the synths underneath. For this process, the vocal is filtered into different bands and the amplitude of the vocal in each band drives the amplitude of the synths in the same band. The keyboard part is made to sing along in tight, machine-like symphony.

Artist: The Smiths
Song: "How Soon Is Now?"
Album: *Hatful of Hollow*
Label: Rough Trade Records
Year: 1984
Notes: Rhythm guitar intro offers tremolo in sixteenth note time, and a sound that The Smiths' fans spot from five miles away — an iconic sound.

Volume

8

"Once upon a time, I was an ocean.
But now I'm a mountain range.
Something unstoppable set into motion.
Nothing is different, but everything's changed."
— "ONCE UPON A TIME THERE WAS AN OCEAN," PAUL SIMON, *SURPRISE*
(WARNER BROTHERS RECORDS, 2006)

For the most part, the signal-processing effects discussed in this book are not available to the music-buying public. These devices are for the professional studio, not the home stereo. Consumers are, however, allowed one of the most important signal-processing devices of all: the volume control. Sound engineers must not overlook the significance of this signal processor. This chapter reviews the important uses of some of the simplest signal-processing devices dedicated to volume: the fader, pan pot, and mute button.

8.1 Volume Controls

The mother of all volume controls is, of course, the fader. Slide it up for more level and down for less volume. Pan pots coordinate two level settings to perceptually place the stereophonic image of a single audio track at or between the two loudspeakers. The mute button expresses level in binary: on or off, level or none.

8.1.1 FADERS

Three types of faders are found in the recording studio: the variable resistor, voltage-controlled amplifier, and digital fader.

Variable Resistor

Electrical resistance is a property of all materials describing how much they restrict the flow of electricity. A high resistance does more to hinder the flow of electrical current than a low resistance. Materials with very high resistance are classified as insulators. Under normal operating conditions, they do not conduct much electricity at all. We appreciate this property when we handle things like power cords.

At the other extreme, devices with very low resistance fall into the category of conductor. Copper wire is a convenient example. The copper within that power cable conducts electricity from the wall outlet to the piece of audio equipment, enabling it to process audio signals, getting the light-emitting diodes (LEDs) to flicker, motivating the meters to twitch, and enabling us to make and record music.

In audio, special use is made of devices whose resistance is *adjustable*, from low to high. The volume knob on a home stereo, electric guitar, or analog synthesizer is (with a few model-specific exceptions) a *variable resistor* (Figure 8.1). Set to a high resistance, electricity has trouble flowing and the volume is attenuated. To turn up the volume, lower the resistance and let the audio waveform through more easily. Lowering the resistance raises the volume. These variable resistors define their own class volume control.

Electrical voltage represents the potential for electrical current to flow, depending on the impedance of the circuit that is made between the two points of contact on the voltage source. The potential is, to those who design and manufacture electrical devices, synonymous with voltage. Accordingly, variable resisters are also called *potentiometers*.

This same device gets yet another name. In order to specifically distinguish it from the voltage-controlled amplifier discussed next, variable resistors

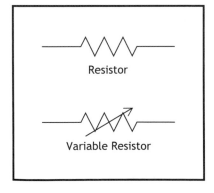

▲ *Figure 8.1 Resistors.*

used in audio are sometimes simply called *audio faders*, or, more emphatically, *true audio faders*. This term indicates that the adjustable resistance of this fader is part of the very circuit that contains the audio. As the discussion of voltage-controlled amplifiers makes clear, this is not always the case.

Voltage-Controlled Amplifier

In the recording studio, it is necessary to look more closely at analog volume controls because there is a second type: the voltage-controlled amplifier (VCA). The idea behind them is simple and clever.

Be forewarned, it risks getting a little philosophical here. For audio circuits using a VCA, the fader that sits on the mixing console is separated from the audio itself by one layer. Instead of having that slider on the console physically adjust the resistance of a component of an audio circuit, it adjusts a control voltage instead. This control voltage in turn adjusts the amount of gain on an amplifier (Figure 8.2). That is, instead of using the fader to directly adjust the amplitude of the audio, the engineer uses the fader to adjust a thing that adjusts the amplitude of the audio. It may seem a tedious distinction at first.

Voltage-controlled amplifiers boost or attenuate an audio signal (that is the *amplifier* part) based not on the position of a fader control, but on the value of a control voltage (that is the *voltage-controlled* part). The result is that an engineer, machine, mix automation system, or compressor can control the level of the signal through a VCA. Without VCAs, the only way to have something other than the engineer adjust the level is to stick a motor on the fader. Motorized faders are certainly available, but they are a pricey, complicated option. But before motorized faders were feasible, machines controlled signal amplitude through VCAs. A machine can output a control voltage as easily as it can output an audio voltage. By using VCAs, faders

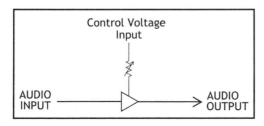

▲ *Figure 8.2 A voltage-controlled amplitier.*

can be manipulated by man or machine. This makes possible a range of effects, notably mix automation (see Chapter 15), compression (see Chapter 6), and expansion (see Chapter 7).

Sitting in front of any analog mixing console for the first time, an engineer is wise to determine (from the manual or the house engineer) whether the faders are VCAs or variable resistors. Variable resistors are comparatively easier to design and manufacture at a low price, with high audio quality. Voltage-controlled amplifiers, on the other hand, are a more sophisticated device, containing active electronics. Noise and distortion rear their ugly heads. A quiet, distortionless VCA is not a given. High-end manufacturers of analog consoles today can be counted on to offer sonically-transparent VCAs. With older devices (consoles from the 1980s and before), and cheaper devices, one needs to listen carefully.

Pay particular attention to low-level details in the signal, such as the imaging of a stereo pair of tracks or the envelopment of a lush reverb tail. Inferior VCAs can collapse stereo images ever so slightly, and they are particularly destructive to stereo reverb, where the reverberation narrows, shortens, and becomes less believable.

Digital Fader

Digital audio offers a third type of volume control. In digital audio, where the audio signal is represented by a long series of numbers, a fader is a multiplier. Multiply by a number less than one to reduce level. Multiply by a number greater than one to increase level. Provided the digital system is functioning properly, has headroom available to accommodate the mathematical result, and dithers the signal properly, it is fair to think of digital volume controls as a third, simpler, kind of level adjuster. The sound quality of a digital fader is driven by the type of math (fixed point or floating point) and bit resolution (storage formats are commonly 16, 24, and sometime 32 bit, but processing bit resolutions often run at much higher resolutions, using 48 bits or more for the math associated with faders, pan pots, signal processors, and mixers).

8.1.2 PAN POTS

The action of the pan pot is so natural and intuitive that one might forget that it is a level control device. In fact, it is really two volume controls. A single user interface — a knob or a slider — adjusts the relative level of two

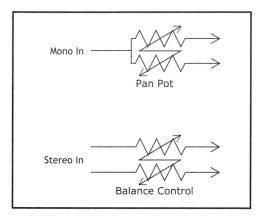

▲ *Figure 8.3 The pan pot is in a one-in, two-out device, while the balance control is in a two-in, two-out device.*

outputs (Figure 8.3). The balance knob is similar, but is a stereo-in, stereo-out device, adjusting the relative volumes of the left and right side, but never mixing the two.

The user, in a single pan pot gesture, turns up one output while turning down the other. Two variable resistors are adjusted by a single control. The variable resistors are wired with opposite polarity. A clockwise motion lowers the resistance of the right output while it increases the resistance of the left output by the corresponding amount. The result, of course, is the perception of sound coming from the direction of the louder speaker. Pan to make a track appear at one speaker, the other speaker, or some point in between.

Autopanners bring the panning capability into an effects device. A single input is fed to a voltage-controlled pan pot, which can be programmed to move in some desired patterns: left to right, right to left, left and right repeatedly, etc. The speed, depth, and shape of the panning are often controllable as well (much like the modulation section of a delay; see Chapter 9). In this way, a machine faithfully repeats a panning pattern, allowing the engineer to set it up once and then focus on other matters.

8.1.3 MUTE SWITCHES

To mute a signal, a simple switch purposefully interrupts the flow of the signal. Using relays or transistors, switches may be controlled or triggered by a machine as well as the engineer.

8.2 Volume Applications

These humble level control devices are profoundly important production tools.

8.2.1 BALANCE

Consider the first step in building a mix. One carefully, systematically, and iteratively adjusts and readjusts the volume of each and every track until the combination starts to make musical sense. When the recording engineer has fine-tuned the level and panning of the core elements of the multitrack arrangement, thoughtfully muting those tracks that undermine the quality of the production, the mix is said to be "balanced."

In the course of overdubbing or mixing a tune, a single song may be played more than a hundred times. Artists will take advantage of multiple passes to refine and hopefully improve their recorded performance. Mixdown is such a complicated process that, in most pop productions, it takes from several hours to several days to mix a $3^1/_2$-minute song.

In the course of just that first playback of the piece, however, the engineer must find the fader levels and pan pot positions that enable the song to stand on its own. The goal is to empower each and every track to make its contribution to the overall music without obliterating other parts of the multitrack production. Flawed tracks are muted. Tracks that are in conflict with each other are carefully evaluated and possibly muted. Balancing a mix is a fundamental skill all engineers must possess.

While there is no single right answer, it would be fair to say that in a *typical* pop mix, the vocal and the snare sit pretty loud in the mix, dead center. Sharing the center, but at lower levels, are the kick drum and the bass guitar.

These four elements — vocal, snare, kick, and bass — are essential ingredients for almost every pop mix. They should always be independently audible, never masking each other (see Chapter 3). The rest of the tracks sit below and beside, never masking these four tracks.

The vocal must reign supreme. Beware of other tracks with strong spectral presence in the middle frequencies. Spectral masking may make it difficult to hear the consonants of the vocals. The words then become difficult to understand. Beware of other instruments playing in the same range as the vocal. A piano, guitar, sax, or cello can render a vocal inaudible if the parts are too similar. Solo sections and instrumental tunes often have a featured solo instrument in lieu of a lead vocal. Give that solo instrument the same priority a lead vocal would get. Find fader positions for the tracks that keep midrange harmony instruments out of the way enough so that the listeners can understand and enjoy the phrasing of the singer or the soloist. Pan the various nonvocal tracks competing in the middle frequencies off to opposite sides, out of the way of the lead vocal and each other. Instruments with similar pitch and spectral content will likely blur together and cloud the enjoyment of each individual instrument. Pan them to different locations for an immediate improvement in clarity and multitrack independence.

While the vocal is typically the most important single track in the entire production, it is not unusual for the snare drum to rival it in loudness. In fact, the snare might even be a bit louder than the lead vocal in some songs. Because the snare is so short in duration, the temporary masking it causes is sometimes (but certainly not always) acceptable. There is a natural tendency to push the snare fader up. A snare that sounds great while soloed may sound disappointing when placed back into the full context of the mix. Keys and guitars will fight the snare in the midrange. Cymbals and distortion (see Chapter 4) can create unwanted masking in the higher frequencies. Bass can rob the snare of low-frequency punch and power. But do not turn the snare up too much; if too loud, the snare loses musical value, becoming a distraction rather than a source of rhythm and energy. The snare drum gets constant attention. Set the fader so it is always an enjoyable part of the multitrack production.

The kick drum and bass guitar may have a low-frequency tug-of-war. Set their levels very carefully. A song with a lot of bass can be thrilling on first listen, but if the bass guitar is too loud, the kick drum gets masked. Choose instruments, tunings, playing styles, musical parts, and recording techniques that aim for at least slightly different frequency ranges, kick versus bass. Use signal processing at mixdown to ensure the result. Typically, the deep tone of the kick drum is lower in frequency than any low emphasis of the electric bass, but there are exceptions. A small kick drum, tuned to an upper low-frequency heartbeat in a production using a five-string bass with the low string tuned down to the B slightly more than three octaves below middle C (and therefore having a fundamental frequency of about 31 Hz) can turn things around. In cases such as these, it might make sense to place the bass guitar below the kick drum, spectrally. In either case, these two instruments must be shown how to get along together. There is a natural tendency for them to fight. The fader positions for these two instruments are given deliberate attention. Raise their levels to points that are clearly too loud and then pull them back. Lower their level until they are too soft, and then push them back up. Somewhere between these two positions lives the most effective musical level. Engineers constantly message the levels to both teach themselves the dangers of losing control of these tracks and to make sure their levels make the most musical sense. The mix achieves new levels of refined clarity when the kick and bass cooperate.

The other parts of the multitrack arrangement (both tracks and effects) fill in around and underneath these most important tracks. If the guitar is louder than the vocals, the band is probably going to sell fewer disks and downloads. If the music fans can not hear the piano when the sax plays, the song loses musical impact (and the engineer probably loses the chance to hire that piano player on the next album). Work the faders hard to find a balance that is fun to listen to, supports the music, and reveals all the complexity and subtlety of the song.

Balancing a mix sounds so straightforward, in concept. Engineers who are early in their career will soon discover that it is not easy. Keeping a mix balanced requires experience, excellent listening skills, patience, focus, and a strong musical opinion. On every session of every project, the balance of the mix directly drives everyone's opinion of the quality of the project. It also influences their ability to get their job done. In order to play, they have to be able to hear. The engineer is on the spot, expected to balance the tune and keep it balanced at all times.

Yet on any given session, there are other points of focus. The drummer needs attention during basics. The singer is everything during vocal overdubs. The compressors and reverbs are complex devices that can distract an engineer during mixdown. No matter what other session priorities are present, no matter what other difficulties and distractions arise, the engineer is always required to keep the production balanced. During a guitar overdub, the guitar might be allowed to be overly emphasized in the mix. As soon as the session advances from guitar to piano overdubs, however, that guitar must be turned down and tucked into the mix so that the vocal, snare, kick, bass, and all other supporting elements can be enjoyed.

This first step of a mix session is really a part of every session. For tracking and overdubbing, the players can not play, the engineer can not hear, and the producer can not produce until the signals from all the live micro-phones, all the tracks previously recorded, and all the effects are brought into balance. Relying almost entirely on volume controls, balancing a mix is one of the most important skills an engineer must master.

8.2.2 IDEAL LEVEL

Every track of the multitrack production begins its life at that critical moment when it is recorded to tape or disk. If a microphone captures the music, a microphone preamp is needed. Of course, microphone preamplifiers are nothing more than volume devices. An engineer has got to set the microphone preamp level just right when recording to tape or hard disk.

Because of the noise floor of all equipment, it is desirable to try to record music at the highest level allowed so that the musical waveforms mask the noise floor. So it seems true that louder is indeed better. The question is how loud. The strategy for recording levels divides into two types, depending on whether the storage format is digital or analog.

Most engineers have heard that, for digital recording (whether to tape or hard disk), the goal is to "print the signal as hot as possible, without going over." This deserves a closer look.

Digital recording systems convert the amplitude of the music signal into a string of numbers. Pressure in the air becomes voltage on a wire (thanks to the microphone), which then becomes numbers on tape or disk (thanks to the analog-to-digital converter). When the digital audio recorder has reached the largest number that it can handle, it maxes out. As the air pressure of the music increases in the air, the corresponding voltage on the microphone cable increases too. When the input into the analog-to-digital converter is too high, the numbers to be stored by the digital system can not get any bigger. When the signal gets too hot, a digital audio system runs out of the ability to keep up with that signal, sort of like a child not being able to keep counting once they run out of fingers. The child just stops at 10 fingers. The 16-bit system stops at: 1111 1111 1111 1111. The 24-bit system stops at: 1111 1111 1111 1111 1111 1111. No higher value can be stored. When the amplitude of the input forces the digital system to "run out of fingers," the digital data no longer follows the musical waveform (Figure 8.4). This is a kind of distortion known as hard clipping, discussed in Chapter 4. Play back the numbers, and the digital recorder can not recreate the original music. The peaks are clipped off, gone forever. Harmonic distortion favoring odd harmonics is the unmistakable result.

To prevent this kind of distortion, it is important to make sure the analog levels going into the analog-to-digital converter never force the system past

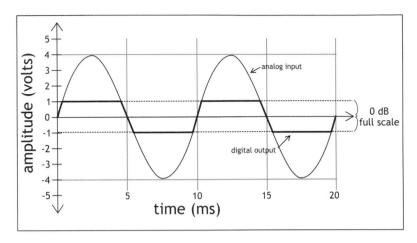

▲ *Figure 8.4 Analog-to-digital converters will hard clip a waveform when overloaded.*

its maximum. Meters are essential here. Digital recorders generally have meters that measure the amplitude of the signal in decibels below full scale. At full scale, the digital system has reached its maximum digital value. Inputs below this level are successfully converted from analog to digital. Inputs above this level will not be stored with appropriately large numbers. Clipping commences. It is the engineer's job to make sure the vocal, snare, or didgeridoo sound being recorded is never so loud that the digital system exceeds full scale and runs out of numbers.

Well . . . mostly. Engineers who are intrigued by the squared-off waveform shown in bold in Figure 8.4, and wonder what it sounds like, might want to overdrive the digital system on purpose. This is certainly allowed, if the engineer is careful. First, monitor at a low level. This kind of distortion is full of high-frequency energy and can melt tweeters. Second, listen carefully. This type of distortion is extremely harsh; it is not a particularly natural or musical effect. Use it sparingly, if at all.

A small amount of digital overload on a snare hit might give it a desirable, aggressive, unnatural sound that fits into some production styles. The instantaneous burst of harmonics associated with hard clipping adds complexity to the timbre of the snare sound during its brief percussive impact. Digital overload on a more continuous sound, such as a vocal or a rhythm guitar, should be reserved as a rare special effect, as it is almost always a rather unpleasant, fatiguing sound.

For analog magnetic recording systems, the typical approach is to record as hot as possible, and occasionally go over and into the red. Unlike digital

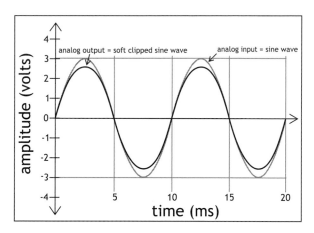

▲ *Figure 8.5 Analog magnetic tape recorders will soft clip a waveform.*

audio, analog audio does not typically hit such a hard-and-fast limit. When the level of the music being recorded gets too high, analog magnetic tape distorts only gradually. This less-abrupt distortion at the peaks is called *soft clipping*, shown in Figure 8.5 and discussed in Chapter 4. At lower amplitudes, the analog magnetic storage medium tracks fairly accurately with the waveform. As the audio signal starts to reach higher amplitudes, the analog storage format can not keep up. It starts to record a signal that is a little lower in amplitude. As analog tape runs out of amplitude capability, it does so gradually and gracefully. Analog magnetic tape, when recorded at too hot a level, distorts, but in a less abrupt way than digital audio systems.

A careful look reveals that overdriving analog tape might have a lot in common with compression (see Chapter 6). In fact, recording at high levels to analog tape is called *tape compression*. No compressor will exactly mimic tape compression, as the way it reshapes the amplitude detail of an audio waveform comes from a kind of low-threshold, variable-ratio compression with infinitely fast attack and release. Tape compression moves in and out on the peaks of the complex waves, affecting the amplitude of each individual peak, whether part of a low-frequency or high-frequency signal.

Augmenting those rack spaces and pulldown menus full of compressors, an engineer can overdrive analog magnetic recorders as an effect. It is not just a storage device; analog tape is part EQ, part distortion, part compression too. Learning to use analog formats to *process* sound is essential for experienced engineers.

8.2.3 IDEAL FADERS

On a split console, engineers dedicate certain input/output modules of the console or digital audio workstation to recording signals to the multitrack, and others to monitoring those signals (see Chapter 2). On an in-line console, every module has two faders: one for the record path (also known as the channel path) and the other for the monitor path (also called the mix path).

Two different forces motivate the engineer when setting the level of the channel faders and the monitor faders. On the record path, the fader determines the level of the signal to the multitrack recorder. Analog or digital, the engineer makes the best use of the dynamic range available. It is primarily a technical decision, balancing signal-to-noise ratio versus headroom.

On the monitor path, the level settings for these faders are almost entirely a creative decision. The engineer chooses the appropriate level of each and every track in the multitrack arrangement so that the desired musical statement is made. Balance the mix, and make it beautiful, sad, exciting, _____ (insert production goal here).

8.2.4 AUTOMATION

In a digital audio workstation, almost every adjustable parameter can be automated. In the hands of a clever engineer, a digital audio workstation can be instructed to wake us to music first thing in the morning (noon), start the coffee maker, check email, and draw a warm bath. While this is all quite useful, automation is almost always used for just two very simple processes: fader rides and mutes. Automation is, in essence, a volume processor. Chapter 15 is dedicated to the subject.

For example, using the humble mute switch, the mix engineer controls the multitrack arrangement: cut the bass in the extra bar before the chorus, pull the flute out of the horn section until the last chorus, etc. These sorts of details can enhance the tune and help keep listeners interested. It is accomplished very simply, using some well-placed automated cuts. This sort of mix move happens throughout pop music. But check out an extreme example by listening to U2's *Achtung Baby*. The album begins with some heavy cut activity as the drums and bass enter at the top of the first tune of the album, "Zoo Station."

Automating fader rides in support of the arrangement is a natural application of automation. Maybe it makes sense to push the guitar up in the choruses, pull the Chamberlain down during the guitar solo, and such. Ideally, the band (maybe with the advice of a producer) gets these dynamics right in their performance. Players know intuitively when to push their level up a little as the song develops. But in the studio, the full arrangement of the song may not come together for several months, as overdubs are gradually added to the tune. Fader rides may be just the ticket to help this assembly of tracks fall into a single piece of music with appropriate expressions of volume.

Volume changes are automated just to keep the song in balance (discussed above) as multitrack components of the song come and go. Typically, it is best to keep these moves quite subtle. These rides are aimed at the musical interpretation of the mix, trying to make the song *feel* right. With few exceptions, it should never sound like a fader was moved. Listeners want to hear the music, not the mixing console. Just a small ride here and another one there will help shape the energy level and mood of the mix.

Another automated volume effect is the automated send. Although the aux send knob is not usually automated, it can be (see Chapter 15). Some very sophisticated mix elements are created with automated sends. In this way, automation is employed to add rich and spacious reverb to the vocal in the bridge only, introduce rhythmic delay to the background vocals on key words, increase the chorus effect on the orchestral strings in the verses, add distortion to the guitar in the final chorus, etc. Automate the additional effect in and back out of the mix wherever it is wanted. The automated send — just another volume effect — offers the engineer a way to layer in areas of more or less effects, using nothing more than straight-forward faders and cuts automation.

8.2.5 CROSS FADES

When editing analog tape, one generally chooses a pronounced sonic moment, such as a snare hit, and cuts the tape at an angle. Placing the edit just before a snare hit helps keep the engineer oriented while editing. The snare hits provide a sonic clue identifying major increments in musical time in the song form. The snare hit also performs a bit of forward masking (see Chapter 3), making the cut in tape that occurs the instant before the snare hit difficult, if not impossible, to hear.

Cutting the tape at an angle makes the transition from one part of the tape to another gentler. Rather than having the pattern of magnetism on tape change suddenly from one performance to another, the angled cut magnetically cross fades between the two edit pieces.

Editing in a digital audio workstation offers more opportunities for rearranging a song. The ability to zoom in on the waveform enables great precision in choosing the edit point. The fact that digital edits may be nondestructive to the audio waveform, amounting only to a difference in the way the computer plays back the data, not an alteration to the digital audio data itself, makes cutting and rearranging an audio waveform as simple as cutting and pasting text. The engineer can try an edit, audition it, undo it, or modify it, ad nauseam. Digital audio workstations have made editing a much more frequent part of the music production process. Physical cuts into analog magnetic tape are more difficult to undo. Zooming in on analog tape is unproductive as the waveform is not visible.

Edits are quite volume dependent. Making edits between the notes, when the instrument is at least briefly silent (i.e., not playing) is a good first strategy. Further, performing all edits at zero crossings, so that the amplitude of the audio at the point of edit is zero, is a good habit. In fact, most digital editors can be persuaded to default to zero crossings for all edits. In this way, the amplitude of the signal at the instant of the edit is as low as practical.

Frequently the goal is to remove any audible artifact associated with the edit so that the listener feels the edited performance is the original performance. In this case, a cross fade is frequently required. The cross fade amounts to a fading out of the first event during the fading in of the second event. There are production situations in which cross fades need to be as short as a few milliseconds, or as long as several thousand milliseconds.

Steady-state signals, such as string pads, piano notes in long decay, long reverb tails, and hiss, can be unforgiving of edits. Cutting from one part of the steady sound to another similar steady sound often causes a noticeable click, thump, or other sonic discontinuity at the edit point. Any abrupt change from one steady-state sound to another, similar sound will reveal even slight differences between them. Editing during a distracting event can mask this. In addition, a cross fade between these similar, but not identical, steady-state sounds might make the edit aurally invisible.

Low-frequency signals, with their relatively slowly cycling waveforms, are intolerant of changes to the progress of those slow cycles. Low-frequency instruments are particularly difficult to edit. Often, it is not enough to work at zero crossings. A brief cross fade will smooth the edit.

8.2.6 FADE OUT

Commercial radio stations can not endure a fade out. The DJ will start talking, the next song will begin, and/or the advertisement will be played before a song is ever allowed to fade to silence. Live concerts make the fade impossible. Yet, many music recordings fade out.

It is a practical solution to a sometimes difficult problem. Composers find it challenging to end a piece, and may not compose an actual ending. Bands find it difficult to stop cleanly; the end can be more difficult than the beginning of a tune. When the band flies through a take, thinking, "This is the one! This is the keeper! We are going to be stars!," the last thing they need to worry about is a tight ending, which, if someone rattles on too long, renders the whole take unusable. It is sometimes deemed better to play several extra bars or an extra repeat of the chorus and fade it in the mix. Sometimes mistakes made by one or more performers may need to be hidden. The fade out covers any number of sins.

The fade is a musical statement, however. If used to simply cover over a problem, the result can be musically weak. Listeners may not be consciously aware of the cause, but will feel the music failed to finish, to take them somewhere, to achieve fruition. The fade out should not be the assumed ending, it should be a deliberate ending. Bands who arrange and rehearse actual endings to their songs are taking advantage of the ending of the tune to have a created, musical effect.

8.3 Selected Discography

Artist: U2
Song: "Zoo Station"
Album: *Achtung, Baby*
Label: Island Records
Year: 1991
Notes: The power of the cut button is displayed in the intro to the song *and* the album as the drums and bass enter.

Artist: The Smiths
Song: "Hand in Glove"
Album: *Hatful of Hollow*
Label: Rough Trade Records
Year: 1984
Notes: The tune fades out, back in, and then out again. Listeners can only wonder what possibly went wrong and had to be hidden.

Section 3
Time FX

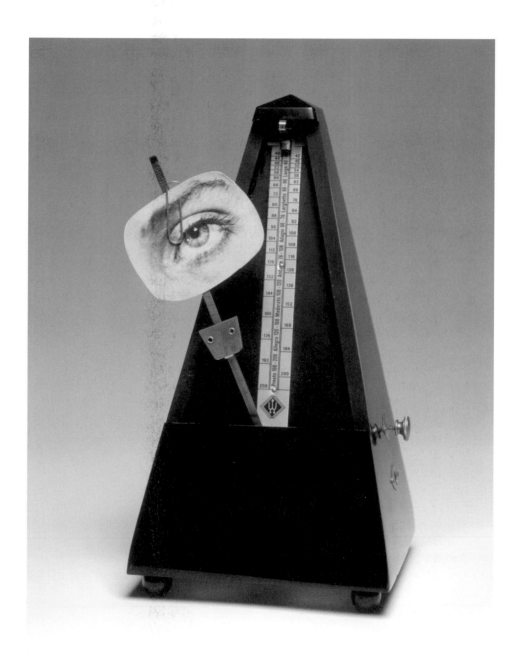

Indestructable Object, 1965. Man Ray (1890–1976). Photo credit: Smithsonian American Act Museum, Washington, D.C./Art Resource, NY. © 2006 Man Ray Trust/Artists Rights Society (ARS), NY/ADAGP, Paris.

Delay

"Nothing ever happens.
Nothing happens at all.
The needle returns to the start of the song,
and we all sing along like before."
— "NOTHING EVER HAPPENS," DEL AMITRI, *WAKING HOURS* (A&M RECORDS, 1989)

The delay line is the single most important signal processor used for manipulating the time axis of an audio signal. In concept, it could not be simpler: delay all signals fed to the device. Audio goes in. It waits the designated amount of time. Audio comes out. Simple on the surface, this process becomes the building block for a vast range of effects, and is a required processor in all recording studio environments.

This chapter begins with some theory, summarizing delay technologies frequently used in recording studios and describing the basic signal flow through a delay processor. This is followed by a detailed look at the many creative applications of delay common in professional music production.

9.1 Source of Delay

The delay in time may be generated in a number of ways — in software or hardware, analog or digital. The digital delay is perhaps the most intuitive source of delay. Once audio is made digital, it is a string of numbers as willing to be stored in random access memory as any word-processing document or web page. All that is needed is a bit of memory management and a user interface. Any computer should be able to do this. Not surprisingly, digital audio workstations are very good at this. Dedicated digital devices — stand-alone units that offer digital delay outside of the computer — are also available.

Purely analog delay lines also exist, though they become increasingly rare as digital audio becomes ever more available, affordable, and capable.

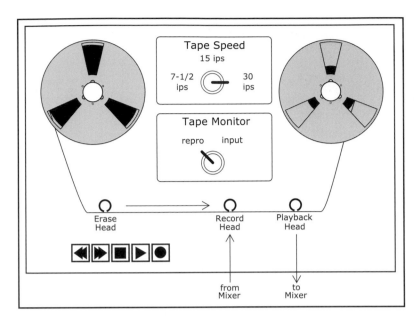

▲ *Figure 9.1 Tape delay.*

Inductors and capacitors are the basic analog circuit components used to introduce some time delay to an analog signal.

An analog tape machine — perhaps sitting underutilized somewhere in your studio — is also a clever source of delay (see Figure 9.1) and is the basis for a whole class of delay effects, discussed in more detail below.

9.2 Signal Flow

Figure 9.2 demonstrates a good starting point for routing audio signals through the console in order to create many of the delay effects discussed in this chapter (see also "Outboard Signal Flow" Chapter 2).

Set up a postfader aux send from the source track to the delay. Return the output of the delay to a separate fader. This sends the desired signal, with all of its carefully thought-out effects like compression (see Chapter 6) and equalization (see Chapter 5), to the delay processor. The delay output is patched into its own mixer input so that the engineer has full control of the return from the delay on its own fader.

Hooking it up as shown in Figure 9.2 and setting the delay time to 200 milliseconds (ms) would add an echo to the source track for the duration of the performance.

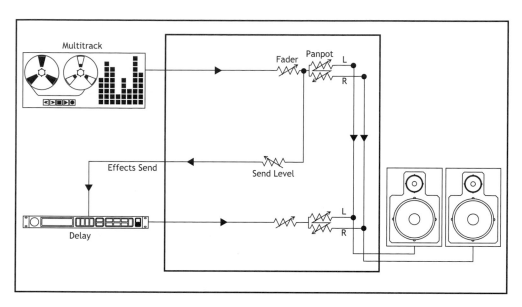

▲ *Figure 9.2 Signal flow at the mixing console.*

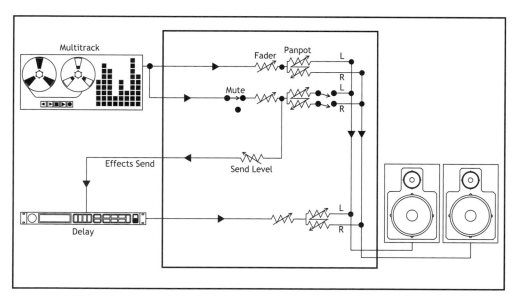

▲ *Figure 9.3 The "automated" effects send.*

To introduce delay at discrete moments rather than for the entire song, patch up an "automated send," shown in Figure 9.3. Here the engineer has a fader and cut button dedicated to the control of the *send* into the delay, not just the return. This echo send might remain cut most of the time. Unmute the send only when the effect is desired.

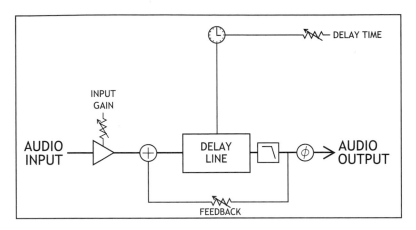

▲ *Figure 9.4 Typical signal flow within the delay unit.*

9.2.2 BASIC FEATURES

What is going on at the delay processor itself? Most delay devices available have the controls shown in Figure 9.4: input and/or output level, delay time, and regeneration control. Input/output levels are self-explanatory. Levels are typically set so that there is unity gain through the unit. Watch the input level to make sure the device is not driven into distortion (unless, of course, distortion is part of the intended effect; see Chapter 4).

The regeneration control, sometimes called the feedback control, sends some of the output of the delay back to the input of the delay. In this way, a delayed signal can be further delayed by running it back through the delay again. This is how an echo is made to repeat more than once.

The simple controls of Figure 9.4 empower the delay to become a fantastically diverse signal processor.

9.2.2 MODULATION

The effect becomes more interesting still when the delay time is allowed to *change*. Many types of signal-processing effects are built on varying delays. Figure 9.5 adds a modulation section to the standard controls on a digital delay device. These controls are an elegant addition to the signal-processing capability of a delay.

Figure 9.6(a) describes a fixed delay time of 100 ms. This creates a slap echo as discussed in detail below. The delay unit takes whatever signal it is sent, holds it for the delay time specified (100 ms), and then releases it. That's it.

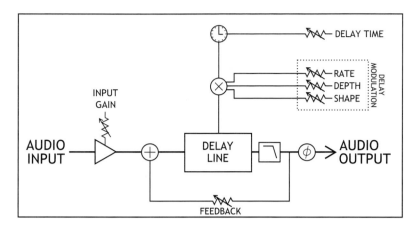

▲ *Figure 9.5 Signal flow for a modulated delay.*

Throughout the song, all session long, the delay time remains exactly 100 ms; all signals sent to it — guitar, vocal, or didgeridoo — experience the exact same amount of delay. That is a delay without modulation.

The graphs of Figure 9.6 may seem peculiar at first, plotting time on the vertical axis versus time on the horizontal axis. Seemingly similar, the two time axes are wholly independent variables. The vertical axis represents the delay time parameter within the delay unit. The horizontal axis represents the typical human experience of time, flying by, if we are having fun. Figure 9.6 (a–d), therefore, plots the delay time parameter as time goes by.

Some beautiful and powerful effects require the use of a changing delay, using the modulation controls: rate, depth, and shape.

Rate controls how quickly the delay time parameter with the effects device is changed. Figure 9.6(b) gives a graphic representation of what happens when this control is changed. Engineers find cases when they want to sweep the delay time imperceptibly slowly, and other times where they need a fast, very audible rate.

Depth controls how much the delay is modulated. It bounds the delay time at the extreme, defining the shortest and the longest delay times allowed. Figure 9.6(c) graphically contrasts two different settings. The original, fixed delay time might be increased and decreased by 5 ms, 50 ms, or more.

Shape describes the path taken by the device as it changes the delay time within the bounds set by the depth control, at the speed determined by the

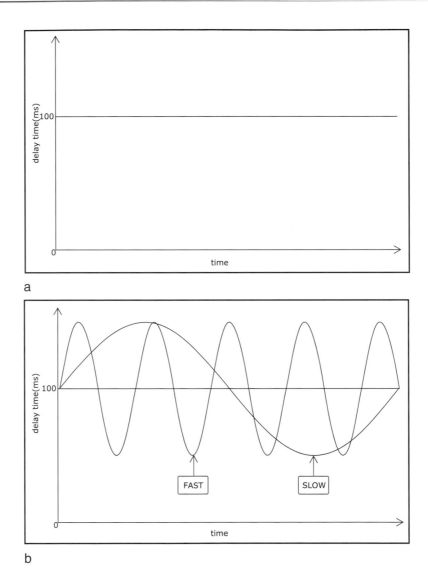

▲ *Figure 9.6 (a) Fixed delay; (b) changing delay: rate; (c) changing delay: depth; and (d) changing delay: shape.*

rate control. As Figure 9.6(d) shows, it can sweep in a perfect, sinusoidal shape back and forth between the upper limit and the lower limit specified (those upper and lower delay limits were set with the depth control described above). Alternatively, there may be a need for a square wave trajectory between delay times, in which the delay time snaps instead of sweeps, from one delay value to the other. Figure 9.6(d) highlights an additional common feature of the shape control: It lets the engineer use a shape that is some mixture of the two — part sinusoid, part square.

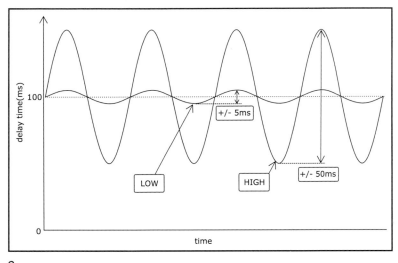

c

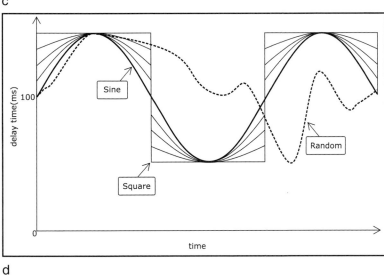

d

▲ *Figure 9.6 Continued*

Beyond these sine and square wave modulation shapes, the delay may have a random setting in which the delay time moves less orderly between the two delay extremes.

Finally, some delay units let you use a combination of all of the above; for example, varying the delay time in a slightly random, mostly sinusoidal general pattern. The shape control may make it possible to mix these shape options and set a unique contour for the delay's motion between its highest and lowest settings.

These three modulation parameters give the recording engineer much-needed control over the delay, enabling them to play it like a musical instrument. They set how fast the delay moves (rate). They set the limits on the range of delay times allowed (depth), and they determine how the delay moves from its shortest to its longest time (shape).

Readers familiar with using a low frequency oscillator (LFO) to modulate signals in some synthesizers, for example, will recognize that the modulation section of a delay unit relies on a simple LFO. Rate is the frequency of the LFO. Depth is the amplitude of the LFO. Shape, of course, is the type of LFO signal. Instead of modulating the amplitude of a signal, as might be done in an A.M. (amplitude modulation) synthesizer, this LFO modulates the delay time parameter within the signal processor.

9.3 Long Delay

Not all delays are created equal. In order to understand the creative sonic possibilities for a delay effect, it helps first to separate them into different categories based on the length of the delay (Figure 9.7). Delays are classified into three broad categories, cleverly called long (greater than about 50 ms), medium (between about 20–50 ms), and short (less than about 20 ms).

Consider the long delay. A delay is classified "long" when the delay time is no less than about 50 ms and as long as . . . well, as long as you are willing to wait around for the music.

Music (available in the mind's ear, or in the studio) will illustrate the musical potential of the long delay quite nicely. Hum or sing along with Pink Floyd's tune "Comfortably Numb" from *The Wall*. There is that all-important first

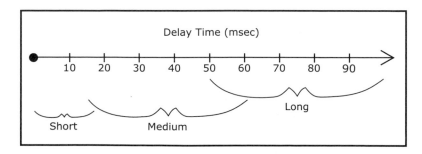

▲ *Figure 9.7 Delay effects divide into three classes based on the delay time.*

line: "Hello . . . hello . . . hello. Is there anybody in there?" That repeating of "hello" is a classic use of a long delay. The dreamy, disturbed, out-of-mind state of our walled-in friend and hero, Pink, is enhanced by (the entire, brilliant rock-and-roll arrangement, including) this repeating, gently fading echo.

How is it done? Perhaps the simplest way is to use a postfader aux send from the vocal to the delay, which is returned on a separate fader (as shown in Figure 9.2). The voice is sent to the delay processor, which when set to a long delay, creates the echo.

But there is a headache here. Hooking it up as shown in Figure 9.2 adds an echo to the entire vocal performance, not just the word, "hello." Each and every word repeats. This leads to a tiresome, distinctly nonmusical bit of vocal chaos. It is hard to find the melody. The lyrics are difficult to understand, obscured by the relentless output from the delay as each and every word happens two, three, or more times. Echo applied to the entire vocal track in this way is rarely useful.

The solution is the automated send shown in Figure 9.3. Now the engineer has a fader and cut button dedicated to the control of the *send* into the delay, not just the return. The echo send remains cut most of the song. Open it up briefly for the single word, "hello," and, presto, that single word starts to repeat and fade. The rest of the vocal line, not sent to the delay, fits nicely around this discrete, one-word echo. No distraction. No trouble understanding the words. The melody is easy to follow and the rest of the musical arrangement is easy to hear.

It may take some practice getting the feed into the delay unit just right. The engineer must unmute the effects send just before the word "hello" and mute it again just after. You may find it helpful to imagine yourself singing the line. With a musician's ear for every nuance of a performance, learn the exact phrasing of this track. Know when the singer breathes, learn how long the singer sustains each vowel, and master the exact rhythm of each syllable.

Getting into the musical performance in this way, the engineer finds the brief spaces between words open up and become easier to find. With practice, this becomes an easy, intuitive bit of hand/ear coordination. Store this unmute/mute gesture into the mixer automation system, and a perfectly tailored delay effect occurs each and every time this part of the tune is played.

9.3.1 CALCULATED DELAY

Approximately 99.9% of the time echoes like these should be set to a time that makes musical sense. The engineer does not simply pick a random delay time. A *musical* delay time is carefully dialed in. Should it repeat with a quarter-note rhythm, an eighth note, a triplet?

Tuning by Ear

One proven way to find the right length of delay relies on the engineer's sense of rhythm. Try using the snare to "tune" a delay — to set a delay time that makes musical sense. Even if the plan is to add delay to the vocal, piano, or guitar, it is usually easiest to use the snare for setting the delay time both because it is a rhythm instrument, and also because it hits so often. So much of pop music has a back beat — the snare falling regularly on beat two and beat four. Send the snare to the delay and listen to the echo. Starting with a long delay time of about 500 ms, adjust the delay time until it falls onto a musically-relevant beat. This can be extremely confusing at first. It may help initially to pan the snare track off to one side, and the delay return to the other. It is pretty jarring to hear a delay fall at a nonmusical time interval. But when it is adjusted into the time of the music, everyone listening will instantly feel it. It is perhaps easiest to find a quarter-note delay, but with practice and concentration, finding triplet and dotted rhythms becomes perfectly intuitive.

After the delay time value appropriate to the tempo of the song is found by ear, pull the snare out of the effects send, and send the vocal (or whichever track is to be treated with echo) to the delay device instead.

Tuning Through Math

Sometimes studio personnel in search of a specific musical time value *calculate* a delay time instead. How is this calculated? It is time for a useful bit of algebra. If the tempo of the song is known in beats per minute (bpm) and the length of a quarter-note delay in milliseconds (Q) is desired, perform the following straightforward calculation:

- First convert beats per minute into minutes per beat by taking the reciprocal: bpm beats per minute becomes 1/bpm minutes per beat.
- Then convert from minutes to milliseconds: 1/bpm minutes per beat × 60 seconds per minute × 1,000 milliseconds per second.

- Putting it all together, the length of time of a quarter note in milliseconds per beat is:

$$Q = \frac{60 \times 1,000}{bpm}$$

$$Q = \frac{60,000}{bpm} \tag{9.1}$$

For example, a song with a tempo of sixty beats per minute ticks like a watch, with a quarter note occurring exactly once per second. Try using Equation 9.1:

$$bpm = 60$$
$$Q = \frac{60,000}{bpm}$$
$$Q = \frac{60,000}{60}$$

$Q = 1,000$ ms per quarter note, or

$Q = 1$ second per quarter note

Double the tempo to 120 bpm:

$$bpm = 120$$
$$Q = \frac{60,000}{bpm}$$
$$Q = \frac{60,000}{120}$$

$Q = 500$ ms per quarter note, or

$Q =$ half a second per quarter note

Milliseconds are used here for two reasons. First, the millisecond is the magnitude most delay units expect. Second, it typically leads to comfortable numbers in musical applications. For the frequencies we can hear, and for the effects we are likely to use, units in milliseconds generate numbers of manageable size — not too many decimal places, not too many digits. Using seconds, minutes, years, or fortnights would still work, theoretically. These are all units of time. But the millisecond is the more convenient order of magnitude.

217

Musical Delay Time

Calculating first the quarter-note delay makes it easy to then determine the time value of an eighth note, a sixteenth note, dotted or triplet values, etc. Table 9.1 can be a useful tool. Use the snare drum approach or the bpm conversion to find the time equivalent of a quarter note. Then use this table and some musical judgment to dial in the right type of echo-based effect.

In Pink Floyd's "Comfortably Numb" example, a dotted eighth-note delay is cleverly used. This is a deliberate production decision, not a happy accident. It is worth transcribing it for some production insight.

The tune is dreamy and lazy in tempo, moving at about 64 bpm. The two syllables of "hello" are sung as sixteenth notes. To count quarter notes, just count, "one, two, three, . . ." To count eighth notes, insert the syllable "and" in between the beats: "one and two and three and. . . ." To count the more complicated sixteenth-note rhythm, stick in two additional syllables to identify all parts of the sixteenth note pattern, "e," which rhymes with "free," and "a," like the a in "vocal." Sixteenth-note time becomes, "one e and a two e and a three e and a. . . ."

To appreciate the perfection in Pink Floyd's dotted eighth-note delay time, consider two other, perhaps more obvious, choices: a quarter-note delay or an eighth-note delay (see Table 9.2).

In Table 9.2, the quarter-note delay strongly emphasizes the time of the song; it is orderly and persistent:

Hello x x hello x x hello . . .

Each repeat of the word falls squarely on the beat. This would make it seem like Pink is being nagged or pushed around, and the very orderliness of the quarter-note repetition takes away from the soporific state intended by this composition.

An eighth-note delay, on the other hand, forces the words to fall immediately and persistently one after the other, with no rest in between:

Hello hello hello hello hello.

This is simply annoying, and stressful, which is not the desired musical effect.

	MILLISECONDS PER NOTE VALUE								
					dotted			triplet	
bpm	quarter	eighth	sixteenth	quarter	eighth	sixteenth	quarter	eighth	sixteenth
40	**1,500**	**750**	**375**	**2,250**	**1,125**	**562.5**	**1,000**	**500**	**250**
41	1,463	732	366	2,195	1,098	549	976	488	244
42	1,429	714	357	2,143	1,071	536	952	476	238
43	1,395	698	349	2,093	1,047	523	930	465	233
44	1,364	682	341	2,045	1,023	511	909	455	227
45	**1,333**	**667**	**333**	**2,000**	**1,000**	**500**	**889**	**444**	**222**
46	1,304	652	326	1,957	978	489	870	435	217
47	1,277	638	319	1,915	957	479	851	426	213
48	1,250	625	313	1,875	938	469	833	417	208
49	1,224	612	306	1,837	918	459	816	408	204
50	**1,200**	**600**	**300**	**1,800**	**900**	**450**	**800**	**400**	**200**
51	1,176	588	294	1,765	882	441	784	392	196
52	1,154	577	288	1,731	865	433	769	385	192
53	1,132	566	283	1,698	849	425	755	377	189
54	1,111	556	278	1,667	833	417	741	370	185
55	**1,091**	**545**	**273**	**1,636**	**818**	**409**	**727**	**364**	**182**
56	1,071	536	268	1,607	804	402	714	357	179
57	1,053	526	263	1,579	789	395	702	351	175
58	1,034	517	259	1,552	776	388	690	345	172
59	1,017	508	254	1,525	763	381	678	339	169
60	**1,000**	**500**	**250**	**1,500**	**750**	**375**	**667**	**333**	**167**
61	984	492	246	1,475	738	369	656	328	164
62	968	484	242	1,452	726	363	645	323	161
63	952	476	238	1,429	714	357	635	317	159
64	938	469	234	1,406	703	352	625	313	156
65	**923**	**462**	**231**	**1,385**	**692**	**346**	**615**	**308**	**154**
66	909	455	227	1,364	682	341	606	303	152
67	896	448	224	1,343	672	336	597	299	149
68	882	441	221	1,324	662	331	588	294	147
69	870	435	217	1,304	652	326	580	290	145
70	**857**	**429**	**214**	**1,286**	**643**	**321**	**571**	**286**	**143**
71	845	423	211	1,268	634	317	563	282	141
72	833	417	208	1,250	625	313	556	278	139
73	822	411	205	1,233	616	308	548	274	137
74	811	405	203	1,216	608	304	541	270	135
75	**800**	**400**	**200**	**1,200**	**600**	**300**	**533**	**267**	**133**
76	789	395	197	1,184	592	296	526	263	132
77	779	390	195	1,169	584	292	519	260	130
78	769	385	192	1,154	577	288	513	256	128
79	759	380	190	1,139	570	285	506	253	127
80	**750**	**375**	**188**	**1,125**	**563**	**281**	**500**	**250**	**125**
81	741	370	185	1,111	556	278	494	247	123
82	732	366	183	1,098	549	274	488	244	122
83	723	361	181	1,084	542	271	482	241	120
84	714	357	179	1,071	536	268	476	238	119
85	**706**	**353**	**176**	**1,059**	**529**	**265**	**471**	**235**	**118**
86	698	349	174	1,047	523	262	465	233	116
87	690	345	172	1,034	517	259	460	230	115

Table 9.1 Converting Tempo into Time

(cont.)

Table 9.1 Converting Tempo into Time—Continued

	MILLISECONDS PER NOTE VALUE								
bpm	quarter	eighth	sixteenth	quarter	dotted eighth	sixteenth	quarter	triplet eighth	sixteenth
88	682	341	170	1,023	511	256	455	227	114
89	674	337	169	1,011	506	253	449	225	112
90	**667**	**333**	**167**	**1,000**	**500**	**250**	**444**	**222**	**111**
91	659	330	165	989	495	247	440	220	110
92	652	326	163	978	489	245	435	217	109
93	645	323	161	968	484	242	430	215	108
94	638	319	160	957	479	239	426	213	106
95	**632**	**316**	**158**	**947**	**474**	**237**	**421**	**211**	**105**
96	625	313	156	938	469	234	417	208	104
97	619	309	155	928	464	232	412	206	103
98	612	306	153	918	459	230	408	204	102
99	606	303	152	909	455	227	404	202	101
100	**600**	**300**	**150**	**900**	**450**	**225**	**400**	**200**	**100**
101	594	297	149	891	446	223	396	198	99
102	588	294	147	882	441	221	392	196	98
103	583	291	146	874	437	218	388	194	97
104	577	288	144	865	433	216	385	192	96
105	**571**	**286**	**143**	**857**	**429**	**214**	**381**	**190**	**95**
106	566	283	142	849	425	212	377	189	94
107	561	280	140	841	421	210	374	187	93
108	556	278	139	833	417	208	370	185	93
109	550	275	138	826	413	206	367	183	92
110	**545**	**273**	**136**	**818**	**409**	**205**	**364**	**182**	**91**
111	541	270	135	811	405	203	360	180	90
112	536	268	134	804	402	201	357	179	89
113	531	265	133	796	398	199	354	177	88
114	526	263	132	789	395	197	351	175	88
115	**522**	**261**	**130**	**783**	**391**	**196**	**348**	**174**	**87**
116	517	259	129	776	388	194	345	172	86
117	513	256	128	769	385	192	342	171	85
118	508	254	127	763	381	191	339	169	85
119	504	252	126	756	378	189	336	168	84
120	**500**	**250**	**125**	**750**	**375**	**188**	**333**	**167**	**83**
121	496	248	124	744	372	186	331	165	83
122	492	246	123	738	369	184	328	164	82
123	488	244	122	732	366	183	325	163	81
124	484	242	121	726	363	181	323	161	81
125	**480**	**240**	**120**	**720**	**360**	**180**	**320**	**160**	**80**
126	476	238	119	714	357	179	317	159	79
127	472	236	118	709	354	177	315	157	79
128	469	234	117	703	352	176	313	156	78
129	465	233	116	698	349	174	310	155	78
130	**462**	**231**	**115**	**692**	**346**	**173**	**308**	**154**	**77**
131	458	229	115	687	344	172	305	153	76
132	455	227	114	682	341	170	303	152	76
133	451	226	113	677	338	169	301	150	75
134	448	224	112	672	336	168	299	149	75

(cont.)

Table 9.1 Converting Tempo into Time—Continued

	MILLISECONDS PER NOTE VALUE								
					dotted			triplet	
bpm	quarter	eighth	sixteenth	quarter	eighth	sixteenth	quarter	eighth	sixteenth
135	**444**	**222**	**111**	**667**	**333**	**167**	**296**	**148**	**74**
136	441	221	110	662	331	165	294	147	74
137	438	219	109	657	328	164	292	146	73
138	435	217	109	652	326	163	290	145	72
139	432	216	108	647	324	162	288	144	72
140	**429**	**214**	**107**	**643**	**321**	**161**	**286**	**143**	**71**
141	426	213	106	638	319	160	284	142	71
142	423	211	106	634	317	158	282	141	70
143	420	210	105	629	315	157	280	140	70
144	417	208	104	625	313	156	278	139	69
145	**414**	**207**	**103**	**621**	**310**	**155**	**276**	**138**	**69**
146	411	205	103	616	308	154	274	137	68
147	408	204	102	612	306	153	272	136	68
148	405	203	101	608	304	152	270	135	68
149	403	201	101	604	302	151	268	134	67
150	**400**	**200**	**100**	**600**	**300**	**150**	**267**	**133**	**67**
151	397	199	99	596	298	149	265	132	66
152	395	197	99	592	296	148	263	132	66
153	392	196	98	588	294	147	261	131	65
154	390	195	97	584	292	146	260	130	65
155	**387**	**194**	**97**	**581**	**290**	**145**	**258**	**129**	**65**
156	385	192	96	577	288	144	256	128	64
157	382	191	96	573	287	143	255	127	64
158	380	190	95	570	285	142	253	127	63
159	377	189	94	566	283	142	252	126	63
160	**375**	**188**	**94**	**563**	**281**	**141**	**250**	**125**	**63**
161	373	186	93	559	280	140	248	124	62
162	370	185	93	556	278	139	247	123	62
163	368	184	92	552	276	138	245	123	61
164	366	183	91	549	274	137	244	122	61
165	**364**	**182**	**91**	**545**	**273**	**136**	**242**	**121**	**61**
166	361	181	90	542	271	136	241	120	60
167	359	180	90	539	269	135	240	120	60
168	357	179	89	536	268	134	238	119	60
169	355	178	89	533	266	133	237	118	59
170	**353**	**176**	**88**	**529**	**265**	**132**	**235**	**118**	**59**
171	351	175	88	526	263	132	234	117	58
172	349	174	87	523	262	131	233	116	58
173	347	173	87	520	260	130	231	116	58
174	345	172	86	517	259	129	230	115	57
175	**343**	**171**	**86**	**514**	**257**	**129**	**229**	**114**	**57**
176	341	170	85	511	256	128	227	114	57
177	339	169	85	508	254	127	226	113	56
178	337	169	84	506	253	126	225	112	56
179	335	168	84	503	251	126	223	112	56
180	**333**	**167**	**83**	**500**	**250**	**125**	**222**	**111**	**56**

1	e	&	a	2	e	&	a	3	e	&	a	the beat
Hel	lo											sung word
X	X	X	X	hel-	lo							**quarter-note delay**
X	X	hel-	lo									**eighth-note delay**
X	X	X	hel-	lo								dotted eighth-note delay
X	X	X	hel-	lo	X	hel-	lo	X	hel-	lo		**with regeneration**
Hel	**lo**	X	hel-	lo	X	hel-	lo	X	hel-	lo		**net effect**

Table 9.2 Evaluating the Musical Timing of Delays

The delay time chosen in the released recording has the effect of inserting a sixteenth-note rest in between each repeat of the word. "Hello" is sung on the downbeat. The echo never again occurs on a down beat. First it anticipates beat two by a sixteenth note, then it falls on the middle of beat two (called the "and" of two, because of the way eighth notes are counted). It next lands a sixteenth after beat three. Finally, it disappears as the next line is sung.

This timing scheme determines that "hello" will not fall squarely on a beat again until beat four, by which time the next line has begun and "hello" is no longer audibly repeating. It is really a pattern of three in a song built on four. This guarantees it a dreamy, disorienting feeling. It remains true to the overall "numb" feeling for the song's atmosphere, giving an uncertain, disconnected feeling. The result is a premeditated creation of the desired emotional effect. And it is a catchy hook — a real Pink Floyd signature.

9.3.2 ECHO

It is a funny idea to add an echo to a singer, or a piano, or a guitar that does not seem to have any motivation based on reality. The only way to hear an echo on the vocal of a song without the help of studio signal processing is to go to a terrible sounding venue (like an ice hockey rink or the Grand Canyon) and listen to music. The sound of an echo across the entire mix that occurs in these places — places not designed for music listening — is quite an unpleasant experience. It is almost always sonically messy and distinctly nonmusical. The echoes found in pop music tend to be used with more restraint. In some cases, the echo is added to a single track, not the whole mix. To keep things from becoming too confusing, the output of the delay is often mixed in at low level, so as to be almost inaudible. As Pink Floyd so ably demonstrates, another valid approach is to apply echo only to key words, phrases, or licks.

9.3.3 SUPPORT

If a constant echo is to be added to an entire track, the echo needs to be mixed in almost subliminally, nearly hidden by the other sounds in the mix. A soft echo underneath the lead vocal can give it added richness and support. This approach can strengthen the sound of the singer, especially when the melody heads into falsetto territory. Pulsing, subliminal echoes feeding a long reverb can create a soft and delicate sonic foundation under the vocal of a ballad (see Chapter 11).

Then there is the vulnerable rock-and-roll singer in front of his mate's Marshall stack. After the last chorus, the singer naturally wishes to scream "Yeaaaaaaaaah!" and hold it for a couple of bars. It is not easy to over-come the guitarist's wall of sound. Help the singer out by pumping some in-tempo delays into the scream.

The best "Yeaaaaaaaaah!" ever recorded in the history of rock and roll (according to your author, based on no data whatsoever) is Roger Daltrey's in "Won't Get Fooled Again" by The Who. The scream occurs right after the reintroduction, when those cool keyboards come back in, and right before the line, "Meet the new boss, the same as the old boss." This scream is a real rock-and-roll icon.

Listen carefully (especially at the end of the scream) and you will hear a set of delayed screams underneath. It is Roger Daltrey, only more so. It is half a dozen Roger Daltreys. It makes quite a statement. Anyone can do this. All they need is a long delay with some regeneration, and young Roger.

9.3.4 SLAP

A staple of 1950s rock is sometimes part of a contemporary mix: slapback echo. Music fans pretty much never heard Elvis without it. Solo work by John Lennon, therefore, often had it. Guitarists playing the blues tend to like it. Add a single audible echo somewhere between about 80 ms and as much as 200 ms, and each and every note shimmers and pulses a bit, courtesy of the single, quick echo. On a vocal, slap echo adds a distinct, retro feeling to the sound. Elvis and his contemporaries reached for this effect so often that it has become a cliché evocative of the period. Pop-music listeners today have learned to associate this effect with those happy days of the 1950s.

On guitar, slap echo makes a performance sound more live, putting the listener in the smoky bar with the band. Most music fans have experienced the music club, with that short echo of sound bouncing of the back wall of the venue. Slap echo can conjure up that specific experience for many listeners.

Before the days of digital audio, a common approach to creating this sort of effect was to use a spare analog tape machine as a generator of delay. During mixdown, the machine is constantly rolling, in record. The signal is sent from the console to the input of the tape machine in exactly the same way one would send a signal to any other effects unit — using an echo send or spare track bus. That signal is recorded at the tape machine and milliseconds later it is played back. That is, though the tape machine is recording, it remains in *repro* mode so that the output of the tape machine is what it sees at the playback head. As Figure 9.1 shows, the signal goes in, gets printed onto tape, the tape makes it's way from the record head to the playback head (taking time to do so), and finally the signal is played back off tape and returned to the console. The result is a tape delay.

The signal is delayed by the amount of time it takes the tape to travel from the record head to the repro head. The actual delay time then is a function of the speed of the tape and the particular model of tape machine in use (which determines the physical distance between the two heads).

Want to lengthen the delay time? Slow the tape machine down. There might be two, maybe three, choices of tape speed: $7^{1}/_{2}$, 15, or 30 inches per second.

None of these delay times seem exactly right? Maybe the tape machine has vari-speed, which lets the engineer achieve tape speeds slightly faster or slower than the standard speeds listed above.

Having trouble making these delay times fit into the rhythm of the song? Now the recording engineer is faced with the rather expensive desire to acquire another analog two-track machine, one with a different head arrangement so that the delay time will be different.

A single tape machine, which might cost several thousand dollars, is capable of just a few different delay settings. A three-speed tape machine used this way is like a really rather expensive effects device capable of only three delay time settings. Tape delay was originally used because it was one of the only choices at the time.

To help out, manufacturers made tape delays. These were tape machines with a loop of tape inside. The spacing between the record and playback heads was adjustable to give the engineer more flexibility in timing the delay. Now, in the twenty-first century, studios have more options. Life is good. Today, engineers can buy a digital delay that is easily adjustable, wonderfully flexible, cheaper than a tape machine, and it either fits in one or two rack spaces or exists conveniently in a pulldown menu on our digital audio workstation.

But those who have a spare open reel tape machine that has perhaps been sitting unused ever since they made the investment in a CD burner have the opportunity to create tape slap. It can even be a cassette deck if it has a tape/monitor switch to allow monitoring of the playback head while recording.

Why bother? Tape delay is more trouble and more expensive than many digital delays. But there is no denying it: Some great old recordings made effective use of it. That is reason enough for some engineers. Retro for retro's sake.

Engineers go to the trouble to use a tape delay when they want that "sound." An analog tape machine introduces it's own, subtle color to the sound. Mainly, it tends to add a low-frequency hump to the frequency content of the signal. The exact frequency and gain of this low-frequency emphasis depends on the tape machine, the tape speed, the tape gauge, and how the machine is calibrated. If one pushes the level to the tape delay into the red, that signature analog tape compression is introduced. At hotter levels still, analog tape saturation distortion results (see Chapter 4).

Tape delay becomes a more complex, very rich effect now. It is not just a delay. It is a delay plus equalizer plus compressor plus distortion device. It can be darn difficult to simulate digitally. It is sometimes the perfect bit of nuance to make a track special within the mix.

9.3.5 EMPHASIS

Adding a long delay to a key word, as in the Pink Floyd example, is a way to emphasize a particular word. It can be obvious, like the "hello" that begins the song. Simulating a call-and-response type of lyric, the delay is often a musical hook. The echo invites others to sing along. Alternatively, it can be more subtle. A set of emphasizing delays hits key words throughout "Synchronicity II" on The Police's final album, *Synchronicity*. The first line of every chorus ends with the word "away," which gets a little delay-based boost. Listen also to key end words in the verses: "face," "race," and, um, "crotch." These are a quick dose of several echoes, courtesy of the regeneration control. The Wallflowers' "One Headlight" on *Bringing Down the Horse* offers a great example of really hiding the delays. Listen carefully to the third verse. The words "turn" and "burn" each get a single, subliminal dotted quarter-note delay. The rest of the vocal track receives no such effect throughout the rest of the tune.

Similar, sparing use of echo occurs on the phrases of slide guitar lines in the middle solo section of "One Headlight". The guitar is panned left. The first echo falls dead center, and the second echo is panned to the right. It is a nice detail in a seemingly straightforward arrangement.

It is not unusual to apply a low-pass filter to these sorts of delays. Attenuating some of the high-frequency content from each repeat of the sound makes it sink deeper into the mix. When sound travels some distance in a room, the high-frequency portion of the signal is attenuated more quickly than the low-frequency part. High-frequency attenuation due to air absorption is an inevitable part of sound propagating through the air. As a result, a well-placed low-pass filter can be used to make any element of the multitrack project seem more distant. The brain seems to infer the air absorption and perceptually pushes the sound away from the listener.

Good delay units provide engineers with this filter as an option (as shown in Figures 9.4 and 9.5). Moreover, there is often the ability to double the delay time on outboard digital delays by pressing a button labeled "X2," meaning "times two." This cuts the sampling rate in half. With half as many samples to keep track of, the amount of time stored in a fixed amount of memory effectively doubles, hence the "times two" label. Halving the sample rate also lowers the upper frequency capability of the digital device. Engineers today know this from their pursuit of higher sampling rates in their productions: 44.1 kHz, 48 kHz, 88.2 kHz, 96 kHz, 176.4 kHz, 192 kHz, and beyond. A key benefit of an increased sampling rate is improved high-frequency response. Even as sampling rates creep up on many studio tools (especially DAWs and multitrack recorders), an engineer may purposefully *lower* the sampling rate on an outboard digital delay device. Doing so lengthens the delay time of the device, and it low-pass filters the signal — an often desirable mix move.

9.3.6 GROOVE

Well-timed delays are an excellent way to fill in part of the rhythm track of a song. Reggae is famous for its cliché echo. Drum programmers have been known to put in an eighth- or quarter-note delay across the entire groove. Guitarists use delay too. U2's the Edge has made delay a permanent part of his guitar rig. A classic example is apparent from the introduction of U2's "Bad" on *The Unforgettable Fire*. The quarter-note triplet delay is not just an effect, it's part of the riff. The Edge has composed the delay element into the song. Used in this way, echo becomes part of the groove, a driving musical force for the tune.

9.4 Short Delay

Delays times can be so short that they are not perceived as echoes. The delayed sound happens too quickly for the brain to notice it as an event separate from the original sound. When delays are short enough, they are perceptually fused with the undelayed sound. The mix of signal plus delayed signal becomes a single entity, with a different sound quality.

So as the delay time falls below about 50 ms, the sound of the delay is no longer an echo. It becomes, well, something else. It is not that the delay is impossible to hear, just that it has a different perceptual impact when the delay time is short. In fact, very short delays have an important *spectral* effect on the sound.

9.4.1 CONSTRUCTIVE AND DESTRUCTIVE INTERFERENCE

Sine waves — with their familiar, faithfully repeating wave shape — are helpful in illustrating the frequency-dependent implications of the short delay. Mixing together — at the same volume and pan position — the original signal with a delayed version of itself might have results like the two special cases shown in Figure 9.8.

If the delay time happens to be exactly the same as the period of the sine wave, constructive interference like that shown with the solid line in Figure 9.8 results. That is, if the delay time of the processor is equal to the time it takes the sinusoid to complete exactly one cycle, then the two signals will combine cooperatively. The combined result is a signal of the same frequency, but with twice the amplitude.

The situation shown by the dashed line in Figure 9.8 represents another special case. If the delay time is set to equal exactly half of a period (half the time it takes the sine wave to complete exactly one cycle), then the original sound and the delayed sound move in opposition to each other

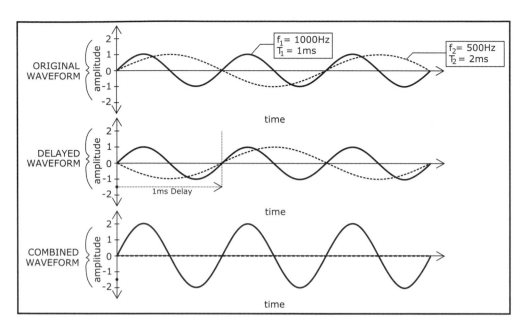

▲ *Figure 9.8 Constructive/destructive interference.*

when combined. For a sound wave in the air, pressure increases meet pressure decreases for a net result of no pressure change. For a voltage signal on a wire, positive voltages meet equal magnitude but negative voltages and sum to zero volts. In a digital audio workstation, positive numbers are cancelled out by negative numbers. The combination results in zero amplitude — silence.

Any audio engineer with access to a sine wave oscillator (either as test equipment or within a synthesizer), can give it a try. 500 Hz is a good starting point. This frequency is not quite as piercing as the standard test tone of 1,000 Hz, yet the math remains easy. The time it takes a pure 500-Hz tone to complete one cycle is 2 ms:

$$\text{Period} = \frac{1}{\text{Frequency}}$$
$$= \frac{1}{500}$$
$$= 0.002 \text{ seconds}$$
$$= 2\,\text{ms}$$

Mixing together equal amounts of the original 500-Hz sine wave and a 2-ms delayed version will create perfectly constructive interference very much like the solid line in Figure 9.8. Lower the delay time to 1 ms — creating the dashed-line situation of Figure 9.8 — and the 500-Hz sine wave is essentially cancelled.

As the bold waveforms on Figures 9.9 and 9.10 show, these doublings and cancellations happen at certain other higher frequencies as well. For any given delay time, certain frequencies line up just right for perfect constructive or destructive interference.

The math works out as follows: For a given delay time (t expressed in seconds, not milliseconds), the frequencies that double are described by an infinite series: $1/t$, $2/t$, $3/t$, etc. These frequencies all possess the feature that, when the delay time t is reached, they are all back exactly where they started, in this case beginning another repetition of their cycle. The frequencies that double in amplitude have cycled exactly once, twice, or some other integer multiple at the instant time t occurs (Figure 9.9).

The frequencies that cancel are $1/2t$, $3/2t$, $5/2t$, etc. These are the frequencies that are one-half cycle shifted when the delay time is reached. They have cycled $1\frac{1}{2}$ times, $2\frac{1}{2}$ times, or any integer-and-a-half times (Figure 9.10).

Using these equations, one can confirm that a signal combined with an equal amplitude 1-ms delay ($t = 0.001$ seconds) of the same signal has spectral peaks at 1,000 Hz, 2,000 Hz, 3,000 Hz, etc. with nulls exactly in between at 500 Hz, 1,500 Hz, 2,500 Hz, etc. This is consistent with the earlier observation in Figure 9.8 that combining a signal and its 1-ms delay can cancel a 500-Hz sine wave. A 2-ms delay has amplitude peaks at 500 Hz, 1,000 Hz, 1,500 Hz, etc. and nulls at 250 Hz, 750 Hz, 1,250 Hz, etc. The math reveals that the peaks and dips happen at several frequencies, not just one. While this pattern theoretically occurs for all frequencies without limit, audio engineers focus on those peaks and valleys that fall within the audible spectrum from about 20–20,000 Hz.

Another way to explore this further would be to set up the mixer so that it combines a sine wave with a delayed version of itself, set to the same amplitude. Sweep the sine wave frequency higher and lower, watch the meters, and listen carefully. With the delay fixed to 1 ms, for example, sweep the frequency of the sine wave up slowly beginning with about 250 Hz. One should hear the mixed combination of the delayed and undelayed waves disappear into silence at 500 Hz, reach a volume peak at 1,000 Hz, fall silent again at 1,500 Hz, reach a louder peak again at 2,000 Hz, and so on. A *delay*, not an equalizer, changes the amplitude of the signal as a function of frequency. A *delay*, not a fader or a compressor, changes

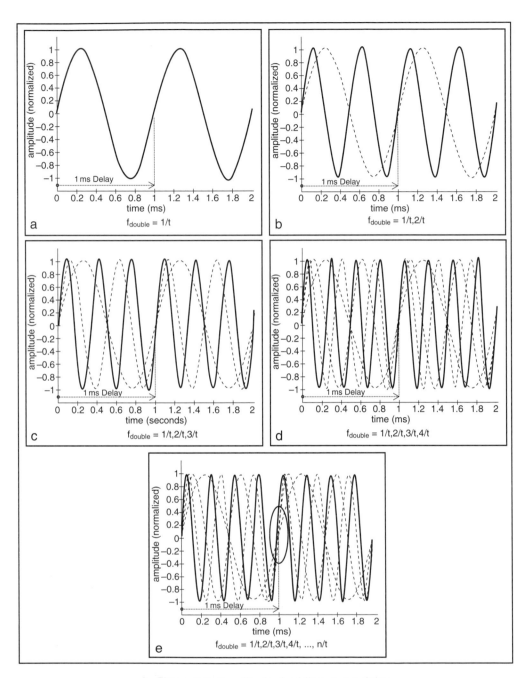

▲ *Figure 9.9 Amplitude doubling: 1 ms delay.*

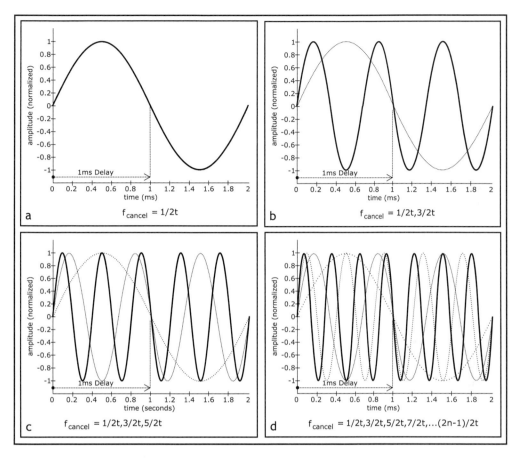

▲ *Figure 9.10 Amplitude canceling: 1 ms delay.*

the loudness of the mix. The connection between time- and frequency-dependent amplitude is an audio surprise to be savored.

9.4.2 COMB FILTER

The constructive and destructive interference associated with short delays clearly leads to a radical change in the amplitude of an audio signal. Is this a silly parlor trick, or valuable music production tool?

To answer this question, one has to get rid of the pure tone (which pretty much never happens in pop music) and hook up an electric guitar (which pretty much always happens in pop music). Run a guitar signal — live from a sampler, or off the multitrack — through the same setup above. With the delayed and undelayed signals set to the same amplitude, listen to what happens.

Is it possible to find a delay time setting that will enable the complete cancellation of the guitar sound? Nope. The guitar sound is not a pure tone. It is a complex signal, rich with sound energy at a range of frequencies. No single delay time can cancel out all the frequencies at once. But mixing together the undelayed guitar track with a 1-ms delayed version of the same guitar track definitely has an audible affect on the sound quality.

It was already observed that a 1-ms delay can cancel entirely a 500-Hz sine wave. In fact, it will do the same thing with guitar (or piano, or didgeridoo). Musical instruments containing a 500-Hz component within their overall sound will be affected by the short, 1-ms delay: The 500-Hz portion of their sound can in fact be cancelled. What remains is the tone of the instrument without any sound at 500 Hz.

But wait. There's more. It was also shown in Figures 9.8 and 9.9 that 1 kHz would double in amplitude when this 1-ms delay was combined with the signal. So for the guitar, the 1,000-Hz portion of the spectrally complex signal gets louder.

Taking a complex sound like guitar, which has sound energy at a range of different frequencies, and mixing in a delayed version of itself at the same amplitude, will cut certain frequencies and boost others. This is called *comb filtering* (Figure 9.11) because the alteration in the frequency content of the signal looks like teeth on a comb.

Combining a musical waveform with a delayed version of itself radically modifies the frequency content of the signal. Some frequencies are cancelled, and others are doubled. The intermediate frequencies experience something in between outright cancellation and full-on doubling. The changes in frequency content associated with combining a signal with a short delay of itself suggest that short delays are less like echoes and more like equalizers (see Chapter 5).

Short delays are too short to be perceived as echoes. In fact they are so short that they start to interact with discrete components of the overall sound, adding some degree of constructive (additive) or destructive (subtractive) interference to different frequencies within the overall sound. Figure 9.8 demonstrates this for a sine wave. Figure 9.11 summarizes what happens in the case of a complex wave — a more typical audio track like guitar, piano, saxophone, or vocal.

Equalization (EQ) is not too far off the mark as a way to think about comb filtering, but it would be nearly impossible to actually do the same with a

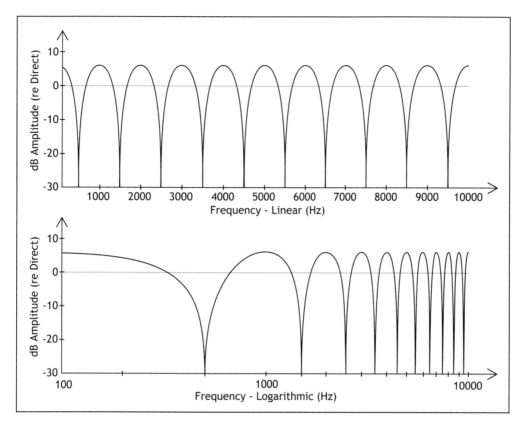

▲ *Figure 9.11 Comb filtering: short delay.*

practical equalization device: a boost here, a cut there, another boost here, another cut there, and so on. In theory one could simulate comb filtering with an equalizer, carefully dialing in the appropriate boosts and cuts. That is the theory. In fact, such an EQ move is unlikely. To fully imitate the comb filter effect that a 1-ms delay creates, the studio would need an equalizer with about 40 bands of parametric EQ (20 cuts and 20 boosts within the audible spectrum), each with its own, unique setting of Q. Such an equalizer would be prohibitively expensive to build, would likely have other side effects (e.g., noise and phase distortion) in terms of quality, and would take a long time to set up. In fact, part of the point of using short delays in this way in music production is to create sounds that can't easily be achieved with an equalizer. A single short delay creates an unbelievably complex EQ-like frequency contour.

Short delays offer a very interesting extra detail: They create mathematical — not necessarily musical — changes to the sound. Study Figure 9.11, comparing the upper curve to the lower curve. Both plots show the same information. But the lower graph presents the information with a logarithmic frequency axis. This is the typical way of viewing music (see "The Decibel," Chapter 1). The keyboard of a piano and the fingerboard of most stringed instruments lay things out logarithmically, not linearly and equally spaced. If one looks at comb filtering with a linear (and nonmusical) frequency axis, as in the upper part of Figure 9.11, one finds that the peaks and dips in the filter are spaced perfectly evenly. In fact, it is only by looking at the spectral result in the linear domain that the name, comb filter, becomes clear. The logarithmic plot on the lower portion of Figure 9.11 would not make for a very useful comb, but it's the more musically and perceptually informative visual representation.

Pure tones, cycling with infinite reliability, would experience the comb filtering effect at any delay time, not just short ones. A delay time of 1 ms has been shown to cause constructive and destructive interference at a precise set of frequencies. The logic still holds when the delay time is 1,000 ms or more. In music production, however, where the target of our signal processing is complex music signals, not pure tone test signals, the effect is practically limited to very short delays.

Two related phenomena are at work. First, as discussed in the section on "Long Delay" above, delays greater than about 50 ms become perceptually separate events. Our hearing system identifies the undelayed and the long delayed signals as two separate events, each with their own properties. Comb filtering relies on the fusing together of the undelayed and the delayed signals into a single perceived event.

Second, music signals do not repeat with the rigid regularity over time that sine waves do. Musical signals change constantly. The singer moves on to the next syllable, the next word, the next line. The guitarist moves on to the next note, the next chord. Long delayed signals appear too late to interact directly with the undelayed signal. Unlike a sine wave, the musical signal has progressed to other things.

Even if the musical signal is relatively unchanging — the whole note of a bass guitar, or the double whole note of a sustained piano chord — it still falls well short of sine wave stability. A bass note held over time still wobbles a bit in pitch. This might be a means of expression, as the performer gently raises and lowers the pitch for musical effect. It might also be a practical inevitability; new strings on a stable instrument hold a pitch more steadily than old strings on a weak instrument. If the pitch has changed, the undelayed and long delayed signals are too different from each other for the mathematically predicted comb filtering to occur. In music productions, the distinct spectral signature of comb filtering is limited only to very short delays of 20 ms or less.

Recording engineers patch up the short delay, creating a comb filter, when they have a special, radical, spectral effect in mind. If a more organic, natural tailoring of sound is desired, reach instead for an equalizer, with its logarithmic, more musical controls.

9.4.3 EARLY REFLECTIONS

It is still fair to ask: Why all this talk about short delays and their effect on a signal? After all, how often do recordists use delays in this way? It is essential to understand the sonic implications of these short delays because, all too often, they simply cannot be avoided. Consider an electric guitar recording session. With the guitar amplifier in the middle of the recording space, on a beautiful wooden floor, the engineer places the chosen microphone a few feet away and tries to capture the natural sound of the amp in the room. This is a good approach, shown in Figure 9.12. The problem is that the sound reflected off the floor and into the microphone will arrive a split second later than the sound that went straight from the amplifier to the microphone. The path is longer via the reflected path, introducing some delay. The result is some amount of comb filtering. Recording a sound and a single reflection of that sound is a lot like mixing a track or sample with a delayed version of itself, as in the discussions above. With any loud reflection representing a delayed copy of a sound, it rapidly becomes apparent that comb filtering is an everyday part of recording.

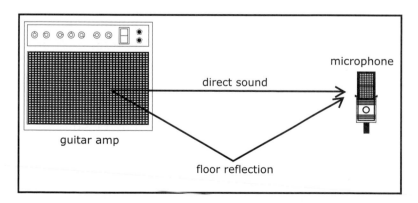

▲ *Figure 9.12 Delay from room reflections.*

Fortunately the sound reflected off the floor will also be a little quieter, reducing the comb filter effect: less-pronounced peaks and shallower notches. If the floor is carpeted, the comb filtering is a little less pronounced at high frequencies. Place a thicker pad of sound-absorbing material at the point of the reflection, and the comb filtering is likely to be even less audible. An important part of the recording craft is learning to minimize the audible magnitude of these reflections by taking advantage of room acoustics when placing musical instruments in the studio and strategically placing sound-absorptive materials around the musical source. This is one approach to capturing a nice sound at the microphone.

Better yet, learn to use these reflections and the comb filtering they introduce on purpose. For example, raising the microphone will make the difference in distance between the reflected path and the direct path even longer. Raising the microphone, therefore, lengthens the acoustic delay time difference between the direct sound and the reflected sound, thereby changing the spectral locations of the peaks and valleys of the comb filter effect.

Of course, other factors are in play. Raising the microphone also pushes the microphone further off-axis of the amp, changing the captured timbre of the electric guitar tone as viewed by the microphone. Another strategy is to raise the amp up off the floor, perhaps setting it on a piano bench or at the edge of a strong table. It is also common to tilt the amp back so that it faces up toward the raised microphone and away from the sound-reflective floor. This is an exercise in acoustically adjusting the amplitude of a short delayed signal. Then again, one can flop the amp on its belly, facing straight down into the floor if that sounds good. Rock-and-roll guitar knows no rules. Delay-induced comb filtering is only part of the equation.

A common approach to recording a guitar amp — and many other instruments, for that matter — is to use a combination of two or more microphones to create the sound, even as it is recorded onto a single track. Consider the session shown in Figure 9.13: two microphones, one track. Here, a close microphone (probably a dynamic) grabs the in-your-face gritty tone of the amp and a distant microphone captures some of the liveness and ambience of the room. An engineer might label the channel fader controlling the close microphone something like "close," and the channel fader governing the more distant microphone "room." The engineer adjusts the two faders to get an attractive mix of close and room sounds and prints that to a single track of the multitrack.

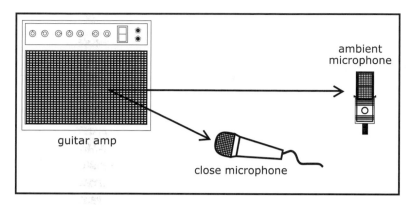

▲ *Figure 9.13 Delay from multiple microphones.*

That is only half the story. As one adjusts the faders controlling these two microphones, not only is the close/ambient mix changed, but also the amount of comb filtering introduced into the guitar tone is affected. These two microphones pick up very similar signals, but at different times. Mixed together, they act very much like the signal-plus-delay scenario just discussed. Moving the distant microphone to a slightly different location is just like changing the time setting on the delay unit. It effectively selects different key frequencies for cutting and boosting using the exact same principles explored in Figures 9.8, 9.9, 9.10, and 9.11. Sound travels a little farther than one foot per millisecond. To lengthen the delay time difference by about a millisecond, move the distant microphone back about a foot. To get a 10-ms delay increase, move the distant microphone back about 10 feet. It is that simple.

Naturally, there is too much to keep track of. Each of these microphones receives reflected sounds from the floor and the ceiling and all the other room boundaries, all in addition to the obvious direct sound from the amp. The engineer tries to orchestrate the complex interaction of the many components of guitar sound radiating out of the amp. The direct sounds into multiple microphones arrive at different times, leading to some amount of comb filtering. The reflections from the various room boundaries into each microphone arrive at a later time than the direct sound, adding additional comb filtering, with peaks and dips falling at different spectral locations. There are an infinite number of variables in recording. Understanding comb filtering is part of how audio engineers master the vast recording process. Recordists rarely do the math; it is enough to know to listen for the comb filter effect.

9.4.4 INEVITABLE COMBINATIONS

Perhaps the recording engineer wants a tough, heavy, larger-than-large guitar tone. Maybe a comb filter–derived hump at 80-Hz is the ticket. Or should it be 60 Hz? That is a creative production decision. Explore this issue by moving the microphones around. Place two microphones on the amp as shown in Figure 9.13. Keep the close microphone fixed and move the distant one slowly. The goal is to introduce a frequency peak at some powerful low frequency that suits the tone of the guitar/amp combination currently performing. It is a great luxury, on sessions like these, to have an assistant engineer. The assistant slowly moves the microphone out in the live room while the first engineer listens to the combined close/distant microphone mix in the comfort and monitoring accuracy of the control room. When one lacks an assistant, record a take onto multitrack while slowly moving the microphone, as quietly as possible. When those comb filter peaks and notches fall into frequency ranges that complement the tone screaming out of the guitar amp, that elusive bit of audio nirvana, the sweet spot, has been found.

Finding the microphone placement that captures the tone that pleases the guitarist simply requires a bit of patience, and an understanding of the spectral implications of short delays.

The art of microphone placement requires mastery of room acoustics, musical acoustics, and psychoacoustics. To predictably achieve good sounding results engineers need recording experience, an understanding of microphone technologies, knowledge of microphone sound qualities, and exposure to the various stereo microphone techniques, among many other topics. Note, though, that an essential tool in microphone placement is the deliberate use of comb filtering to modify the sound being recorded.

Electric guitar responds well to comb filtering. With energy across a range of frequencies, the peaks and dips of comb filtering offer a distinct, audible sound property to be manipulated. Other instruments reward this kind of experimenting. Try placing a second (or third or fourth) microphone on an acoustic guitar, piano, or anything. Experiment with the comb filter–derived signal processing to get a sound that is natural, or one that is unnaturally beautiful.

All engineers one day find themselves in a predicament: The guitar amp sounds rich and full out in the live room, but thin in the control room. Perhaps the problem is that, courtesy of the short delay between two microphones, they have a big dip in frequency right at a key low-frequency region. Undo the problem by changing the spectral location of the frequency notch: move a microphone, which changes the delay, which changes the frequencies being cancelled.

Every time more than one microphone is placed on an instrument, make it routine to listen for symptoms of wanted or unwanted comb filtering. Check out each microphone alone. Then combine them, listening for critical changes in the timbre. What frequency ranges are attenuated? What frequency ranges get louder? The hope is to find a way to get rid of unwanted or less interesting parts of the tone while emphasizing the more unique and more appealing components of the timbre. Good engineers make short delays part of their mixing repertoire. For subtle tone shaping or a radical special effect, the short delay is a powerful signal processor. It takes some trial and error, some experience, but it will lead directly to better sounding recordings.

9.4.5 FLANGER

Dialing in a very short delay time and modulating it via the three delay modulation controls leads to an effect known as *flanging*. The only rule is that the delay time needs to be in that range short enough to lead to audible comb filtering. That suggests a starting delay setting less than about 10 ms, though the effect may be more obvious at delay times closer to 5 ms. This ensures audible comb filtering will occur. Set the delay modulation controls to taste.

That ringing, whooshing, ear tickling sound that is created by a flanger comes from the simple comb filter effect enhanced by these modulation controls. Consider two different short delay settings. One delay time causes the peaks and valleys in the frequency content of the combined signal to occur at one set of frequencies. A different delay setting results in a different set of peaks and valleys (Figure 9.14). This suggests that when a continuously modulated delay value sweeps from one delay time to another, the comb filter bumps and notches sweep also. Figure 9.14 shows the result: flanging.

Sometimes there is no holding back — the entire mix gets flanged. "Life in the Fast Lane," on *Hotel California* by the Eagles presents a classic example at the breakdown near the end of the tune.

In other applications, the effect is applied to a single instrument — just the drum kit or just the rhythm guitar. The audio engineer might flange just a

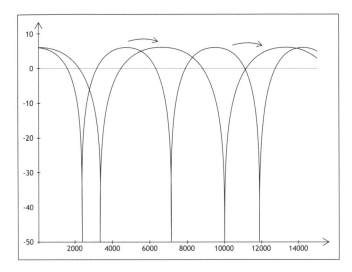

▲ *Figure 9.14 Swept comb filter.*

single track. Or they might limit the effect to just one section of that track (e.g., only on the bridge). Audio experience gives the engineer the maturity and judgment needed to know when, how, and how much of effects such as this to use. It becomes a matter of personal opinion and musical taste.

Pop music is full of examples of flanging. "Then She Appeared" from XTC's last effort as a band on the album *Nonesuch* offers a good case study of a gently sweeping flange. Notable from the first time the words "then she appeared" occur, a bit of your traditional flange begins, courtesy of a set of short vocal delays being slowly modulated. In this example the flange comes and goes throughout the song, offering us a good chance to hear the vocals with and without the delay treatment. A more subtle approach occurs on Michael Penn's "Cover Up" from the album *Resigned*. A bit of flanging appears on the vocal for the single word "guests" near the end of the second verse. That is it. No more flanging on the vocal for the rest of the tune. It is just a pop-mix detail to make the arrangement that much more interesting. The flange effect actually softens the rather hard sounding, sibilant, and difficult to sing "sts" at the end of the word "gue*sts*."

The simple effect that comes from mixing in a short, modulated delay offers a broad range of audio effects. Flanging invites your creative exploration.

9.5 Medium Delay

In the above discussion on long delays (those greater than about 50 ms), we saw how they are used for a broad range of echo-based effects. Short delays of about 20 ms or less create the radical comb-filtered effect that, especially when modulated, we call flanging. What goes on in between 20 and 50 ms?

This middle range of delay times does not cause echoing or flanging. The delay is too short to be perceived as an echo. It happens too fast. But the delay is too long to lead to audible comb filtering for most musical signals.

Try a 30-ms delay on a vocal track for a good clue about what is going on. A medium delay sounds a little bit like a double track — like two tracks of the same singer singing the same part.

9.5.1 DOUBLE TRACKING

It is a common multitrack production technique to have the singer double a track. The engineer, producer, and musician work hard to capture that elusive "perfect" take. It can require several tries. It might even require several hours. It can get a little grim, but it might even take several different sessions over several weeks or months before everyone is happy with the vocal performance. Once this is finally accomplished, have the singer record the part again, on a separate track. She sings a new track while listening to the just recorded "perfect" track of herself. The resulting sound is stronger and richer. It even shimmers a little.

If you are unfamiliar with the sound of doubled vocal tracks, a clean example can be found at the beginning of "You Never Give Me Your Money" on the Beatles' *Abbey Road*. Verse one begins with solo vocal. On the words "funny paper," the doubling begins. The vocal remains doubled for the next line and then the harmonies commence. Naturally, each harmony part is doubled, tripled, or more.

Roy Thomas Baker is famous for, among other things, pushing doubling to the hilt. Check out the harmonies, doubled and tripled (and beyond), throughout The Cars' first album, *The Cars*. For example, listen to the first harmonies on the first song, "Good Times Roll," when they sing the hook, "good times roll." It sounds deep and immense; the vocals take on a slick, hyped sound.

The deep layering of multiple vocal tracks by the same vocalist goes back to 1951 when Les Paul and Mary Ford covered "How High the Moon." Mary Ford sings lead vocal. Mary Ford sings multipart harmonies. And most ear-tingling of all, Mary Ford doubles and triples any and all parts. Pop music has always found occasion to be more than a little over the top.

This layering of tracks borrows from the tradition of forming instrumental sections in orchestras and choirs. Consider the sound of 1 violin. Then imagine the exact same piece of music with 12 violins playing the same line, in perfect unison.

The value of having multiple instruments play the same musical part is almost indescribable. Adding more players does not just create more volume. The combined sound is rich and ethereal. It transports the listener.

A crystal clear application of doubling can be found on Macy Gray's "I Try" from the album *On How Life Is*. Typically, double tracks support the vocal, adding their inexplicable extra bit of polish. They are generally mixed in a little lower in level than the lead vocal, reinforcing the principle track from the center or panned off to each side. The Macy Gray tune turns this on its head. At the chorus, where pop production tradition calls for a good strong vocal, the vocal track panned dead center does something quite brave: It all but disappears.

The chorus is sung by double tracks panned hard left and right. It is brilliantly done. Rather than support the vocal, they *become* the vocal. The chorus does not lose strength. The tune does not sag or lose energy at all. The doubled tracks — panned hard but mixed aggressively forward — offer a contagious hook that invites the listeners to sing along, to fill in the middle.

While pop vocals, especially background vocals, are often doubled, any other instrument is fair game. A common track to double is rhythm electric guitar. The same part is recorded on two different tracks. On mixdown, they might appear panned to opposite sides of the stereo field.

The two parts are nearly identical. Change just one variable in the recording setup: switch to a different guitar, a different amp, or a different microphone, or slightly change the tone of the doubling track in some other way. Maybe the only difference between the tracks is the performance. As no two performances are ever identical, the resulting pair of guitar tracks will vary slightly in timing. The "chugga chugga" of the left guitar track is slightly early in one bar and slightly late in the next. The result is that, through the interaction of the two guitar tracks, the ears seem to pick up on and savor these subtle delay changes. At times, the two tracks are so similar they fuse into one meta instrument. Then one track pulls ahead and gets noticed on its own. Then the other track pulls ahead in time and temporarily draws the listener's attention. Then, for a brief instant, they lock in together again. And so it goes. Doubled guitars are part audio illusion and part audio roller coaster; they are an audio treat.

Layering and doubling tracks can be simulated through the use of a medium delay. If it is not convenient, affordable, or physically possible to have the singer or guitarist double the track, engineers reach for a medium delay. Some amount of delay modulation is introduced so that the doubled track moves in time a little relative to the source track. This helps it sound more organic, not like a cloned copy of the original track. The addition of a bit of regeneration creates a few additional doubling layers of the track underneath the primary one.

Some delay units have the ability to simultaneously offer several delay times at once (called *multitap delays*), each modulated at it's own rate. Use several slightly different delay times in the 20- to 50-ms area and synthesize the richness of many layered vocals. Using pan pots, spread the various medium delay elements out left to right and front to back (in surround applications) for a wide wall of vocal sound. As with all studio effects, make sure the sonic result is appropriate to the song. The solo folk singer doesn't usually benefit from this treatment. Neither does the jazz trumpet solo. But many pop tunes welcome this as a special effect on lead vocals, backing vocals, keys, strings, pads, and so on.

9.5.2 CHORUS

An alternative name for this use of modulated medium delays is *chorus*. The idea is that, through the use of several different delays in the 20 to 50 ms range, one could upgrade a single vocal track into a simulation of the sound of an entire choir. Thus the term, *chorus*.

Naturally, stacking up 39 medium delays with a single vocalist will not convincingly sound like a choir of 40 different people. Think of it instead as a special electronic effect, not an acoustic simulation. And it is not just for vocals. John Scofield's trademark tone includes a strong dose of chorus (and distortion, a sweet guitar, and brilliant playing among other things). It is not uncommon to add a bit of chorus to the electric bass. This medium-delay concoction is a powerful tool in the creation of musical textures.

To see how "out there" the effect can be made, reach for the album *Throwing Copper* by Live and listen to the beginning of the tune "Lightning Crashes." It's difficult to impossible to know exactly what kind of signal-processing craziness is going on just by listening. The guitar sound includes short and medium delays, among the panning, distortion, and phase shifting effects going on. A fair summary is that the delay is being modulated between a short, flanging sort of sound (around 10 ms) and up to a longer, chorus sort of delay time (around maybe 40 ms). Note especially the sound of the guitar in the second verse, when the effect and the relatively clean sound of the guitar are mixed together at similar volumes. This presents a good taste of chorus. Such amazing projects are a great inspiration to the rest of us to do more with the effect.

9.6 Experimental Sessions

With all this discussion of delays, it is essential to hear it for yourself.

9.6.1 EXPERIMENTAL #1

Patch up a sampler loaded with a variety of sounds or find a multitrack tape with a good variety of tracks. On your mixer, set up a delay fed by an aux send that returns to your monitor mix at about the same volume as the synth or original tracks. Pan both the source audio and the return from the delay dead center. Your assignment is to carefully listen as you mix each source sound with the output of the delay.

Start with a bass line. Check out the combination of the bass with a long delay, maybe 500 ms. Yuk. It is a blurry, chaotic mess. Now start shortening the delay: 200 ms, 100 ms, 80 ms, 60 ms, 20 ms, 10 ms, 5 ms, down to 3 ms and below. As you do this, occasionally mute the delay return so you can remind yourself of the sound of the unprocessed audio signal.

What the heck is going on? The long delay is just an echo, and it probably does not help the music much. The very short delays (15 ms and lower) sound strange, sometimes hollow and sometimes boomy. At one short delay setting you may find extra low end, then at a slightly different delay time, a lack of low end.

This mix of a bass sound with a very short delay sounds like it is been equalized. Gradually lengthen the delay time and listen for the point at which it starts to sound like a distinct echo again. Depending on the bass sound and type of performance, you may hear the delay separate from the bass into an echo somewhere between about 60 and 80 ms. Staccato performances reveal an echo at a lower delay time — closer to about 60 ms — than long, sustained, double whole-note performances that may not have an audible echo until 80 ms or more. In between the very long and the very short delay times, well, it is hard to describe. More on the middle zone later.

Okay. Next try a snare sound. Again start with a long delay and gradually pull it down to a short delay. Again you should hear a distinct echo at long settings. The delay introduces a strange timbral change at short delays and something tough to describe as it transitions between the two. While we are here, do the same experiment with an acoustic or electric guitar track, or a string patch on the sampler.

Welcome to the real world of delays. They are not just for echoes anymore. When delays become shorter than about 50 or 60 ms (depending on the type of sound you are listening to, as demonstrated above), they are no longer repeats or echoes of the sound. The same device that delays a signal starts to change the color, texture, and spectral content of the signal.

9.6.2 EXPERIMENTAL #2

Ride the faders in the following experiment. On your mixer, patch things up so that one fader has a sine wave at 500 Hz, panned to center. The same sine wave should also be sent to a delay unit, set to a delay time of 1 ms. Another fader controls the return from this delay, also panned to center. Start with both faders down. Raise the fader of the source signal to a reasonable level. Now raise the second fader. As you *increase* the level of the delayed signal, your mix of the two waves gets *quieter*. As you add more of the delayed sine wave, you get more attenuation of the original sine wave. This is the phenomenon shown by the dashed line in Figure 9.8. The mix reaches its minimum amplitude when the two signals are at equal amplitude.

9.7 Summary

A simple delay unit offers a broad range of audio opportunity, representing a nearly infinite number of sound qualities. Short delays create that family of effects called flanging. Medium delays lead to doubling and chorusing. Finally, when the delay is long enough, it separates from the original signal and becomes its own perceptible event: an echo.

Take a quick tour of all of the above with a single album: *Kick* by INXS. To hear a terrific use of flange, listen to "Mediate." This is a textbook bit of flanging. For a true doubling, listen to "Sweat" and those hard-panned questions: "How do you feel? What do you think? Whatcha gonna do?" Finally, the same album demonstrates a classic application of chorus to an electric guitar. Check out the rhythm guitar in "New Sensation" and the steely cool tone the chorusing adds.

Flange, chorus, and echo are three very different kinds of effects that come from a single kind of effects device: the delay.

9.8 Selected Discography

Artist: The Beatles
Song: "You Never Give Me Your Money"
Album: *Abbey Road*
Label: Parlophone/EMI
Year: 1969
Notes: From single-track vocals to double- and triple-tracked vocals and harmonies. A good case study. The layering begins after the words, "funny paper."

Artist: The Cars
Song: "Good Times Roll"
Album: *The Cars*
Label: Elektra Records
Year: 1978
Notes: This is double tracking taken as far as it can reasonably go. Listen to the vocals when they sing the title "Good Times Roll." One imagines they filled every available track on the multitrack at this point.

Artist: The Eagles
Song: "Life in the Fast Lane"
Album: *Hotel California*
Label: Asylum Records
Year: 1976
Notes: Over-the-top flanging of the entire stereo mix in the chorus repeats into the coda at 3:38.

Artist: Les Paul and Mary Ford
Song: "How High the Moon"
Label: Capitol Records
Year: 1951
Notes: Sets the pace for doubling and tripling multitrack performances.

Artist: Michael Penn
Song: "Cover Up"
Album: *Resigned*
Label: Epic Records
Year: 1997
Notes: One word, and one word only, gets flanged: "guests" at the end of the second verse.

Artist: The Police
Song: "Synchronicity II"
Album: *Synchronicity*
Label: A&M Records
Year: 1983
Notes: Ghost of an echo on most of the lead vocal.

Artist: U2
Song: "Bad"
Album: *The Unforgettable Fire*
Label: Island Records
Year: 1984
Notes: Not the only example, by far. The Edge uses delay as part of the guitar riff.

Artist: The Wallflowers
Song: "One Headlight"
Album: *Bringing Down the Horse*
Label: Interscope Records
Year: 1996
Notes: The lead vocal is wholly without delay, for the entire song, until the third verse. The words "turn" and, to a lesser extent "burn," get an almost subliminal dotted quarter-note echo. The slide guitar solo gets a thoughtful quarter note delay treatment that comes and goes, mid-solo. Solo is panned left, first echo is panned dead center, and the second repetition is panned hard right.

Artist: The Who
Song: "Won't Get Fooled Again"
Album: *Who's Next*
Label: Polydor
Year: 1971
Notes: It's not just the, "Yeaaaahhhhh!" It's the layers of echoes underneath.

Artist: XTC
Song: "Then She Appeared"
Album: *Nonsuch*
Label: Caroline Records
Year: 1992
Notes: A modern update of a psychedelic 1960's flanger effect. The title is the hook, and gets the effect.

Pitch Shift

"I'll tell you something.
I am a wolf,
But I like to wear sheep's clothing."
— "TEMPTATION WAITS," GARBAGE, *VERSION 2.0* (ALMO SOUNDS, 1998)

Use a playback sample rate that is higher than the original sample rate used when the audio was recorded, and the pitch goes up. Play an audiotape at a slower speed than intended and the pitch of the recording goes down. Somewhere in this simple principle lies an opportunity for audio exploration and entertainment.

This chapter explains the fundamentals of pitch shifting and initiates an inventory of the production potential.

10.1 Theory

The principle of shifting pitch is straightforward. The basis for the effect is the delay. The same device stretched to such creative limits in Chapter 9 has still more applications.

10.2.1 PITCH RECALCULATION

To appreciate the elegance of pitch shifting, a bit of math is in order (follow along with Figure 10.1). Figure 10.1(a) shows one cycle of a simple sine wave with a chosen period of 4 milliseconds (ms) and, therefore, a frequency of 250 Hz. This sine wave is seen to complete exactly one cycle every 4 ms, and the frequency calculated courtesy of the following familiar equation:

$$f = \frac{1}{T} \qquad (10.1)$$

where f is the frequency (hertz), and T is the period (seconds), which is the time to complete precisely one cycle (see Chapter 1). Using Equation 10.1 for Figure 10.1(a):

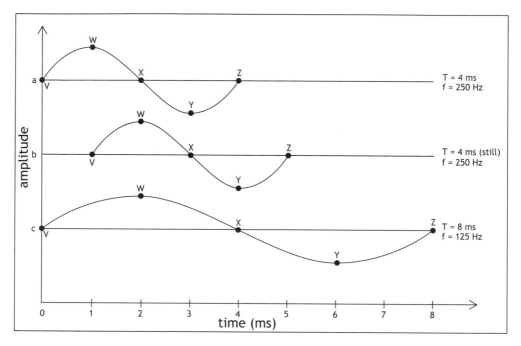

▲ *Figure 10.1 Pitch shifting through a variable delay.*

$$f = \frac{1}{0.004}$$

$$f = 250 \text{ Hz}$$

Figure 10.1(a) labels some key landmarks during the course of a single cycle, using the letters V, W, X, Y, and Z. It all starts at point V. Here, at time equals zero, the sine wave has an amplitude of exactly zero and is increasing. It reaches its positive peak amplitude at W, taking exactly 1 ms to do so. It returns again to an amplitude of zero at the halfway point (time equals 2 ms), labeled X. The amplitude looks at this instant a lot like point V. While the amplitude is the same, notably zero, it is decreasing through point X. Y is the point of maximum negative amplitude and occurs 3 ms after the beginning of the cycle. At Z, the sine wave returns 4 ms later to what looks exactly like the starting point: the amplitude is zero, and increasing. Follow these points as some signal processing is applied.

10.1.2 FIXED DELAY

Run this sine wave through a fixed delay of 1 ms, and the situation described in Figure 10.1(b) results. Visually, one might observe that the sine wave appears to have slipped along the horizontal time axis by 1 ms. Looking point by point, Table 10.1 shows what happens. Point V originally occurred

Table 10.1 Follow Landmark Points Into and Out of a Fixed Delay (in Milliseconds)

	V	W	X	Y	Z
Phase	0°	90°	180°	270°	360°
Input	0	1	2	3	4
Delay	1	1	1	1	1
Output	1	2	3	4	5

Table 10.2 Follow Landmark Points Into and Out of a Changing Delay (in Milliseconds)

	V	W	X	Y	Z
Phase	0°	90°	180°	270°	360°
Input	0	1	2	3	4
Changing Delay	0	1	2	3	4
Output	0	2	4	6	8

at a time of 0 ms. Inserting a 1-ms delay on this signal, point V now occurs at time equals 1 ms. The other points follow. Point Y, for example, occurred undelayed at a time of 3 ms. After it has been run through a fixed, unchanging delay of 1 ms, point Y is forced to occur at a time of 4 ms.

10.1.3 ACCELERATING DELAY

Here's the mind bender. What happens if the delay isn't fixed? What if the delay sweeps from a starting time of 1 ms and then increases, and increases, and increases? Table 10.2 summarizes. Here the delay changes at a rate of 1 ms per millisecond. That can seem a bit confusing at first. For every millisecond that ticks by during the course of this experiment, the delay gets longer by 1 ms. At one instant the delay might be 10 ms. Whatever audio is fed into the delay at this instant will appear at the output in 10 ms; 5 ms later, the delay time has swept up to 15 ms. Audio entering the delay unit now won't appear at the output for 15 ms.

The simple act of increasing the delay by 1 ms each millisecond has a surprising result. The *pitch* of the audio is changed. Table 10.2 shows the time location of the sine wave landmarks both before and after the introduction of this steadily increasing delay. Point V initially occurs at a time of zero. At this time, the delay time setting within the delay line is also zero. Point V then remains unchanged and occurs at time zero. Skip to point

X. It originally occurs at a time of 2 ms. By this time, the delay has increased from 0 ms to 2 ms. This delay of 2 ms leads point X to finally occur 4 ms after the beginning of this experiment. Doing the math point by point leads to the sine wave of Figure 10.1(c).

The key landmarks are identified. The result is clearly still a sine wave. But because it takes longer to complete the cycle, the pitch is known to have changed. Return to Equation 10.1. Looking at the new sine wave in Figure 10.1(c), one can calculate its frequency. The sine wave in Figure 10.1(c) takes a full 8 ms to complete its cycle:

$$f = \frac{1}{T}, \text{ where } T = 8\text{ms or } 0.008 \text{ seconds}$$

$$f = \frac{1}{0.008}$$

$$f = 125\,\text{Hz}$$

The constantly increasing delay caused the pitch of the signal to change. A 250-Hz sine wave run through a delay that increases constantly at the rate of 1 ms per millisecond will be lowered in pitch to 125 Hz. An increasing delay lowers the pitch. It is also true that a decreasing delay raises the pitch. That is, start with a delay of 50 ms and decrease it by, say, 0.5 ms per millisecond, to 49.5 ms, 49 ms, etc. The result is a waveform exactly one octave higher.

In addition, the case studied above found that increasing the delay at a rate of 1 ms per millisecond raised the pitch by an octave. It is also possible to change the pitch by two octaves, or a minor third, or a perfect fifth — whatever is desired. One need only change the delay at the correct rate.

The underlying methodology of pitch shifters is revealed. A pitch shifter is a device that changes a delay in specific controlled ways so as to allow the user to affect the pitch of the audio.

Naturally, there are some significant details that must be addressed. Return to the example in which the 250-Hz sine wave was lowered by an octave through a steadily increasing delay. If this effect were applied to an entire $3\frac{1}{2}$-minute tune, not just a single cycle of a sine wave, then one would find it necessary to increase the delay from a starting point of 0 ms to a final delay time of 210,000 ms ($3\frac{1}{2}$ minutes equals 210,000 ms). That is, from the start of the tune an increasing delay is needed: 1 ms, then 2 ms, and so on. By the end of the tune, the delay time has reached 210,000 ms. This highlights two problems.

First, it appears a delay device capable of a very long delay time is needed. Many hardware delays start to run out of memory closer to a 1-second delay (1,000 ms). The more capable, more expensive delay lines might go up to maybe 10 seconds of delay. But a delay of hundreds of thousands of milliseconds is a lot of signal processing (i.e., memory) horsepower that is not typically available outside of a computer.

Second, the song that used to be $3\frac{1}{2}$ minutes long doubles in length to 7 minutes as the pitch is lowered by one octave. Consider the last sound at the very end of the song. Before pitch shifting, it occurred $3\frac{1}{2}$ minutes (210,000 ms) after the beginning of the song. By this time, the pitch-shifting delay has increased from 0 ms to 210,000 ms. Therefore the final sound of the pitch-shifted song occurs at 210,000 ms (original time) plus 210,000 ms (the length of the delay). That is, the song now ends 420,000 ms (that's 7 minutes) after it began. The $3\frac{1}{2}$-minute song is lowered an octave, but doubled in length.

Simply increasing the delay forever as above is exactly like playing a tape back at half the speed it was recorded. The pitch goes down and the song gets longer. Pitch-shifting signal processors differentiate themselves from tape speed tricks in their clever solving of this problem. Digital delays can be manipulated to always increase, but also to reset themselves. In the sine wave example, what happens if the digital delay increases at a rate of exactly 1 ms per millisecond but never goes over 50 ms in total delay? That is, every time the delay reaches 50 ms, it resets itself to a delay of zero and continues increasing from this new delay time at the same rate of 1 ms per millisecond. The result is pitch shifting that never uses too much delay, and never makes the song more than 50 ms longer that the unpitch-shifted version. After all, the analysis above showed it was the *rate* of change of the delay that led to pitch shifting, not the *absolute* delay time itself. Any delay time that increases at rate of 1 ms per millisecond will lower the audio by one octave. Any delay time. So why not keep it a small delay time?

The devil is in the details. Getting the pitch shifter to reset itself in this way without being noticeable to the listener is not easy. It is a problem solved by clever software engineers who find ways to make this inaudible. Older pitch shifters "glitched" as they tried to return to the original delay time. Today, those glitches are mostly overcome by intense signal processing. Software algorithms can evaluate the audio and find a strategic time to reset the delay time, applying cross fades to smooth things out. Another approach might be to use two delay lines, sweeping them at the same rate but resetting them at different times. If one cross fades between them at opportune times, one avoids the moment when either delay is reset. Digital signal-processing approaches have become more advanced still, with the result that convincing pitch shifting is a staple effect available in any well-equipped studio.

10.2 Applications

Pitch-shifting effects are common in multitrack production — sometimes subtle, other times obvious; sometimes accidental, other times deliberate.

10.2.1 SIDE EFFECTS

Before any discussion of effects deliberately using pitch shifting, it is worth noting that pitch shifting is a natural part of some effects already studied in this book. Recall the chorus effect that comes from adding a slowly modulated delay of about 20–50 ms (see "Medium Delay," Chapter 9). Chorus is difficult to describe in words. To study it, one must listen to it. A careful, critical listen to the richness that the chorus effect adds to a vocal or guitar reveals a subtle amount of pitch shifting. Beyond that blurring of things in time through the use of perhaps several medium delays, pitch shifting is a fundamental component of that effect known as chorus. Since a chorus pedal relies on a *modulating* delay, it introduces a small amount of pitch shifting. As the delay time sweeps up, the pitch is slightly lowered. As the delay time is then swept down, the pitch is then raised, ever so slightly. One can not have chorus without at least a little pitch shifting.

10.2.2 SPECIAL EFFECTS

In Chapter 12, use is made of a common effect built, in part, on pitch shifting: the spreader. Here is a quick summary of this effect. The spreader is a "patch" that enables the engineer to take a mono signal and make it a little more stereolike. A single track is "spread" out by sending it through two delays and two pitch shifters, each hard panned left and right. The delays are kept short, each set to different values somewhere between about 15–50 ms. If they are too short, the effect becomes a flange/comb filter; if they are too long, the delays stick out as distinct audible echoes. Delay left might be 17 ms, while delay right is 22 ms.

In using a spreader, the return of one delay output is panned left while the other is panned right. The idea is that these quick delays add a dose of support to the original monophonic track. In effect, these two short delays simulate some early sound reflections that one would hear if the sound were performed in a real room. The "spreader" takes a single mono sound and sends it to two slightly different, short delays to simulate reflections coming from the left and right.

That is only half the story. The effect is taken to the next level courtesy of some pitch shifting. Shift each of the delayed signals ever so slightly, and the mono source material becomes a much more interesting loudspeaker creation. Detune each delay a nearly imperceptible amount, maybe 5–15 cents. This is not a significant pitch change. An octave is divided into twelve half steps, representing adjacent keys on a piano or adjacent frets on a guitar. Each half step is further divided into 100 equal pitch increments, called cents. The pitch shifting called for in the spreader, then, is just 5 to 15% of a half step — all but imperceptible except to the most trained listeners. The goal of the spreader is to create a stereo sort of effect. As a result, one seeks to make the signal processing on the left and right sides ever so slightly different from each other. Just as unique delay times are selected for each side of this effect, choose different pitch shift amounts left and right as well — maybe the left side is shifted down 8 cents while the right side is shifted up 8 cents.

Like so much of what is done in recording and mixing pop music, the effect has no basis in reality. When adding delay and pitch shifting, the engineer is not just simulating early reflections from room surfaces anymore. The spreader makes use of common studio signal-processing equipment (delay and pitch shifting) to create a wide stereo sound that only exists in loudspeaker music. This sort of thing does not happen in symphony halls, opera houses, stadiums, or pubs. It is a studio creation, plain and simple.

Take this effect further and the sonic result might be thought of as more of a "thickener." There is no reason to limit the patch to two delays and two pitch changes. Ample signal-processing horsepower in most digital audio workstations makes it trivial to chain together eight or more delays and pitch shifts. Strategic selection of unique delay times, pitch shift increments, and pan locations control the fullness, width, and coloration of the effect. It is likely to sound unnatural when used heavily, but a light touch of this effect on vocals, guitars, or keyboard parts can help those tracks sound larger, fuller, and more exciting. Modulate each of those delays like a chorus, and more complex pitch shifting is introduced. Added in small, careful doses, this densely packed signal of supportive, slightly out-of-tune delays will strengthen and widen the loudspeaker illusion of the track.

10.2.3 SURGICAL EFFECTS

Pitch shifting is also used to zoom in and fix a problematic note. Maybe this sounds familiar: It is 5 a.m. It's the fiftieth take of the song. It's a great take. Then on the fifth repeat of the last chorus, the singer — tired from working all night long — drifts flat on the key word, the title word of the song, held for two bars: "Gaaaaaaaarlic." The song simply does not work if the last time the listeners hear the title of the song, "Garlic," the line is flat. No problem.

In the old days of multitrack production (and the reader is encouraged to try this approach), the sour note was sampled. It was then manually tuned using a pitch shifter, and the engineer's sound musical judgment. It was raised or lowered to taste. Finally, the sampled and pitch shifted note was re-recorded back onto the multitrack. With the problematic note shifted to pitch perfection, no one was the wiser.

Alternatively, and more frequently, the engineer reaches for clever digital signal processes that can automatically shift pitch into tune. Such an effects device can monitor the pitch of a vocal, violin, or didgeridoo. When it detects a sharp or a flat note, it shifts the pitch automatically by the amount necessary to restore tuning. The engineer typically watches over this process, choosing when the pitch shifting is used and when the original performed pitch is to remain. These pitch-correction tools are very effective, but one has to be careful not to overuse these devices.

First, engineers should not overpolish their productions. Pitch shifting everything into perfect tune is not always desirable. Vibrato is an obvious example of the musical detuning of an instrument on purpose. And, if Bob Dylan had been pitch shifted into perfect pitch, where would folk music be now? There is a lot to be said for a musical amount of "out-of-tuneness." Remove all the bends and misses, and we risk removing a lot of emotion from the performance.

Second, producers should not expect to create an opera singer out of a lounge crooner, or a pop star out of a karaoke flunky. There is no replacement for actual musical ability. If the bass player can not play a fretless, give them one with those pitch-certain things called frets. If the violin player can not control their intonation, hire one who can. Do not expect to rescue poor musicianship with automatic pitch correction. Use it to add to a stellar performance, not to create one. Musical sense and good judgment must motivate everything that is done in the recording studio. People generally want to hear the music, not the effects rack.

That said, Cher's title track from the album *Believe* puts wholly unnatural, machine-driven pitch shifting front and center and makes it one of the hooks of this pop tune. The lifespan of this effect is likely short, but variations on this effect are there for the creative engineer to explore.

10.2.4 OBVIOUS EFFECTS

Pitch shifting need not be subtle, as shown in the many examples that follow.

Leslie

Hammond B3 organs, many blues guitars, and even vocals are often sent through a rather unusual device: the Leslie cabinet. The Leslie sound is a hybrid effect built on pitch shifting, volume fluctuation, and often a good dose of tube overdrive distortion. The Leslie cabinet can be thought of as a guitar or keyboard amp in which the speakers sound as if they rotate. A two-way system, the high-frequency and low-frequency parts work in slightly different ways. The high-frequency driver of a Leslie is horn loaded. The driver is fixed, but it fires into a rotating horn. The perceived location of sound is the end of the horn, which moves toward and then away from the listener as the horn spins.

It would be very difficult to spin the large, low-frequency driver to continue the effect at low frequencies. Instead, the woofer is enclosed inside a drum. The drum has a few large holes in it. While the woofer conveniently remains fixed, the drum rotates, opening and closing the woofer sound. The result is a low-frequency approximation of what the Leslie is doing with the horn at higher frequencies.

In addition, the spinning system has three speeds. The rotating horn and drum may be toggled between off (the amp stays on, but the rotating mechanisms stop), slow, and fast during the performance.

The sound of the Leslie is fantastic. With the drum and horn rotating, the loudness of the music increases and decreases — tremolo or amplitude modulation (see Chapter 7). With the high-frequency horn spinning by, a Doppler effect is created: The pitch increases as the horn comes toward the listener/microphone and then decreases as the horn travels away.

The typical example used in the study of the Doppler effect is a train going by, horns ablaze. That classic sound of the pitch dropping as the train passes is based on this principle. Sound sources approaching a listener with any appreciable velocity will increase the perceived pitch of the sound. As the sound source departs, the pitch similarly decreases.

The high-frequency portion of the Leslie sound is heard through a horn. The perceived location of that sound source is the bell of the horn. So while the driver sits fixed within the Leslie cabinet, the high-frequency sound source is moving. The Doppler effect results.

The low-frequency driver, housed within a spinning drum, does not experience a pronounced pitch shift. As the holes within the drum rotate by, the low-frequency signal gets slightly louder. Continued rotation of the drum causes the sound of the woofer to be again attenuated, when holes have not opened up the woofer to the listener/microphone. Amplitude modulation without pitch bending is the signature low-frequency sound of a Leslie. The net result of the Leslie system then is a unique fluttery and wobbly sound.

The Leslie effect is common wherever B3s and their ilk are used. Switching between the three available speeds of rotation, one can create a fast Leslie and a slow Leslie effect or no spinning Leslie effect, as well as the acceleration or deceleration in between. Listen to the single note organ line at the introduction to "Time and Time Again" on The Counting Crows' first record, *August and Everything After*. The high note enters with a fast rotating Leslie. As the line descends, the speed is reduced. Listen carefully throughout this song, this album, and other B3-centric tunes, and the Leslie pitch-shifting vocabulary that keyboardists love will be revealed. Of course, sound engineers can apply Leslie to any track they like — guitar, vocal, and oboe — if you have the device, or one of its many imitators or simulators.

Big Shift

The straightforward pitch shifting that is the basis for the spreader and the thickener can be used in a more forward, don't-try-to-hide-it way. The hazard with an obvious pitch shift is that it can be hard to get away with musically. Special effects — in movies and on some records — where a vocal is shifted up or down by an octave or more can have a comedic effect. If it is too low, the pitch-shifted vocal conjures up images of death robots invading the mix to eat entire villages. If too high, the singer becomes a gerbil-on-helium sort of creation.

In the hands of talented musicians, aggressive pitch shifting really works. Prince famously lowers the pitch of the lead vocal track and takes on an entirely new persona in the song "Bob George" from *The Black Album*. The effect is obvious, and an incredible story results.

No effort was made to hide the deeper than typically found in nature bass line of "Sledgehammer" on Peter Gabriel's classic *So*. The entire bass track seems to include the bass plus the bass dropped an entire octave. The octave-down bass line is mixed right up there with the original bass. There is nothing subtle about it.

The pitch-shifting effect can be used to add two-, three-, or four-part harmony if the engineer is so inclined. Get out the arranging book though, because the pitch shifter makes it easy to inadvertantly add a dissonant interval. Each pitch-shifted note is a fixed interval above whatever note occurs in the source track. Only the octave will stay in tune with the harmony of a song. The perfect fourth and perfect fifth, tempting as they are, lead to a couple of nondiatonic notes when applied as a fixed interval above all the notes of the scale. The major third, applied rigidly to every note in a scale, will lead to dissonance and confusion; many notes will be out of tune with the key of the song.

A bit of additional signal processing is needed in the pitch-shifting device if it is to use minor thirds or major thirds and stay in the appropriate key. Digital signal processors offer this ability. The pitch-shift interval is variable, as needed to add harmony to a line.

The pitch shifting can be tied to Musical Instrument Digital Interface (MIDI) note commands enabling the engineer to dictate the harmonies from an MIDI controller. The pitch shifter is processing the vocal line on tape or disk according to the notes played on the keyboard. The result is a harmony or countermelody line with all the harmony and dissonance desired.

This production tool can reach beyond harmonies. One can use pitch shifting to turn a single note into an entire chord. String patches can sometimes be made to sound more orchestral with the judicious addition of some perfect octave and perfect fifth pitch shifting (above and/or below) to the patch.

It does not stop with simple intervals. Chords loaded with tensions are okay too when used well. Progressive rockers, Yes, put it front and center in "Owner of a Lonely Heart" on the album *90125*. Single-note guitar lines are transformed into something more magical and less guitarlike using pitch shifters to create the other notes.

Stop Tape

A final obvious pitch-shifting effect worth mentioning is the stop tape effect. As analog tape risks extinction, this effect may soon be lost on the next generation of recording musicians. When an analog tape is stopped, it does not stop instantly; it takes an instant to decelerate. Large reels of tape, like two-inch 24 track, are pretty darn heavy. It takes time to stop these large reels from spinning. If one monitors the tape while it tries to stop (and many fancy machines resist this, automatically muting to avoid the distraction this causes during a session), one will hear the tape slow to a stop. Schlump. The pitch dives down as the tape stops. This sometimes is a musical effect. It is not just for analog tape as Garbage demonstrates via a digital audio workstation effect between the bridge and the third chorus of "I Think I'm Paranoid" on their second album, *Version 2.0*.

Start Tape

The stop tape effect can be turned around, at least in the analog domain. Have the performer start playing before the tape recorder is rolling. Go into record immediately as the tape machine gets up to speed. The result, on playback at full speed, is a high-pitched descent into proper pitch. While the tape machine was coming up to speed, the signal was being recorded at an improperly slow speed. Playback at proper speed raises the pitch of that portion of the signal, and a unique pitch shift results.

The intro to "Synchronicity II" on the album, *Synchronicity* by The Police demonstrates this effect quite clearly. The squealing electric guitar in feedback pops into the mix with a sharp pitch bend courtesy of some coordinated start tape effects.

10.3 Selected Discography

Artist: Cher
Song: "Believe"
Album: *Believe*
Label: Warner Brothers Records
Year: 1998
Notes: The reference for just how far blatant pitch shifting can be pushed.

Artist: Counting Crows
Song: "Time and Time Again"
Album: *August and Everything After*
Label: Geffen Records
Year: 1993
Notes: This album abounds in good Leslie examples, but the intro of this song isolates the Leslie on the B3 on the right channel. It enters with fast rotation, and slows before the lead vocal enters.

Artist: Peter Gabriel
Song: "Sledgehammer"
Album: *So*
Label: Geffen Records
Year: 1986
Notes: Pitch shift that bass, one octave down. Don't hide it. Mix it in with the original bass. Larger than life.

Artist: Garbage
Song: "I Think I'm Paranoid"
Album: *Version 2.0*
Label: Almo Sounds
Year: 1998
Notes: The pitch dives as if analog tape were slowly stopped between the bridge and the third chorus, though this is most likely a digital effect simulating the tape stop.

Artist: The Police
Song: "Synchronicity II"
Album: *Synchronicity*
Label: A&M Records
Year: 1983
Notes: Intro reveals start tape effect. Electric guitar is made to feedback first. Then the tape machine is punched into record from a standstill. It ramps up to speed while recording. Played back at speed, the formerly slow bits are now pitched up. The sound of the machine achieving full speed in record becomes a pitch dive when played at uniformly correct speed.

Artist: Prince
Song: "Bob George"
Album: *The Black Album*

Label: Warner Brothers Records

Year: 1983

Notes: Tracked at a fast speed, the protagonist vocal is low and ominous when played back at regular speed. This all-analog pitch-shifting effect supports a convincing performance by Prince.

Reverb

11

"Nobody knows where you are,
how near or how far."
— "SHINE ON YOU CRAZY DIAMOND," PINK FLOYD, *WISH YOU WERE HERE*
(EMI RECORDS, 1975)

In the grand opera houses and symphony halls of the world, reverberation is an integral part of the music listening experience. The reverberant sound of a space is intimately bound to the sound of the music being performed within. The acoustic design of any music performance venue is directly influenced by the intended programming for the space. Musical theater, opera, classical music, chamber music, worship music, jazz — all types of music expect a specific sonic contribution from the hall.

Recorded music also places its own unique demands on reverberation. This form of music, importantly, is enjoyed through loudspeaker playback. This music is typically recorded in a sound studio, using natural and synthesized reverberation that flatters, alters, enhances, or otherwise refines the aesthetic value of the loudspeaker playback performance.

A wide variety of pop-music listening environments exist: living rooms from high-rise condos to country farm houses, automobiles from two-seater sports cars to family-of-five minivans. Unlike symphony halls, the acoustics of these listening spaces is rarely influenced by music because these rooms serve other more utilitarian priorities. Unable to rely on help from the acoustics of the loudspeaker playback venue, recorded music seeks to sound pleasing in any space. For recorded music, the most important reverberation exists within the recording, not the playback space. In the creation of music recordings, reverberation is selectively added, avoided, and/or manipulated to suit the creative needs of the music. That reverb — a fixed part of the recording — then follows the recording to every playback venue. The reverberation in the recording is thus even more intimately bound to the music than the natural reverberation of a symphonic hall is to the sound of the orchestra.

This chapter tours the technologies used to create the reverb used in recording studios and illustrates the broad range of studio effects it generates.

11.1 Reverberation

Imagine listening to the sound of someone's voice as they sing in the peaceful outdoors. Now imagine the sound of her voice as she sing in Paris' Notre Dame Cathedral. What changes? It is well known, even to those who have not been lucky enough to hear music performed at Notre Dame, that the cathedral adds a rich, immersive, heavenly decay to every note sung. That augmentation of the music by the resonance of the space is reverberation.

When one sings outside, there are no walls to reflect the sound back to the audience, no roof to hold the sound energy in. Listeners hear the voice directly. When the singing stops, the sound stops immediately. In a cathedral (or symphony hall, or shower, or any mostly-enclosed, sound-reflective space), listeners hear the voice directly plus the sound of the multitudinous reflections of that original sound bouncing off all the surfaces of the room.

Ignore for a moment the direct sound from the singer and focus only on the reflected sound energy from the building enclosure. Each sound reflection in that reverberant decay is both lower in level (due to all that traveling and bouncing) and later in time (it takes time for sound to travel, about 1 millisecond (ms) for every foot, as discussed in Chapter 1) than the original sound that was sung. The size and shape of the room and the geometric complexity and material make-up of the walls, floor, ceiling, and furnishings, drive the number, amplitude, and timing of the reflections. There are generally so many reflections arriving so close together that none of them can be heard distinctly. The countless reflections fuse into a single continuous wash of energy arriving steadily after the original sound. The particular blur of reflections that is associated with every sound in a specific space provides the sound signature of that space, its reverberant character.

11.1.1 KEY PARAMETERS

Fans of classical music are well aware of the contributions made by a hall to the sound of the orchestra's performance. Performances are sought out based on the repertoire, the orchestra, the conductor, and the hall.

Experienced recording engineers know that even slight changes to a studio reverb setting can have a significant effect on the overall recording. It is perilous to try to reduce reverb to a few numerical quantities. Art defies such a distillation. Imagine trying to describe Jimi Hendrix' "Little Wing" with just a few numbers.

Something as complicated as reverberation cannot be reduced to a handful of numbers. The field of architectural acoustics employs a vast range of measurements, tests, and analyses in an attempt to predictably improve the sound within a space. The broad field of room acoustics is understood by studying many dozens of books and participating in many years of experience and experiments. Nevertheless, engineers focus on a short list of measurable quantities to summarize the quality and quantity of any kind of reverb. The most important such parameters available to recording engineers working on popular recorded music are reverb time, bass ratio, and predelay.

Reverb Time (RT_{60})

The perceived liveness of a hall is measured objectively by *reverb time*, RT_{60}. Perhaps the most noticeable quality of a room's acoustics, reverb time describes the duration of the reverberant wash of energy. More specifically, it is the length of time it takes the sound to decay by exactly 60 decibels (see "Decibel" in Chapter 1).

Allow sound to play in the hall. It could be music or pink noise. Abruptly cut off this sound. The hall does not instantly fall silent. It takes a finite amount of time for the hall to return to silence again. RT_{60} is the standard measure of this length of time.

Bass Ratio (BR)

The reverb time measurement procedure just described is very much dependent on the spectral content of the signal used to initiate the reverb. Pink noise is a good choice because it contains sound energy distributed evenly across all the audible of octaves. If the playback system can handle it (and this is often a challenge, particularly at the lower frequency ranges), the hall is energized at all frequencies of interest to an engineer, nominally 20–20,000 Hz (see Figure 3.1, Chapter 3). The decay observed in this way represents the decay across the entire range of frequencies relevant to music.

Imagine the reverb is caused by a music signal instead of pink noise. This requires a choice to be made as to the type of music. Is the music selection bass heavy? Is it overly bright, having a strong high-frequency emphasis? Then one must decide the moment when the music must be stopped and the duration of the resulting decay measured. When is the best time to stop the music? During a chorus? After a snare hit? After a kick drum? While the vocal is singing?

Reverb time is better understood across a range of frequencies. Rather than relying on a single number to describe reverberation, it is helpful to find a low-frequency, mid-frequency, and high-frequency reverb time. In fact, measuring reverb time as a function of octave bands is the preferred approach.

The generation of reverb remains the same: use full bandwidth music or pink noise, abruptly stopped. The resulting decay is then band-pass filtered into the octave bands of interest. Finally, the time it takes each individual band to decay by 60 dB is found. Measuring reverb time as a function of frequency makes moot the issue of the spectral content of the test signal. The frequency-dependent decay of a space is a function of the size, shape, and materials used in making the space. The test signal only governs which frequency ranges are energized, and which are not.

RT_{1000} describes the length of time it takes sound energy in the one-octave band centered on 1,000 Hz to decay by 60 decibels. RT_{500} describes the 60-dB decay time one octave below, centered on 500 Hz. As the audio window spans some 10 octaves, this more refined method of calculation suggests that a space is better described by 10 reverb time measurements! In truth, even 10 frequency-dependent decay times do not come close to fully describing reverb — more and different measurements are needed to more meaningfully describe reverb. On the other hand, a set of 10 numbers is too much information to keep up with in the recording studio, with the engineer responsible for so many other aspects of the recording besides reverb. Further simplification is called for.

Bass ratio offers a single number comparison of lower octave reverb times to middle frequency reverb times. Specifically:

$$BR = \frac{RT_{125} + RT_{250}}{RT_{500} + RT_{1000}}$$

where BR = bass ratio, and RT_X = 60-dB decay time in the octave band centered on the frequency, X.

If the lower octave reverb times are longer than the middle frequency reverb times, the bass ratio will be greater than one. The perceived overall warmth and low-frequency richness of a performance space is very much influenced by its bass ratio, and ratios slightly greater than unity are often the design goal for a hall expecting to play romantic orchestral music.

Due to the importance of reverb time as a function of frequency, RT_{60} stated alone is generally understood to be a middle frequency reverb time (usually 1,000, maybe 500 Hz). Bass ratio adds required extra context.

Bass ratio is a term borrowed from the field of architectural acoustics. Measuring real symphony halls is a challenge, even with the advanced measurement equipment available today. Low-frequency measurements are the most difficult of all. Getting a portable transducer to reliably create significant sound pressure level in the bottom octaves remains impractical. Making reliable measurements in the bottom two octaves (with center frequencies of 31.25 and 62.5 Hz) is difficult today, and was essentially impossible until recently. Low-frequency behavior was best viewed at these, the third and fourth octave bands in human perception, centered at 125 Hz and 250 Hz, respectively. There are certainly exceptions, but in large, well-behaved spaces, it might reasonably be assumed that the behavior at frequencies below 125 Hz and 250 Hz is a reasonable extension of what is observed at these more accessible frequencies.

Predelay

In addition to the duration of the reverberant decay and the relative duration along the frequency axis (low versus mid), every recording engineer must understand a third reverb parameter, *predelay*. A gap in time exists between the arrival of the direct sound straight from the sound source to the listener and the arrival of any sound reflections or the reverberant wash of energy that follows. Predelay is the difference in time of arrival between the direct sound and the subsequent first associated reflection.

The size and shape of the performance space is the key determinant of predelay time. In orchestra halls, it is often the sidewalls that create the first reflection. Reflections off of the ceiling or the rear walls arrive much later. Therefore, a narrow hall is likely to have a shorter predelay time than a wide hall. In smaller spaces, the ceiling may be the closest reflecting room partition. With reverb-generating signal processors, of course, predelay is simply an adjustable parameter almost without limits.

11.1.2 REFERENCE VALUES

The very idea of reverb for music comes from real spaces, such as symphony halls and houses of worship. The reverb used in recording studios is typically generated by signal-processing devices. These user-adjustable pieces of equipment are wonderfully — and sometimes frustratingly — independent of the physics of sound constrained by architecture. The total freedom to synthesize any kind of reverberant sound is at times paralyzing for the novice engineer, and has been known to bog down even veteran engineers. It is useful to bracket the range of studio reverb parameters based on the architectural acoustics of classical performance venues. An engineer is welcome to venture beyond physically-realizable reverb properties, but clever engineers knows when they have done so.

The symphony halls most adored by conductors, orchestras, critics, and enthusiastic music fans represent perhaps the highest form of achievement in reverb, specifically reverb for romantic orchestral music. Three halls are consistently rated among the best halls in existence today and serve as our reference point in the recording studio:

1. Boston Symphony Hall, Boston, MA, United States
2. Concertgebouw, Amsterdam, The Netherlands
3. Musikvereinssaal, Vienna, Austria

Detailed analysis of these halls and other halls approaching their quality leads to a useful set of representative reverb values (Table 11.1). Setting up a studio reverb so that it's parameters fall within these preferred values does not guarantee success. This quality of reverb may not sound appropriate for the music production at hand. These values have proven themselves appealing for the live performance of romantic orchestral music, not all forms of music. Also, a hall that falls within the desirable ranges for these three values might fail on other fronts. As mentioned above, something as complex as the reverb within a hall is not fully defined by so few

Table 11.1 Preferred Ranges for Three Key Reverb Parameters, in Halls for Romantic Orchestral Music (from Beranek, 1996)

Parameter	Value
Reverb Time	1.8 to 2.0 s
Bass Ratio	1.1 to 1.45
Initial Time Delay Gap	20 ms or less

It is often a recording goal to minimize this leakage. This influences the audio engineer's placement of musicians in the room, and the selection and orientation of directional microphones around the instruments. Clever use of high-transmission-loss sound barriers between players within highly absorptive rooms is also part of this approach. The result is a further reduction in the recording of room ambience in every session.

The desire for isolation between and among the musicians contributes additional motivation to use close-microphone techniques as described above. The sound engineer places microphones closer still to the instrument being recorded, shifting the relative levels of the target instrument versus the "leaking" instruments decidedly in favor of the target.

11.2.3 CREATIVE MICROPHONE TECHNIQUE

When microphones zoom in close to the instruments they record, they find an unusual "view" of the instrument. Good microphone practice requires the engineer to be able to capture a natural sound of an instrument despite such proximate microphone placement. Moreover, the varied, unusual, and at times unnatural sounds that exist in various locations so close to an instrument provide the creative recordist with the chance to record sounds that have a distinct sonic flavor, exaggerated sound feature, extreme intimacy, or heightened timbral detail. Microphones are brought in close to the instruments ultimately in search of ways to benefit the recorded end product. Recording engineers take advantage of this, moving beyond a goal of accuracy in pursuit of more creative sounds. For all recorded instruments, any characteristic sound qualities to be enhanced are identified and emphasized. Unwanted portions of the sound are strategically avoided. The engineer is, therefore, further motivated to place the microphone(s) quite close to the instruments being recorded.

11.3 Sources of Reverb

With so many forces conspiring to deemphasize recorded reverberation, the multitrack production process often requires reverberation to be added back into the loudspeaker music at a later time. At the mixdown session, all individual tracks have been recorded and the technical and creative process of combining them into the single loudspeaker listening experience occurs. To add reverb, the engineer has two choices: record natural reverberation separately (i.e., onto separate tracks of the multitrack recorder, often called *room tracks*) at the time of the original music performance, or employ reverberation devices that create the desired spatial, ambient, or other qualities at mixdown via signal-processing effects.

11.3.1 ROOM TRACKS

While recording overdubs with close-microphone techniques, the engineer may also place distant microphones in the studio in order to capture the room's liveness. These ambient signals are recorded as separate audio tracks so that they later appear as fully adjustable elements of the multitrack arrangement. While a single mono room track has value, stereo productions will record room tracks in stereo pairs. Surround productions will record room tracks in four or five track sets. During mixdown, these room tracks can themselves be adjusted and processed in any way the engineer chooses. Some high-end recording studios are prized for the sound of their live rooms that have proven particularly effective in recorded music. Architecturally much smaller than opera houses and symphony halls, these live rooms offer particularly supportive early reflections with a dose of relatively short reverberant decay.

The history of recorded music documents the value of recording even popular music in reverberant spaces. Many of the most important works of recorded art were recorded in studios much larger than is common today. Studios existed in converted churches and giant loft spaces. The talented engineer was able to capture recorded ambience at the tracking stage that worked for the production when finally mixed and enjoyed over loudspeakers. Few engineers today get the chance to record in such large, live spaces.

While it does not happen very often, multitrack music productions will sometimes go to the trouble to record elements in very reverberant spaces. The session will leave the recording studio and track drums in a church or string section in a hall, and background vocals in a solid concrete basement. Room tracks are a critical part of such sessions.

11.3.2 ACOUSTIC REVERBERATION CHAMBERS

Clearly it is impractical to utilize an orchestra hall as an effects device. Such buildings are rented at great expense and cannot be proximate to every recording studio. As a result, the natural reverberation of a large hall is rarely part of a pop or rock multitrack music production. Knowledge of room acoustics — beginning with the work of Wallace Clement Sabine in the 1890s — can still be employed directly in the recording studio, however. Sabine's well-known equation illustrates the concept of the reverb chamber.

$$RT_{60} = \frac{0.049V}{A}$$
(11.1)

RT_{60} is the length of time in seconds required for the reverberant sound level to decay by 60 decibels, V is the volume of the room (cubic feet), and A is the total sound absorption of the room (square feet).

The reverb time is directly proportional to the volume of the room, and inversely proportional to the room's total sound absorption. For a long reverb time, seek out a large room volume. As large halls are impractical for recording studio work, multitrack productions must generally make do with a much smaller space. If the room volume cannot be large, use a room with little sound absorptivity. A small space is utilized and equipped with loudspeaker(s) and microphones (Figure 11.1). To achieve a long reverb time, the lack of cubic volume in a reverb chamber (Figure 11.2) is overcome through high sound reflectivity. Plaster, concrete, stone, and tile are typical materials for such a space.

Sabine's equation (Equation 11.1) assumes a fully diffuse sound field. It is unlikely that a space as small as a reverb chamber generates a truly diffuse reverberant field within the reverb time window of interest, but the equation still provides informative guidance about using reflectivity to overcome small volume. It follows that the surface treatment within a chamber may

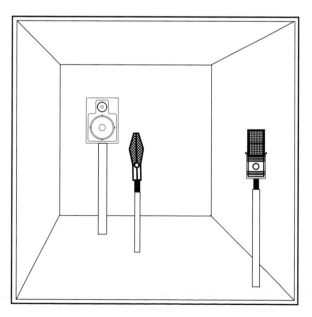

▲ *Figure11.1 Reverb from a small, highly sound-reflective room volume.*

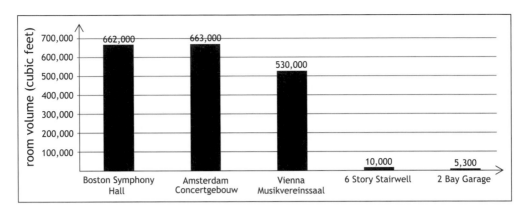

▲ *Figure 11.2 Chambers must generate reverberation within a significantly smaller room volume than symphony halls.*

benefit from being highly irregular to maximize diffusion, increase modal density, and decrease audible coloration within the relatively small space.

There is no such thing as a "typical" reverb chamber. Recording studios install them wherever they can find affordable space. Custom-built rooms just for reverb are an expensive endeavor for recording studios, which are often located in urban areas (e.g., New York, Los Angeles, London, Tokyo) where real estate costs are substantial. Practicality has motivated recording studios to award the lofty title of "reverb chamber" to such unlikely spaces as multistory stairwells (CBS on 7th Avenue, Avatar/Power Station); attics (Motown, Capital Studios on Melrose Avenue); bathrooms (project studios around the world); and basements (CBS 30th Street, Capital Studios on Hollywood and Vine, and Abbey Road).

Some trends in the architecture of the proven chambers may be observed. Highly sound-reflective materials finish theses spaces. Concrete, stone, brick, and tile are common. Frequently, thick layers of paint cover the naturally-occurring pours in the concrete and masonry for minimum sound absorption. For reasons of tradition and superstition as much as science, irregular shapes and sound-diffusing elements are frequently sought out. Stairs, columns, pipes, highly articulated surfaces, and nonparallel walls are the norm.

Selection and placement of the loudspeakers and microphones are critical to the sound of the chamber. The loudspeakers are often selected for their dynamic range capabilities. To energize the chamber and overcome any extraneous noises within, the ability to create very loud sounds is desired. Horn-loaded, sound-reinforcement speakers make a good choice, with efficient ability to handle high power. Practicality influences the loudspeaker selection too. Last year's control room monitors are frequently repurposed as this year's chamber speakers. As coloration is so prevalent in chamber reverbs, often being deliberately sought out, loudspeakers with a flat frequency response are not strictly required. Nonflat frequency response becomes a creative variable. Clever engineers turn the spectral imperfections of the loudspeakers in combination with the frequency biases of the space into a chamber reverb with a distinct and hopefully pleasing flavor.

Microphones are placed very much in the same way room microphones are placed in a recording session. Engineers develop intuition about where in a room a microphone might sound best. Experimentation follows, and the microphone placement is revised as desired. Condenser microphones are most common, ranging from large diaphragm tube condensers to small diaphragm electret condensers. While omnidirectional microphones may seem best for fully picking up the sound of the chamber, directional microphones are common too as they reward the engineer for even slight changes in location and orientation.

Generally, a line of sight between the loudspeakers and microphones is avoided. The loudspeakers are placed so as to energize the space generally. The microphones are placed to capture the subsequent reverberation. The direct sound from speaker to microphone amounts to an acoustic delay only, whose level is likely to be much higher than the reverberation that follows. Avoiding that direct sound maximizes the reverb that is captured by the chamber microphones and returned to the mixing console. Reverb is the intended effect, after all. Directional loudspeakers (horns) and microphones, though not required, help the engineer achieve this.

Because reverb chambers are so small, a certain amount of modal coloration is unavoidable. It is hoped that the coloration this causes will sonically flatter any "dry" sounds sent to it. Equalization (EQ; see Chapter 5) is a signal processor purpose-built to introduce coloration to the sound. Engineers reach for reverb generated in a chamber when they feel the chamber's coloration will lead to a desirable change in sound quality. Otherwise, the strong coloration must be avoided (no chamber reverb is added); ignored (other elements of the multitrack arrangement mask the coloration problem); or perhaps deemphasized with some EQ on the send and/or return.

The typically small dimensions of a chamber lead to a predelay that is on the order of just a few milliseconds, well short of the 20 ms associated with large halls. Typically a delay line (see Chapter 9) is inserted on the send to the chamber so that all reverb is appropriately delayed when it is combined with the other elements of the multitrack production.

11.3.3 SPRING REVERB

A simple spring can be used to create a kind of reverberation. A torsional wave applied to a spring will travel the length of the spring. Upon reaching the end of the spring or encountering any change in impedance along its length, some of that twisting wave reflects back down the length of the spring from which it came. In this way, the wave bounces back and forth within the spring until the energy of the wave is converted into heat and dissipated through friction. Analogous to reflections between just two walls, a spring mechanically emulates the sound of a reverberant space — a *one-dimensional* space. Combine several springs of different lengths, thicknesses, and spring constants into a network of coupled springs and the pattern of reflections can be made more complicated.

The spring reverb has a unique sound but falls well short of simulating a real, physical, reverberant space. Its most important shortcomings in this regard are:

- The frequency response of the driver-spring pick-up system has limitations in the audio band.
- The number of reflections in even a multispring system provides only a finite imitation of the nearly infinite collection of reflections in a real room.
- The modal density of so simple a system is insufficient to prevent strong coloration.
- The build-up of reflection density does not grow exponentially with time as happens in a real room.

The spring reverb is simply not a room simulator. Yet it remains in (limited) use today. Affordable, analog, and portable, it is a common reverb in many electric guitar amps. Artists and engineers have locked onto it for its unique, if unreal, sound. Measuring one example of this device in the same way that acousticians measure a space offers a better understanding of its reverberant behavior.

The impulse response of Figure 11.3 shows an extraordinarily long reverb time. Reverb times vary from about 2 seconds in the upper octaves to more than 5 seconds in the 250-Hz octave band (Figure 11.4).

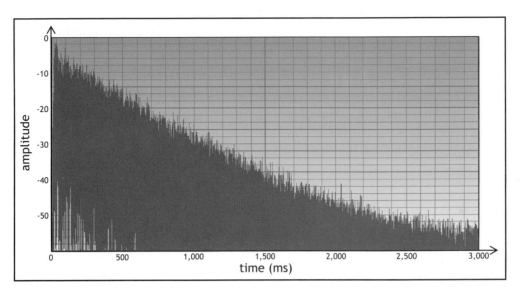

▲ *Figure 11.3 Impulse response of a spring reverb.*

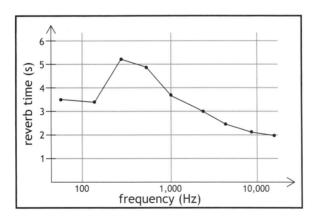

▲ *Figure 11.4 Reverb time by octave band of a spring reverb.*

The frequency response of this mechanical system very much favors the upper octaves, with an approximately 6 dB per octave tilt up to the 4-Hz octave band (Figure 11.5). This helps give the spring reverberator its characteristic metallic sound. Yet the late decay becomes distinctly modal, leading to further coloration of any sound sent to the spring reverb (Figure 11.6).

While the spring does not closely resemble the sound of a hall, it is informative to compare its performance to the three referenced orchestra halls.

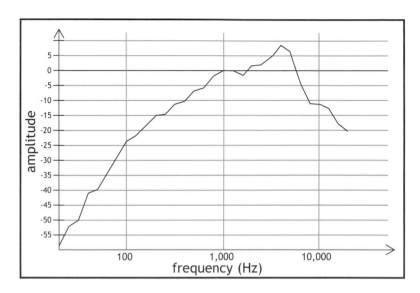

▲ *Figure 11.5 Frequency response of a spring reverb.*

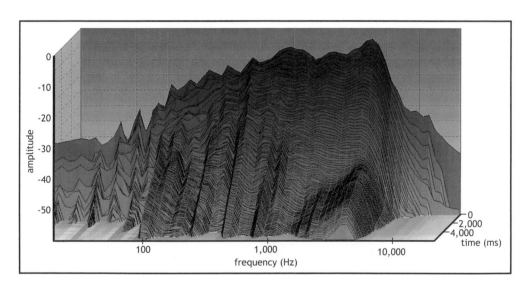

▲ *Figure 11.6 Spectral decay for a spring reverb.*

The spring reverb is compared to Boston Symphony Hall, the Concertgebouw, and the Musikvereinssaal in Figures 11.7 and 11.8. The reverb time of the spring is clearly much longer than that of any desirable hall. The bass ratio of the spring reverb might seem to suggest that the spring could to do an adequate job of creating the warmth associated with good low-frequency reverb. This is not at all the case. The bass ratio metric alone is misleading.

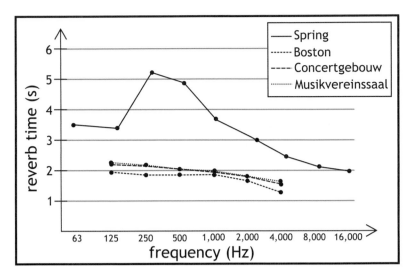

▲ *Figure 11.7 Reverb time, spring reverb versus benchmark halls.*

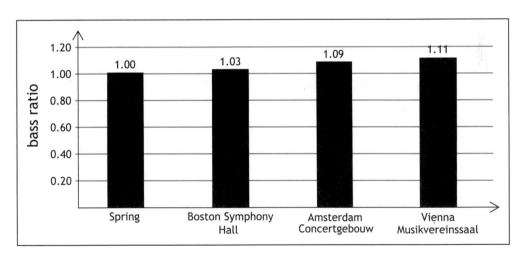

▲ *Figure 11.8 Bass ratio, spring reverb versus benchmark halls.*

Refer again to the reverb time plot of Figure 11.4 and note the extreme difference in reverb times in the two low octave measurements that are a part of the bass ratio calculation, 125 Hz and 250 Hz. While these two octave bands share the same average reverb times as the two mid-band octaves, there is a nearly two-second difference in reverb times between the two lower octaves. Such extreme reverb time variation in the lower octaves does not lead to the same quality of warmth that a more consistent low-frequency device or space would possess.

Though differences certainly exist, it might be surprising to any reader who has used spring reverb how similar the spring reverb measurements are to benchmark room measurements. However, these metrics provide an incomplete picture. The sonic signature of a spring reverb, apparent to experienced sound engineers, is not fully explained by this analysis. Sound engineers are aware (sometimes only intuitively) that, among other things, a spring's initial density of echoes as well as the rate of change of the echo density over time are different from that of an actual reverberant space. Resonance in a space builds and decays as the sound propagates in three dimensions. A spring offers a single dimension of propagation for the torsional wave within. This contributes to a unique and distinctly unnatural-sounding decay. In the right musical and sonic context, engineers use spring reverb precisely because of this different sound quality.

11.3.4 PLATE REVERB

The concept of the spring reverb is improved upon through the use of a metal plate instead of a spring. This upgrades the mechanical reverb to a two-dimensional design. A thin plate of metal has attached to it a driver that initiates a bending wave. This bending wave propagates through the plate, reflecting back at the edges, leading to an accumulation of reverblike energy. A pick-up transducer captures this reverberated signal. Reverberation on this plate behaves in a way very much analogous to a room with two pairs of opposing walls (but without a floor or ceiling).

Because the plate is excited with bending waves, not compression waves, the speed of propagation is determined by the elasticity and mass distribution of the plate and by the plate's thickness and suspended tension. Plate reverb designs can, therefore, slow the speed of propagation down to one thousandth the speed of sound in air. A mechanical device much smaller than a hall can be used to generate reverberation similar in duration to that of a large hall.

The plate offers further advantages over the spring. By applying any of various means of damping (e.g., placing liquids or porous materials against the plate), the reverb time is adjustable across a useful range, offering the sound engineer highly desired production flexibility.

The plate reverb, like the spring reverb, also fails to accurately simulate a real, physical, reverberant space. Its most important shortcomings versus acoustic reverberation are:

- The frequency response of the driver-plate pick-up system has limitations. There is a lack of response in the lower portion of the audio band (measurements are provided in the discussion that follows). In addition, even with damping, there is too much high-frequency resonance, leading to an unnaturally bright sound.
- The modal density of even a large plate is still insufficient to prevent strong coloration.
- While the build-up of reflection density does grow exponentially with time, the two-dimensional, rectangular, mechanical reverberator does not grow at the same rate as a real, three-dimensional space with complex shape.

As with the spring reverb, the plate reverb is simply not a room simulator. However, plates are extremely popular in multitrack production even today. Artists and engineers have found the quality of its sound to be useful for many applications (discussed below). A plate reverb, damped to a commonly used short reverb time, is measured as if it were a reverberant performance space.

The impulse response (Figure 11.9) shows a very steep slope. Reverb times (Figure 11.10) measured in octave bands all fall below about 0.5 seconds.

Though it varies a great deal from unit to unit, and depends very much on tuning and maintenance, the frequency response of a well-tuned plate

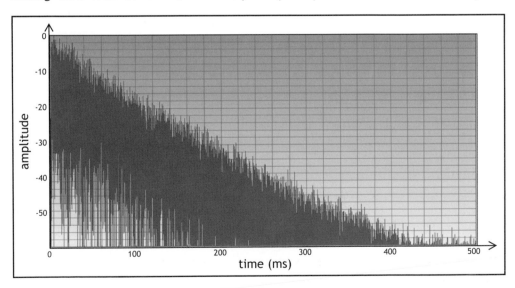

▲ *Figure 11.9 Impulse response of a plate reverb.*

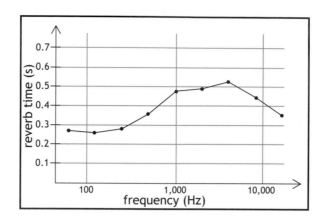

▲ *Figure 11.10 Reverb time by octave band of a plate reverb.*

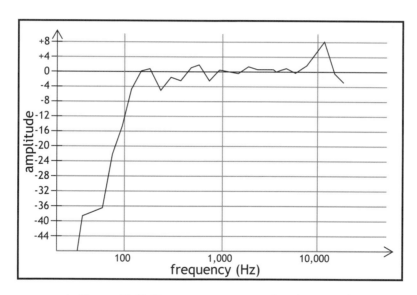

▲ *Figure 11.11 Frequency response of a plate reverb.*

reverb can be made remarkably flat (Figure 11.11), at least above about 100 Hz. These devices are tuned very much like a musical instrument, with the tension of the plate adjusted until a pleasing sound is created. For the most part, the plate is made as tight as practical for maximum undamped sustain, with uniform tension around the entire perimeter for a less clouded sound.

The relatively flat frequency response, from about 100 Hz up through the rest of the audio band, is preserved throughout the decay of the reverb (Figure 11.12).

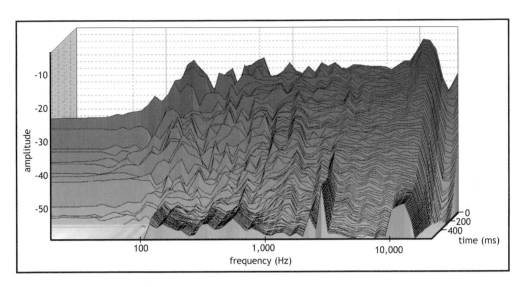

▲ *Figure 11.12 Spectral decay for a plate reverb.*

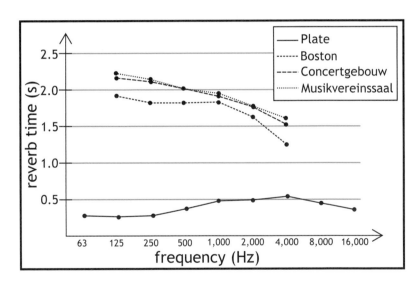

▲ *Figure 11.13 Reverb time, plate reverb versus benchmark halls.*

Comparison of this plate reverb to the benchmark halls reveals noticeable differences. The plate's reverb times (Figure 11.13) are markedly lower than these halls at all octave bands measured.

The bass ratio comparison in Figure 11.14 highlights the lack of warmth in the reverberation of this mechanical device. However, creative recording engineers have used this to advantage, finding successful applications for

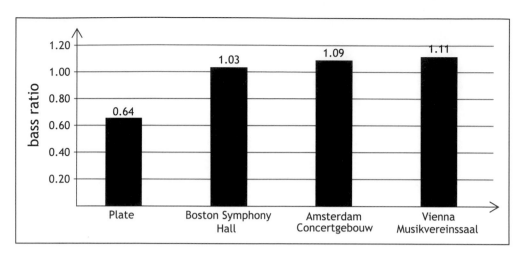

▲ *Figure 11.14 Bass ratio, plate reverb versus benchmark halls.*

the thinner sounding reverb. Popular music has little in common with the orchestral music typically played in these benchmark halls after all.

As with the spring reverb, the full sound quality of the plate reverb is not fully captured by these measurements. Sound engineers rely on intuition, experience, and critical listening when choosing to work with this effect.

11.3.5 DIGITAL REVERB

While the synthesis of reverberation through math requires an intense amount of calculation horsepower, the resources are certainly available to do so. In fact, digital reverberators have been the most common tools for adding reverb in popular music recordings for some 20 years.

Countless algorithms exist for generating reverb digitally. Below is a discussion of the basic building blocks that are a part of the digital reverb.

Infinite Impulse Response Reverberators

Digital signal-processing algorithms, based on recirculating delay systems forming a combination of comb filters and all-pass filters, form a class of infinite impulse response (IIR) reverberators, which are very much in use in recording studios today.

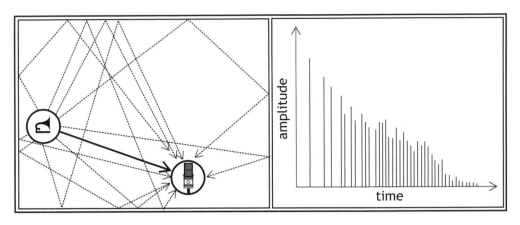

▲ *Figure 11.15 Reverb is a very large collection of chosen delays.*

Comb Filter

Reverberation is a wash of sound reflections. It follows that each reflection could be modeled using a well-chosen digital delay. Figure 11.15 simplifies this approach, beginning to trace out all available reflected sound paths propagating from a sound source in one location to a sound receiver in another. Map out the time of arrival and amplitude of each reflection, and the concept for a machine generating reverb through delay begins to emerge. A key question arises. How many delays are needed?

In a time of significantly less digital signal-processing horsepower, Manfred Schroeder famously explored simple ways to fabricate reverberation from a "dry" signal. Back in 1962, when the Beatles were singing, "She loves you, yeah, yeah, yeah," Schroeder first proposed creating reverberation through the use of a single delay with regeneration, the comb filter (Figure 11.16a). The time response to an impulse shows the reverblike quality of the filter, with its exponentially decaying reflections. The frequency response of a comb filter (see Chapter 9) reveals its spectral shortcoming as a source of reverb.

When the delay time is set to a high value, flutter echo is unmistakable. When the delay time of the comb filter is set to a low value, a dense set of echoes is satisfyingly produced. However, the frequency response will exhibit a wide spacing of peaks and notches. Strong coloration results. It is only these frequency peaks that are reverberated, and the spectrum in-between decays quickly.

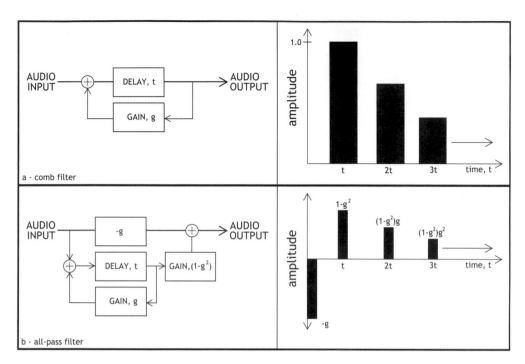

▲ *Figure 11.16 Reverberation from (a) a comb filter, (b) an all-pass filter.*

All-Pass Filter

Schroeder immediately improved on his reverberator. Through feed-forward and scaling, the comb filter can be made into an all-pass filter (Figure 11.16b) creating a series of exponentially decaying reflections with flat frequency response. This flat frequency response is based, mathematically, on an infinite time integration. Of course, human hearing does not wait that long. The result, unfortunately, is that in the short term, the frequency response is not necessarily flat, some coloration remains, and the sound is dependent on the delay time selected and can sound quite similar to the comb filter discussed above.

Hybrid Combs and All-Pass Filters

Combining multiple comb filters and all-pass filters is productive. Perceived coloration can be reduced if the peaks of a comb filter are made more numerous (increasing modal density). Flutter echo is diminished if the number of delayed signals is increased (increased echo density). Comb filters combined in series are ineffective because only the frequencies in the peaks of one filter are passed on to the next. For a signal passing through several comb filters in series, only frequencies within the peaks common to all comb filters will appear at the output. Multiple comb filters, when used, are placed parallel to each other.

All-pass filters, on the other hand, may be connected in series freely, as the net output will still be all-pass. Using multiple all-pass filters will, therefore, increase echo density while preserving the (long-term) frequency response. Use of multiple comb filters in parallel, with strategically chosen delay values that have no common factors, can create a frequency response with peaks distributed throughout the frequency range of interest, increasing modal density and decreasing perceived coloration. Parallel application of multiple comb filters also increases the echo density as the output is simply the sum of the individual comb filters, with a corresponding increase in echo density.

Schroeder founded an industry of digital reverberation when he offered his hybrid combination of parallel comb filters feeding a series of all-pass filters back in 1962, as shown in Figure 11.17. Countless permutations exist for choosing any number of parallel comb filters and serial all-pass filters. Designers are also free to adjust in complex ways the variables, such as delay times and gain factors, for each of the individual comb and all-pass filters. Furthermore, the introduction of low-pass filters within the comb filter simulates air absorption and makes for a more realistic reverberation. Limitless options are available to the creative digital signal-processing designer.

A common, high-performance digital reverb based on this approach, with parameters set to emulate a large orchestral hall, is measured. The measured impulse response (Figure 11.18) is analyzed to reveal a mid-frequency reverb time of about 2.5 seconds. Octave band calculations of reverb time

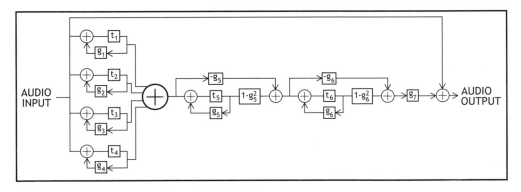

▲ *Figure 11.17 Reverberation from parallel comb and serial all-pass filters.*

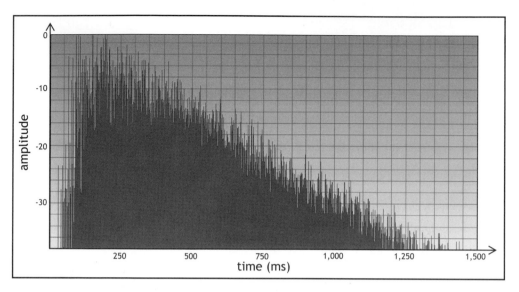

▲ *Figure 11.18 Impulse response of a digital reverb (large hall).*

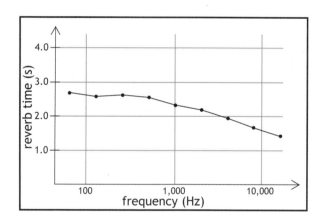

▲ *Figure 11.19 Reverb time by octave band of a digital reverb (large hall).*

(Figure 11.19) demonstrate behavior consistent with an idealized real room. No single octave band dominates, minimizing unpleasant coloration. Additionally, there is a clear trend for high frequencies to decay more quickly, mimicking the air absorption of a real hall.

The frequency response (Figure 11.20) shows a natural roll-off of high frequencies consistent with the performance expected in a highly diffused, air-filled, real hall. The digital algorithm decays naturally, without pronounced modes, but with lingering low-frequency reverberation adding desirable warmth (Figure 11.21).

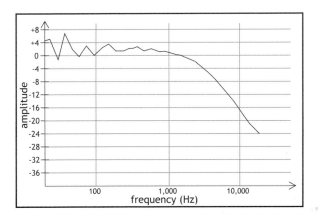

▲ *Figure 11.20 Frequency response of a digital reverb (large hall).*

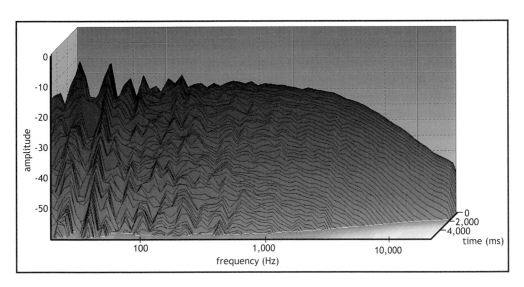

▲ *Figure 11.21 Spectral decay for a digital reverb (large hall).*

Of course, digital reverbs like this one have many adjustable parameters that enable the recording engineer to modify this particular measured response. The reverb can be significantly altered. Nevertheless, it is informative to make comparison to the three benchmark halls.

Reverb time comparison (Figure 11.22) shows this popular reverb algorithm is consistently longer than the most desirable halls across all octave bands. Keep in mind that the recording engineer has many tools to control the perceptual impact of this reverb (discussed below), with the result that a very long reverb can be integrated into a music recording without the

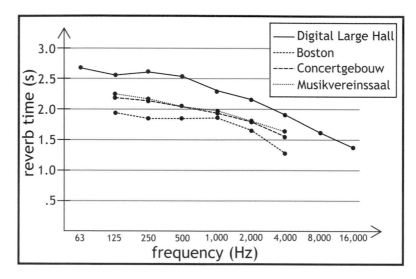

▲ *Figure 11.22 Reverb time, digital reverb (large hall) versus benchmark halls.*

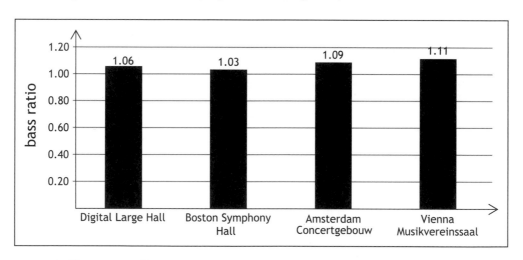

▲ *Figure 11.23 Bass ratio, digital reverb (large hall) versus benchmark halls.*

expected clutter and obscuring that a similarly performing room would have; the digital reverb in this common configuration creates a longer reverb than might be expected to be pleasing.

The trend, octave to octave, of the digital reverb parallels that of the benchmark halls. The highly sought-after warmth that comes from a bass ratio slightly above unity is clearly built in to this reverb algorithm (Figure 11.23).

Convolution Reverberators

If there were a way to capture some sort of blueprint that describes the detail (both the time of arrival and the amplitude) of all the reflections that make up reverberation, and apply it to other sounds, reverb could be synthesized. Convolution reverbs do exactly that.

The impulse response that drives the convolution might be measurement data from an existing space, calculated data from room modeling analysis, or wholly synthesized data fabricated by the creative sonic artist.

Convolution Defined

Earlier in this chapter, comparison was made of a voice singing outside, and then singing inside the Notre Dame Cathedral. Consider a handclap instead of singing: "Pop!" The hands colliding make a sound of very short duration. Outdoors the single handclap is over as soon as it began. Inside the cathedral, the handclap is followed by a burst of reverberant energy that is made up of a very specific set of reflections — handclaps that have been attenuated and delayed according to the size, geometry, and material properties of the space in which the sound bounced around. The way the space converts a single spike of a sound into a more complicated wash of decaying sound energy is a defining characteristic of the space. Impulsive sounds are sounds of very short duration, like the handclap, a gun shot (blanks only, please), a balloon pop, a clave, a snare drum, etc. The way the room responds to an impulsive sound is called its *impulse response*. The impulse response contains the signature of the space as it augments every bit of every sound with its own specific pattern of reflections (see again Figure 11.15).

Convolution is the process of applying that impulse response to other sounds, like a vocal, an orchestra, a guitar, a didgeridoo, to create the wash of energy the space would have added to such audio events if they had occurred there.

It helps to simplify the complicated world of a real space to one that is easier to understand: a space that impossibly offers only two reflections (Figure 11.24). A single impulse within this space would always be followed by this exact pattern of two reflections, arriving exactly at the time and amplitude described by the impulse response. A louder impulse would trigger a correspondingly louder pair of reflections.

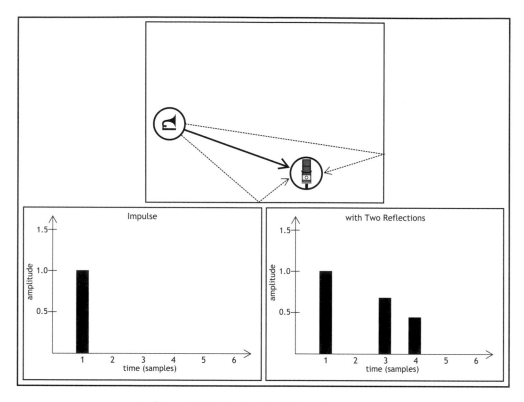

▲ *Figure 11.24 A simplified impulse response.*

If two impulses are played one after another in the space, each would trigger this impulse response and they would combine into a more complicated sound (Figure 11.25).

Convolution is the word that physicists and mathematicians use to describe this math. As a concept, it is fairly straightforward. As a calculation, though, it gets complicated fast.

Consider a more complicated sound — a snare drum, for example (Figure 11.26). Zooming in on the digital waveform to the individual sample level reveals the sound to be nothing more than a stream of pulses. For a sample rate of 44,100 Hz, the audio is made up of 44,100 such pulses of varying amplitudes each second. Let each one of these trigger the impulse response individually. Combine new sounds with those sounds still lingering. Keep track of the decay associated with each sound so that it can be combined with future sounds. Add them all up. The overwhelming data crunching rewards us with convolution reverb.

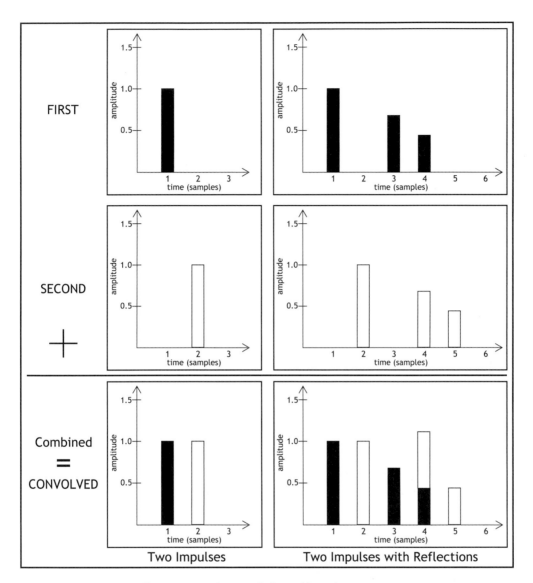

▲ *Figure 11.25 Accumulation of impulse responses.*

If one convolves the complicated impulse response of a real space with the close-microphone recording of a snare drum, the result is a sound very much as it might have occurred if the snare had been played in the real space. Knowing what a space does to a simple impulse is enough to predict what it would have done to almost any signal, as long as there is a machine willing to keep up with all the associated calculations.

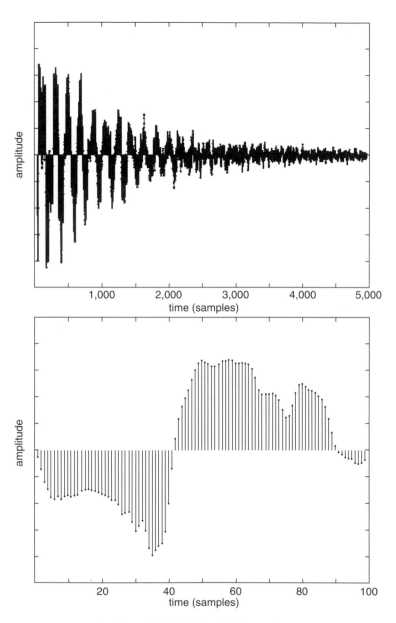

▲ *Figure 11.26 A snare hit, up close.*

This is the basis for the family of reverb devices variously known as sampling, room modeling, and convolution reverbs. The math is so daunting that while theory suggested it many decades ago, engineers could not take advantage of it until computers got fast enough. In this, the twenty-first century, home computers are more than capable of implementing convolution in the recording studio.

To better appreciate the math, consider a symphonic hall with a two-second reverb. At a sample rate of 44,100 Hz, the two-second impulse would be made up of 2 × 44,100, or 88,200 samples. For stereo, two impulses are needed: that seen by the left ear and that received at the right. The hall is then characterized by a stereo impulse response that has 2 × 88,200, or 176,400 pulses. When this is convolved with an audio waveform like that snare drum that is also digitized at a 44,100-Hz sample rate, each one of those samples triggers 176,400 numbers that must be stored and added to the variously decaying elements of all the other pulses of the audio. That is, 44,100 times per second, the system gets another wash of 176,400 numbers to keep track of. Each second of audio requires 3,889,620,000 samples of reverb to be stored and included in the calculations. Raise the sample rates in those calculations from Redbook CD to the middle-of-the-road high-resolution sample rate of 96 kHz, and some 18,432,000,000 samples are part of the calculation for every second of source audio that generates stereo reverb. It takes a fair amount of CPU horsepower to do this in anything close to real time. This horsepower was not readily available in off-the-shelf computing until about the year 2000. These are good times to be creating audio.

Summing up convolution:

- Pulse code modulation converts audio into a series of pulses.
- Each audio pulse triggers an associated impulse response.
- Each impulse response is scaled by the height and polarity of the audio pulse that triggers it.
- Keep track of the amplitude, polarity, and time of arrival of all associated pulses and add them up.

Convolution Parameters

Engineers who have been using those legacy digital reverbs not based on convolution have grown accustomed to the process of selecting a reverb patch and then tweaking it into submission. That is, they select a "hall" or a "medium room" program as a starting point and then adjust parameters until the quality of the reverb fills the needs of the multitrack production.

Convolution reverbs do not offer quite as much flexibility. Their strong basis in reality (it is a theoretical representation of what actually happens in a real room after all) makes them difficult to modify. To select the type of space, simply choose an appropriate impulse response — one from a hall or a medium room, for example. So far, so good.

With nonconvolution reverbs, adjustable parameters beyond reverb time, predelay, bass ratio, and such are freely available for modification. Such parameters are not always available in a sampling reverb. In a real room, the reverb time gets longer when the space gets larger or more acoustically reflective. This implies that to lengthen the reverb, one has to move the walls or change the sound absorptivity of the ceiling, remeasure the resulting impulse response, and then execute the convolution. In other words, to lengthen the reverb time in a convolution engine, one must select a different impulse response entirely, one with a longer reverb time (Figure 11.27).

This causes headaches. The reverb time of the Concertgebouw in Amsterdam, a much-loved hall, is fixed. There is simply not a knob to adjust reverb time in the space. Therefore, there is not an easy way to get that Concertgebouw sound at 1.5 seconds, 2.0 seconds, and 2.5 seconds. The Concertgebouw sound is what it is. If it sounds gorgeous on the lead vocal, then the engineer is satisfied and gets on to the next task. If it sounds good, but the reverb time is a bit long, the engineer is going to have trouble shortening it without altering the many subtle qualities that lead to the selection of this impulse response to begin with.

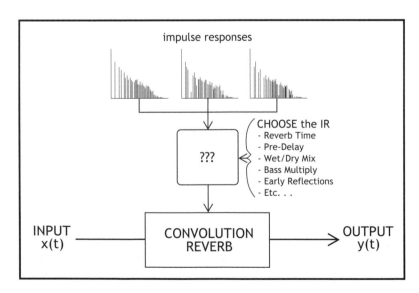

▲ Figure 11.27 Changing the reverb quality generally means changing the impulse response.

The impulse response contains not just the reverb time of a space, it contains the many other properties of a space that give it its character. While any two impulse responses may look quite similar, the ears can extract quite subtle differences between them. Human hearing systems perform a miraculous amount of analysis on the sound quality of a space that is only partially explained by parameters such as reverb time, predelay, and bass ratio.

Consider two different halls with the exact same reverb time. Even though they both might take exactly 2 seconds to decay by 60 dB, they will likely sound different. One hall might be a bit brighter, adding airiness and shimmer as the sound decays within. The other hall might be warmer, adding low-frequency richness to the sounds within. The frequency response, not just the reverb time of the hall, is very much captured in the complex detail of a room's impulse response.

Achieving reverb through convolution is terrific for emulating a known space. However, working in the recording studio, production demands may ask the engineer to modify the sound quality of that known space. It is common to find, for example, the sound of the convolution reverb on the vocal is directionally correct, but needs to be a little shorter or a little brighter. With limited adjustability, convolution can back the engineer into a signal-processing corner. The sound is perfect, almost. No small tweak is available that can bring it in line with what the engineer really needs. On a nonconvolution reverb, engineers simply adjust the appropriate parameter. On a convolution reverb, engineers most commonly hunt around for another impulse response and hope it sounds similar, only shorter. That is a different process. Hunting for different impulse responses and auditioning them is clumsy enough to interfere with the creative process.

Increasingly, convolution reverbs are adding adjustable parameters. Three signal-processing targets present themselves. The quality of the convolution reverb can be manipulated by processing the audio signal before it is sent into the convolution engine, the reverberation coming out of the convolution engine, and, perhaps most intriguing of all, the impulse response used by the convolution engine (Figure 11.28). Reverb time can be made shorter if the impulse can be shortened. The trick is shortening the actual impulse response of the Concertgebouw without robbing the reverb of other sonic features that make the Concertgebouw so desirable. If the engineer wants a shorter reverb overall, and a bass ratio that favors low frequencies a little more, it is difficult for the convolution reverb to respond and still preserve the original sound quality of the defining impulse response. Engineers must

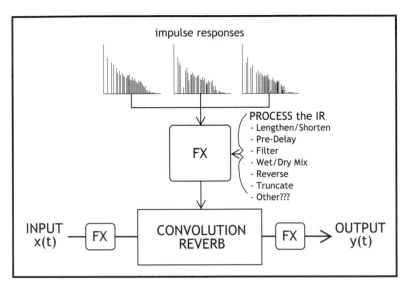

▲ *Figure 11.28 Effects processing can be implemented on the input, the output, and the impulse response.*

listen very carefully when modifying convolution reverbs in this way. The history of digital signal processing makes clear, however, that audio professionals can look forward to continued dramatic improvements. Tomorrow's reverbs will likely be better than today's as the software engineers create better algorithms, and the computer develops more signal-processing power.

Convolution Caveats

All of this is well and good as long as the convolution theory discussed above is true. Skeptical readers may be wondering: Is the impulse response really a perfect picture of a space that can be applied to any signal? It is worth pointing out the limitations so that engineers can make better use of this kind of technology in their productions.

Reverb Quality = Impulse Response Quality
The convolution reverb is only as good as the impulse response measurement. That is, a bad measurement (noise, distortion, poor frequency response, undesirable measurement positions, etc.) is going to flow through the convolution calculation to create similarly bad-sounding reverberation.

When a performance hall is measured, high-quality gear is required. The measurement system requires both a loudspeaker and associated electronics to produce the test signal (which might be an impulse, noise burst, swept sine, pseudorandom noise, or other) and a microphone/microphone preamplifier system to capture the measurement data. This entire measurement system has a number of demands placed upon it. The system needs to have a flat frequency response so that the equipment itself does not color the measured spectral content of the hall. The test equipment needs to have a full-range frequency response — reaching as low and high as practical — to engage and capture the hall's full bandwidth, low- to high-frequency characteristics. The impulse response measurement system needs to be omnidirectional to trigger and collect the contributions from all the room's boundaries. The measurement gear must have great dynamic range — low noise and high level before distortion — so that the resulting impulse response is itself low in noise and not clipped.

In order to measure spaces and add to the collection of available impulse responses, one needs very high-quality gear. Impulse response measurements are approached like an important location-recording gig. The same mindset that leads to great classical recordings captured at the hall leads to great impulse responses for convolution reverbs. Discipline, practice, a certain amount of paranoia, and the highest quality equipment can lead to clean and beautiful impulses.

Engineers do not just shop for the convolution reverb product that meets their needs. Impulse responses for many of the famous spaces are available for purchase too. Acquiring impulse responses is a different kind of gear acquisition, perhaps without precedent in the professional audio industry. Behind every impulse response considered, someone else gained access to the space, chose the measurement gear, placed the speakers and microphones, and captured the measurement. The art of measuring a hall is not unlike the art of recording drums. Spaces, like drum kits, are big, vague instruments that respond well to good microphone selection and placement. They are also unforgiving of bad recording craft. The simple fact is some measurements are better than others. The quality of the reverb coming out of the convolution engine depends on their skills in capturing the impulse response. When a recording engineer uses a convolution reverb in their multitrack production, he is also using the recording skills of another recording engineer, the one who orchestrated the impulse response measurement.

One Space = Countless Impulse Responses
No space is completely defined by a single measurement. It depends on the location of both the sound source and the listener (Figure 11.29). A

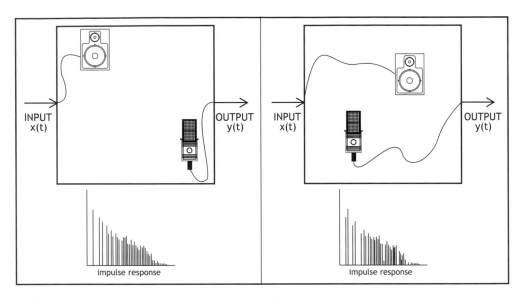

▲ *Figure 11.29 The impulse response changes when the source and/or reciever position changes.*

measured impulse response changes whenever the measurement position changes. After all, the hall sounds different in the front row than it does on the balcony. Moreover, it sounds different at the left ear compared to the right ear. For good sounding convolution reverberation, one seeks to measure in locations that "sound good." Recording experience, technical know-how, creativity, and intuition influence these decisions.

But this is not just about microphone placement. The impulse response changes whenever the position of the sound source is changed. That is, a singer standing center stage has a certain reverberant sound. But a singer standing at the rear right of the stage creates a different impulse response. Any single impulse response is only representative of a single location for the source and a single location for the listener. A 2,000-seat auditorium needs 2,000 stereo impulse response measurements for each stage location to even begin to tell the whole story of that performance space. A large stage capable of holding an orchestra of 100 plus a chorus of 100 leads to a sobering realization: One must choose which of the 800,000 choices of impulse response will be used.

The engineer first envisions the type of reverb needed for the multitrack project — perhaps a little Carnegie Hall on the snare drum. This is not enough. Decisions must be made about the location of the snare drum in Carnegie Hall. Is it on the stage or not? Sure, use the stage. But exactly where on stage will sound best? In addition, the engineer must consider the seat in which the listener is to be virtually placed. When these decisions are made, one then finds the correct impulse response — where the measurement engineer puts the measurement loudspeaker at the desired snare drum location and the measurement microphones at the preferred seat in the house.

In the course of the mix session, an engineer may be inspired to try the snare drum on the balcony, with the listener on stage. Good idea. Tough luck. They will need an impulse response with that exact configuration. Such ideas cannot be explored with the push of a button or turn of a knob.

Impulse Response$_{NOW}$ ≠ Impulse Response$_{THEN}$

The impulse response is really only representative of the reverberation of a space if the hall's performance doesn't change over time. Just as the impulse response is specific to a given source and receiver location, it is also somewhat specific to that instant in time.

Does the sound of a hall change over time? Surprisingly yes — ever so slightly. We look here, not at how the symphony hall sounds different now than it did in 1917, but at how the hall sounds different at the end of the song than it did at the beginning of the song. That distinct pattern of reflections that the impulse response documents does change slowly with time. If the walls move, the patterns of reflections in the impulse response must change. But even if the room partitions are not moving (which is generally the case), the impulse response can still change. These changes are caused primarily by the typical air drafts that occur whenever a group of people occupies a space.

We warm-blooded humans generate heat. Hot air rises, and does so inconsistently. Air conditioning fights back with an occasional breeze of cool air. As the air speeds up, slows down, and changes direction within these thermal drafts, it shifts ever so slightly the time of arrival of each reflection. The speed of sound is about 1 ft/ms *plus* the speed of the medium. If the air is moving with the sound, it propagates a bit more quickly. If the air is moving into the sound, it slows its progress ever so slightly.

This problem is most apparent at shorter wavelengths where a slight breeze might push the sound wave by a larger portion of a cycle. That is, a large wavelength (low frequency) pushed by this small convection current might vary a degree or two of phase sooner or later when it arrives at the listener during the song — likely imperceptible.

The same breeze would create a relatively larger phase shift for the shorter wavelengths (high frequencies) — possibly more perceptible. A single impulse response, therefore, might be thought of as being reliably representative of a hall at low frequencies. But it provides only a snapshot in time of the high-frequency portion of the hall's reverberant character. An instant later, the actual impulse response of a room would drift into a slightly different pattern.

The slight time variations that occur to the reverberation of a real space are not captured by convolution of a static impulse response.

Sampled Gear ≠ Actual Gear
The time variability of the impulse response turns out to be a problem in a surprising way. With the convolution reverb's ability to measure and reproduce the sound of a real space, it did not take recording engineers long to decide they could also sample the expensive racks of artificial reverb gear that the fancy studios have and that most engineers can't afford.

Need an actual plate reverb? Sample one. Missing that top-of-the-line $15,000 digital MegaSweetVerb? Measure one. No access to unobtainable, vintage, first generation, weird sounding, early digital reverbs? Capture one.

Not so fast. Unfortunately, many of these other reverbs exhibit time variation. In fact, the best sounding digital reverbs that do not rely on convolution (e.g., those terrific sounding high-end reverbs by folks like Lexicon and TC Electronic) have gone to great lengths to make the reverb change over time. It is built into the algorithms of these devices. The designers of this equipment figured out pretty quickly that the reverb they synthesized sounded infinitely more natural and beautiful if the algorithm had some dynamic processing within, not unlike the modulation of a typical delay line set to sweep ever so gently.

Sample a top-of-the-line digital reverb to drive a convolution algorithm and the resulting reverb may have noticeably inferior sound quality. More than natural spaces, infinite impulse response digital reverbs strategically shift things in time. They cannot be convincingly defined by a single impulse response.

Convolution is not mathematically valid as a way to describe a system that changes over time. That is why simple convolution will not let us model other devices like compressors, wah wah pedals, distortion boxes, autopanners, chorus effects, etc. These devices cannot be fully described by any impulse response. Their response to an impulse changes over time. With no overarching impulse response, convolution has nothing useful to convolve.

On the other hand, those devices that do not change over time can be sampled, such as equalizers. Measure the impulse response of a reputable EQ and the convolver just became an equalizer. Trouble is, the impulse response is different for each and every knob position on the equalizer. So while the vintage Pultec tube EQ is tempting to capture with a convolution reverb, it will need several hundred impulse responses to even start to emulate it. Oh, and by the way, some of the appeal of that old EQ is its tubes — the way they distort so sweetly. Trouble is, since distortion depends on how hot the signal is going through the tube-based gain stage, and that amplitude changes over time, convolution does not apply.

In the final analysis, all of these reverb technologies — chamber, spring, plate, IIR, and convolution — are valid today. Each possesses unique advantages that the informed engineer knows can be strategically leveraged as needed on any production.

11.4 Reverb Techniques

This single device, reverb, offers not just one but a vast range of effects, helping engineers solve technical problems and pursue creative goals. It takes many years to master the wealth of reverb-based studio effects, but the work pays big dividends. Exquisite use of reverb sets apart the truly great productions from the merely average ones.

Reverb effects, almost without exception, are parallel processes using effects sends, not serial processes using inserts (see "Outboard Signal Flow" in Chapter 2). Patch up aux sends into the reverbs so that a single effect can be accessed by any and all parts of the multitrack production. The reverb outputs are patched to effects returns or spare monitor paths feeding the mix bus.

11.4.1 SOUND OF A SPACE

For many applications, evoking the sound of a space (and all the feelings, memories, and social importance invested in that space) is the goal of artificial reverberation. Three-dimensional space, as it exists in a hall, church, club, or canyon, can be a useful analogy for the sound engineer who seeks to create a sonic landscape between or among loudspeakers. Among all of the reverb effects discussed here, simulating the sound of a real space is the primary motivation for adding reverb to a multitrack production.

Specific Real Space

Discrete elements of a multitrack production, or possibly the entire mix, might be processed to convert close-microphone studio recordings into a sonic illusion that the instruments were played in a real space. Reverb devices synthesize a pattern of early reflections and a reverberant wash evocative of an actual space.

Specific spaces, such as Boston Symphony Hall or Carnegie Hall are sometimes the goal. Engineers, faced with a need to add reverberation to a recording, may wish to stay true to the original recording venue. A recording made in Carnegie Hall is often required to have the reverberant signature of Carnegie Hall. The end of a movement often reveals noise within the hall: people coughing, traffic just outside, air conditioning, etc. Sometimes the recording is edited immediately at the end of the movement, and artificial reverb replaces the acoustic decay at that critical moment. The artificial reverb provides a clean, realistic musical ending free of extraneous noises. Any additional reverb added in the recording studio must, in this case, be perfectly consistent with the sound of — or our memories of the sound of — Carnegie Hall. Reverb devices are selected and the parameters are adjusted to achieve this.

In video postproduction, it is often necessary to replace the dialog. The dialog is recorded at the time the film is shot. Problems arise: some words are difficult to understand, the acting is not quite what was desired, the script changes, a distracting and unwanted noise happens off camera, the field-gathered sound quality is poor, etc. A single word or the entire scene may need new speech recorded. The actor wrestles with getting a great per-formance while syncing their timing with the original take. The engineer must worry about matching the ambience information in the replaced dialog. If the scene takes place in a church, taxi, or on another planet, the reverberation surrounding the dialog must sound exactly like that church, that taxi, or that particular planet.

Matching studio reverb to natural reverb elsewhere in the recording is a fine skill that the great engineers possess. An IIR reverb patch is selected, and its various parameters adjusted, until a satisfactory match is obtained. A convolution reverb is the ideal reverb for matching an existing space, if a quiet impulse response consistent with the original recording approach can be obtained.

Creative opportunities specific to convolution reverb await the adventurous engineer. If it can be measured, it can be convolved. When one wants the sound of the snare in the car, all that is needed the impulse response of the car. For the engineer who cannot help but notice how awesomely thunderous it is when someone slams a door in the concrete stairwell at the parking garage at work, who likes the sound of an empty swimming pool, who sings in the shower — all of these spaces can become reverb patches as soon as an impulse response is captured. Many convolution reverbs come with this very desirable ability. Take loudspeakers, microphones, and the convolution reverb anywhere, capture its impulse response, and then use the signature sound of that space — any space — in a future recording.

Moreover, thoughtful and generous people are sharing and swapping their own impulse responses with other engineers. Reverb transactions occur: "I'll trade you my parking garage stairwell for that church in your neighborhood." Every engineer has access to these opportunities.

Generalized Real Space

A realistic space might be evoked through loudspeakers for purely creative reasons. Relieved of any burden to simulate a specific place, the goal of the reverb signal processing might simply be to create a believable sound of a likely, surprising, or otherwise appealing space. Here the engineer seeks a reverberant character that sounds appropriate for the music and evokes an architecturally-grounded image of a realistic space. It is not necessarily Boston Symphony Hall. It is some kind of hall, convincing and realistic. Maybe it is a little smaller than the real hall. Maybe it is a little darker (sonically or visually or both). The creative engineer seeks to stimulate the imagination and sonic memory of the listeners and let them fill in some of the details themselves.

Evoking realistic spatial qualities is not limited to symphony halls and opera houses. Production goals might lead the engineer toward the sound of a canyon with its long echoes, a gymnasium with its shorter echoes, or a shower with its distinct resonance. Many of the same digital reverberators that evoke real halls can also be adjusted into configurations evocative of other real-sounding spaces. In these applications, the engineer has more freedom to adjust parameters. The space must be realistic, but not specific, allowing the engineer to be quite creative.

Spatial Adjustments

Less obvious spatial applications are common in popular music recordings. The lushness of a reverberant symphonic hall is easily heard, but more subtle approaches have value.

Close Microphone Compensation

Much of the audio in popular music is recorded using close-microphone techniques. Placing the microphone very near the musical instrument leads to tracks full of intimate detail. Done well, this can certainly be a pleasure to listen to; it is a proven technique for many important and successful recording artists. It is very much part of what makes the recorded music sound "better than real." However, a multitrack project consisting entirely of close-microphone tracks may be too intense, too intimate, too exaggerated, or too unnatural for some styles of music. On any given individual track of the multitrack arrangement, the close-microphone intimacy may not suit the creative goals of the art.

In this case, audio engineers use a reverb device with a good dose of early reflections (generally a simple cluster of strategically chosen delays) and very short decay time (digital or plate reverb), mixed in at a just-noticeable level, with the goal of diminishing this close-microphone immediacy. The sonic result is that the recording sounds as if the microphone were further away from the musician at the time the recording was created. To the less-critical listener, the sonic result is a believable and pleasing reduction in the intimate detail of the particular track. It sounds more realistic, less exaggerated, and less intense.

This type of effect might use any of a number of reverb technologies, such as a plate or a digital reverb. Even if a "large hall" type of reverb setting is employed, the large sound of a large space is very much suppressed here. The reverb is used only to diminish and blur the spectral detail of a music track to better assemble a loudspeaker performance that is consistent with the artist's goals.

Liveness

Employing reverb signal processing to simulate a collection of early reflections only, without the wash of late energy typical of a real space, can perceptually diminish the close-microphone signature and contribute to the illusion that the audio track was performed in a real space. A direct sound followed by a volley of reflections indicative of a real room geometry can make for a more compelling stereophonic illusion of location (left to right). In addition it can help evoke in the listener's mind a real location such as a room or stage.

Springs and plates do not create discrete early reflections. These mechanical devices initiate dense reverberation very quickly. For sparse early energy, digital devices are needed. The spacing, in time, of these early reflections is typically adjustable. The pattern of the early reflections — their relative amplitude and time of arrival — is determined in a real space by the room geometry and acoustic properties of the materials used to construct the space. The overall timing of these delays is determined by the scale of the space. Larger rooms will have these reflections spread out in time more than smaller rooms. Many digital reverbs emulate and make this property adjustable (Figure 11.30).

Application of reverb in this way adds the pattern of early reflections of any space, small to large, to a recorded signal with an adjustable amount of the associated late decay of a reverberant room. With digital reverb, it is possible to have early reflections evocative of a space with certain size, without any of the reverberant energy such a space would physically be required to create. The audio track achieves a more precise placement within a be-lievable, real space without the reduction in clarity and intelligibility that the late reverberant energy would typically cause.

Note that the creation of the sonic illusion of a large or small space without the associated reduction in clarity and intelligibility caused by reverb benefits not only the individual track being processed but also the many other elements of the mix seeking to be heard clearly. A long, reverberant wash on the snare drum can mask some speech frequencies (see Chapter 3) and diminish the intelligibility of the lead vocal. The decaying energy of a real space can mask other important elements in the multitrack production. A reverb consisting of early reflections without this late energy does not contribute as much to this masking.

Multitrack projects are often crowded with 24, 48, and often more tracks filling the arrangement simultaneously, fighting to be heard. It requires great care on the part of the mixing engineer and the arranger to prevent the mix from sounding cluttered when played back over loudspeakers. Long reverb times can be the bane of clarity. Unburdened by room acoustics,

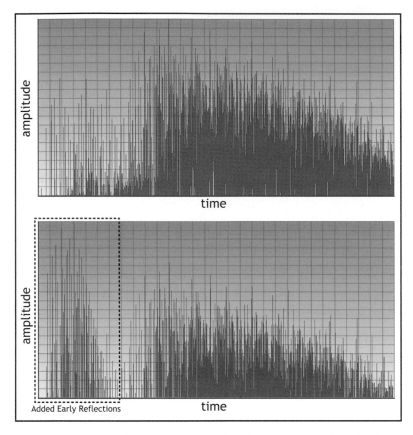

▲ *Figure 11.30 An IIR reverb with and without added early reflections.*

engineers regularly remove the late energy and use only the early energy, creating a live, but intelligible sound. In this manner, recorded music is made to sound as if it were occurring in an actual space, while the perceptual detail of the tracks remains unnaturally, unbelievably vivid.

Width

Through reverb, any single track of the multitrack production can be made to sound as wide as the loudspeakers are placed, perhaps a little wider. Augmenting a sound with reverberant energy that is clearly driven by the source sound (but provides each ear with even slightly different signals) can create a thrilling effect. Subtle differences at the two ears can lead to a perceived widening of the apparent width of the sound source, and can create a more immersive feeling for the listener. Surround sound offers the engineer the chance to make this effect more pronounced and more robust.

Ensemble

An additional hazard of the close-microphone craft is that even sections of musicians (a string section, a horn section, a choir) might be tracked in sonic isolation. That is, each member of the section has an individual, independent track on the multitrack recorder and thereby its own, isolated sound. It is not blended with the other members of the section by the acoustics of a stage or a room. The close microphone picks up the individual member with very little acoustic energy from the other players in the section. Valid, practical motivations push pop music recording in this direction. The result is that sometimes reverb is added back later, when desired, to reassociate the players into a single ensemble.

Multitrack production gives artists the freedom to have possibly every single element of the multitrack arrangement fall into its own, unique ambient environment. This freedom is not always exercised. Using a single kind of reverb effect on many instruments helps to reattach these isolated tracks of audio back into a single section again. A collection of isolated point sources is merged into a single, broad section of players coming out of the loudspeakers with the unifying sonic signature of a single ensemble. Any reverb type is appropriate to this application, but digital reverbs set to "hall" or "medium room" are a good starting point.

Unreal Space

Beyond creating a space that is at least conceptually born in architecture, reverb devices are also used for synthesizing a reverberant character that may not exist in nature. Wishing to enhance the music with lushness or some other form of beauty and held only to a standard that it "sound good," an engineer might dial in settings on a reverb device that violate the physics of room acoustics. For example, an engineer might collect a pattern of early reflections typical of a small room, combined with a late decay typical of a large hall. In addition, the decay might be brightened (high frequencies emphasized) for a pleasing timbre. Digital reverbs make these parameters adjustable. Such an amalgamation of acoustic elements is gathered despite the implications that there now exists a small room geometry (early re-flections) inside a large hall (long decay) in which air absorption has been miraculously overcome (the bright quality).

Such a reverb can be made to sound glorious coming out of loudspeakers. When reverberation is synthesized in a device, it is not burdened by the physics of room acoustics. It is a signal processing creation. Sound engineers regularly abandon any connection to room acoustics and stretch the capabilities of a reverb device to whatever limits necessary in pursuit of improved sound supporting the art of music, not the physics of acoustics.

Reverberation as described above to evoke spatial qualities is part of the signal processing — along with level settings, panning, EQ, and other effects — used to assemble an auditory scene. Within the sound field created by stereo or surround monitoring, this processing begins to make adjustable the subjective spatial parameters of depth and distance, the immersive attributes of discrete sound sources, groups of sound sources, and the size and quality of the performance and playback environment. All the spatial variables are accessible and changeable for the sound engineer.

11.4.2 NONSPATIAL APPLICATIONS

Reverb devices have their roots in room acoustics. The human experience in reverberant spaces motivated scientists and engineers to invent reverberant devices to "fix" tracks recorded in the highly absorptive rooms of recording studios, using close-microphone recording techniques. This historical cause (room acoustics) and effect (reverb devices) does not limit reverb devices to this single use, however. Engineers will apply reverb signal processing for other reasons: for some other practical benefits or simply for the aesthetic change it brings to the production. In this way, reverb devices take on many of the attributes of musical instruments. A given reverb might be sought out for its own subtle, subjectively beautiful sonic character. Another reverb might be utilized for its ease of use, its playability. Engineers select and adjust reverbs to create timbres and textures that support the music, to influence the audibility of the various elements of the multitrack arrangement, and to synthesize completely new sounds; all of these results are not spatial in nature. Reverb signal processing finds other functions.

Timbre Through Reverb

In some situations, reverb is used specifically because of the coloration it brings to a sound. Their frequency response, density of reflections, and resonant nature make springs, plates, and chambers useful devices for coloration. Productions that use these devices (and the digital devices that seek to simulate them) in this way do not seek to mislead anyone into believing that a real space exists around the instruments. There is a reverberant wash of energy. The character of reverb is certainly there. However, evoking an illusion of a symphony hall or an opera house is beyond the capability (or, indeed, the modern day intent) of these devices. Yet, despite the existence of many powerfully effective digital signal processors that can reliably evoke illusions of real spaces, preceding technologies with all their shortcomings are still used in popular music.

State-of-the-art digital signal processors provide the ability to simulate real spaces with presets labeled "hall," "medium room," and "cathedral." It is interesting to also note that they simulate other reverb processors, with patches called "chamber" and "plate." While springs, plates, and chambers were invented out of necessity to add reverberant character back to dry recordings, pop music engineers (and less consciously, pop music listeners) came to like their sound qualities even as more realistic and natural reverberation technologies were developed.

The coloration of a chamber comes in part from its small size. The highly reflective surfaces of the chamber have indeed lengthened the reverb time per Sabine's equation. The modal density is not sufficient, however, especially at the lower end of the frequency range, to prevent obvious resonances. The room is simply so small that the existence of resonances may be audibly obvious. In a hall, this is a well-known problem to be avoided. In popular multitrack production, the engineer creatively matches this resonant behavior with tracks of music that are flattered by this coloration.

In this way, reverb devices are used as an alternative to equalizers and filters (see Chapter 5), which directly alter the frequency content of the signal by design. The resonance of a chamber, plate, or spring might add some sort of glow to the track that sounds pleasing.

Beyond resonance, reverbs can influence the perceived timbre of a signal through frequency-dependent reverb time differences. A bass ratio target value of 1.1 to 1.5 is known to contribute to the perceived warmth of the instruments that play in the space. Likewise, employing a digital reverb that creates reverberation without modeled air absorption, high-frequency reverb times are often stretched longer than mid-frequency reverb times. This adds high-frequency energy to the loudspeaker music, leading to a perceived airiness, sparkle, shimmer, or other pleasing high-frequency quality.

The frequency response of a spring reverb (see Figure 11.5), sloping steeply upward with frequency through about 4 kHz, influences the perceived timbre of the audio track being processed. The mixture of an audio track with reverberation like this colors the sound, ideally in a flattering way. A steel string acoustic guitar might be made to sound a bit brighter still through judicious use of spring reverb.

The plate reverb offers a different spectral adjustment (see frequency response in Figure 11.11), offering more uniform output from just above 100 Hz to just above 10,000 Hz. Brightness with midrange complexity can be created by adding some plate reverberation to an audio signal.

In this way, reverbs are used to subtly shape the perceived timbre of the audio tracks being processed. Chambers, springs, plates, and radically manipulated digital reverbs are well suited to this approach.

Texture Through Reverb

The unique sonic character of some reverbs, especially springs and plates, leads to their use for elements of texture. Reverb is added not for the spatial attributes it evokes, but for the quality of the sound energy it offers. This approach enables a sound engineer or sound designer to add to a signal perhaps a pillowlike softness, a sandpaperlike roughness, a metallic buzziness, a liquid stickiness, etc.

Musical judgment motivates this application, and it requires experience and a good understanding of reverb and synthesis because there is no reverb patch labeled "pillowlike softness." There is, however, the capability to make a track take on a soft, pillowlike texture if the sound engineer makes clever use of the reverb devices in the recording studio. Much as the disciplines of architecture and acoustics must collaborate to create a great sounding space, the fields of sound engineering and music intersect to find musical, aesthetically appropriate reverb applications.

Examples abound. A reverb might be used to slightly obscure a particular part of the multitrack arrangement giving those instruments an ethereal, veiled texture. The metallic sound of a spring reverb might be used to help a track take on a steely, industrial texture. That pillowlike soft texture can be created through use of a long (reverb time exceeding 1.25 seconds) plate reverb with a long (about 120 ms) predelay, and a gentle roll-off of the high frequencies (starting at about 3–4 kHz).

When a composition is orchestrated, a horn chart arranged, or a film scored, elements of texture are a part of what motivates the creative thinking. A similar approach can be applied to reverb to help create textures and their associated feelings in support of the music.

Overcoming Masking Through Reverb

The multitrack arrangement contains musical elements of style, melody, harmony, counterpoint, and rhythm plus the sonic elements of loudness, frequency, timbre, texture, and space, generally with the overarching intellectual element found in the lyrics. All of these are explored, created, composed, and adjusted to values that are appropriate to the music. Such musical and audio complexity must somehow be communicated through loudspeaker playback. The difficulty of fitting so much detail into a pair of loudspeakers is captured by the common saying in audio circles that, "Mixing is the process of making each track louder than all the others." Of course, this circular reasoning makes no logical sense. But it resonates with every pop music recording engineer who has tried to mix a complicated multitrack production. A given guitar performance might sound terrific alone. Add in the horn parts, and the guitar performance likely becomes more difficult to hear and enjoy; its musical impact is diminished. A talented mixing engineer figures out ways to introduce the horns to the mix and yet maintain the important musical qualities of the guitar. Reverb is one of the tools used by the engineer to extract selective clarity, size, richness, etc. from the various elements of the multitrack production.

Contrast Through Reverb

Taste and style might dictate that the entire multitrack production have the same reverb signal processing, putting the entire mix in a single space. The accumulated overdubs, instrument by instrument, are processed with the goal of placing each of them in the same, single space spread out between and among the loudspeakers. The lead vocal, drums, piano, and hand percussion are all treated with reverb so as to create the illusion that all these players are in the same room together. That the players played separately, at different times, possibly in different studios, does not necessarily diminish this illusion of a single space. Signal processing is applied with the goal of uniting these instruments together in a single fabricated sonic space. While less likely, a similar global approach may be used with any of the nonspatial applications of reverb as well.

However, as a long list of rebellious rock-and-roll musicians will testify, there are no rules in music. While the creation of a single unifying space may be a goal for many pieces of music, it is perfectly reasonable to pursue contrasting spatial qualities among various musical tracks in a multitrack production instead. The vocal might be made to sound as if it is in a warm symphony hall with one reverb, while the drums may appear, sonically, to be in a much smaller and brighter room courtesy of another reverb, and the tambourine gets some high-frequency shimmer from yet another reverb. Applying different processing to different elements of a multitrack production reduces masking, making each sound or group of sounds treated with a unique reverb easier to hear (see Chapter 3). This enables a broad range of spatialities and effects to coexist in a single, possibly crowded multitrack recording. In popular and rock music, this approach is the norm. Even simple productions commonly run three or four different reverb units at once, each creating very different reverb qualities. A globally applied, single space or effect is the exception.

As so many pop productions have shown, the sound of a voice in a hall coexisting with the drums in a small room does not lead to any mental dissonance. Listeners are not troubled by the fact that such sounds could never happen naturally in a live performance. Listeners are motivated by what sounds beautiful, exciting, intense, and so on. Listeners are drawn to each of the different reverberant effects, making the variety of musical tracks within the multitrack production easier to hear, and therefore easier to enjoy.

Scene Change

Similarly, the reverb effect used can change during the course of the song. Typical approaches include a scene change from the intro to the first verse. Perhaps the introduction occurs in a lush, highly reverberant soundscape. When the first verse begins, the reverb vanishes, leaving a more intimate setting.

This approach knows no limits. The chorus might be accompanied by a transition from intimacy to a live concert feel, and so on. Active manipulation of the reverb choices and parameters along with advances in the song form is a cliché pop music effect.

Reverb Extrema

Applying a given type of reverb to isolated elements of the multitrack arrangement enables extreme reverb to be employed. Every instrument in the orchestra is treated to the same reverb because they play together in a single hall. Multitrack productions are able to add extraordinarily long reverberation in part because that reverb can be applied to selected tracks only, not the entire mix. Generally, a rock-and-roll drum kit would not be very satisfying to listen to with a reverb time in excess of 2.5 seconds. A noisy wash of reverberant energy would make it difficult to hear any details in the drum performance. But the vocal of a ballad — the vocal alone — may soar toward the heavens with such a long reverb. The drums, meantime, might be sent to a different, much shorter reverb to better reveal their impulsive character.

The multitrack production process makes it possible not only to apply a given reverb to any single track, but also to apply reverb to a single phrase or single note of a performance. The reverb effect can be turned on and off, instant by instant, during the song. It is all or nothing for an orchestra in a reverberant hall. Any degree of fine control is allowed in a multitrack mix using reverb devices.

In this way, it is not unusual for a reverb — when applied to isolated multitrack elements — to climb to reverb times in excess of 10 seconds. It is not uncommon for predelay to range from some 60 ms to beyond 100 ms, or even more. Bass ratio moves freely from a value 0.5 to beyond 2.0. High-frequency reverb times are routinely allowed to rival mid-frequency reverb times even though this could never happen in a symphony hall. The recording engineer reaches for whatever reverb settings support the sonic art they seek to create. Radical, physically impossible reverb settings are applied at any time, to any part of any track in search a better sound.

To be clear, many styles of music are not allowed so liberal an approach to reverb-based treatment. Some pop, most folk and jazz, and nearly all forms of classical symphonic music are presented through realistic recordings. This realistic approach looks for a space — a real, believable space — recorded in an actual room and/or synthesized by signal processors that creates a single convincing performance location. In the world of popular music, however, there is typically great freedom to apply spatial attributes and special effects to any element of the multitrack production, unburdened by the realities of room acoustics. Limits are pushed to the extreme, often with thrilling artistic results.

Unmasking the Reverb

While some reverb techniques strive to make the audio tracks easier to hear, other reverb signal-processing approaches seek to make the *reverb* easier to hear. The contrasting reverbs and the reverbs with extreme settings of parameters can serve to unmask the reverb itself. Predelay, for example, pushes the reverberant wash of energy further back in time, separating it from the audio track that causes the reverb to be generated. Classify the unprocessed audio track as the masker, and the output from the reverberator the signal to be detected (see Chapter 3). This point of view reveals predelay to be a key tool for overcoming temporal masking. A loud individual word within a vocal performance, as might occur in the chorus of a pop song, might easily mask the reverb effect that has been so carefully added to it. Predelay separates the vocal from the reverb making it easier for listeners to hear both.

Unmasking the reverb has an additional benefit. As reverb (or any other effect) is unmasked, its level within the mix may be reduced without removing the perceptual impact of the effect. Attenuating the reverberant energy within the mix without diminishing its aesthetic impact will leave more room sonically for the other elements of the multitrack production. The task of mixing becomes a bit easier. The clarity and precision of even a complicated mix improves.

Synthesis Through Reverb

Reverb signal processing is often the basis for the synthesis of wholly new sounds, acting very much like an electronic music synthesizer, played very much like a musical instrument. As with sound design and sound synthesis, the options are limited only by the imagination and creative force of the musician and the technical capability of the equipment. Some examples among limitless options of using a reverb device as a synthesizer are discussed below.

Gated Reverb

Signal processing aggressively violates laws of room acoustics in this application. A reverberant wash of energy is sent through a compressor (see Chapter 6), followed by a noise gate (see Chapter 7) that radically alters the decay of the reverb, resulting in an abrupt, double-slope decay.

A gated reverb system is connected (Figure 11.31). The outputs from the reverb device are sent to a stereo (two-channel) compressor, whose outputs in turn are sent to a stereo noise gate, whose outputs are sent to the mix via the mixing console. The gate is keyed open reliably by the close-microphone snare track.

An illustrative example begins with a snare drum recording. Reverb is added to this sound (Figure 11.32a). The resulting reverb is sent to the compressor. The compressor, by attenuating the louder portion of the reverberant wash, flattens out the initial decay of the reverb, giving it a more gradual slope implying a longer reverb time (Figure 11.32b). The gate then cuts off the decay with a slope indicative of a much shorter reverb time (Figure 11.32c).

The result is a concentrated burst of uncorrelated energy associated with each strike of the drum. This reshapes the sound of the snare drum into a more intense and more exciting sound. The effective duration of each snare hit is now stretched longer, making it easier to hear (see Chapter 3).

This effect can be made prevalent in the mix, for all to hear. It is unmistakable. It is unnatural. Recordings from the 1980s elevated this effect to a cliché. If one uses it today, a bit of 1980s nostalgia is attached to the production.

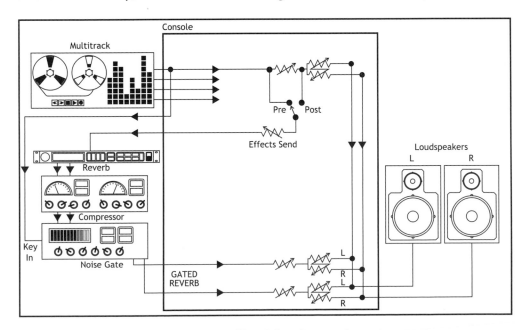

▲ *Figure 11.31 Signal flow for gated reverb.*

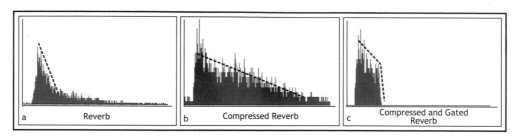

| a | Reverb | b | Compressed Reverb | c | Compressed and Gated Reverb |

▲ *Figure 11.32 Gated reverb.*

However, gated reverb is also used in more subtle ways. A valid philosophy is to make the gated effect so subtle that it is barely noticeable, if at all, to the untrained listener. The goal is to add a bit of sustain to the sound so that the track becomes easier to hear without having to turn it up. In the case of snare drum, the additional harmonic complexity and stereo width are also welcome (see Chapter 13).

Reverse Reverb

Another sound synthesis technique based on reverb requires the temporary reversal of time in the recording studio. The goal is to have the reverb occur before the sound that causes it.

Consider a snare drum back beat within a pop song, falling on beats two and four of a measure (Figure 11.33). The typical addition of reverb to such a track is shown in the lower portion of the same figure. Note the lower waveform is the reverb from a snare drum, not an impulse response. This is the typical use of reverb for such a track. Reverse reverb takes a different approach.

First the snare track itself is played backwards in time (Figure 11.34). This can be done on an open reel analog tape machine by turning the tape over and playing it upside-down. Alternatively, within a digital audio workstation, the audio track is selected and the computer is instructed to calculate the time-reversed waveform. Reverb is then added to this time-reversed snare, creating the signal shown in the lower part of Figure 11.34 and recorded to an available track on the multitrack recorder. Finally, the original snare track and the reverse reverb track are reversed in time (by flipping the tape back, or executing another time-reversal command). This restores the snare track to its original place in time. The reverb derived from the time-reversed snare is found to occur before the snare drum sounds (Figure 11.35).

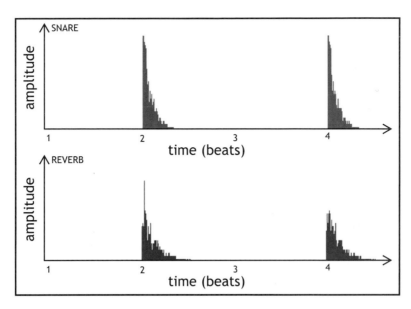

▲ *Figure 11.33 Adding reverb to a snare back beat.*

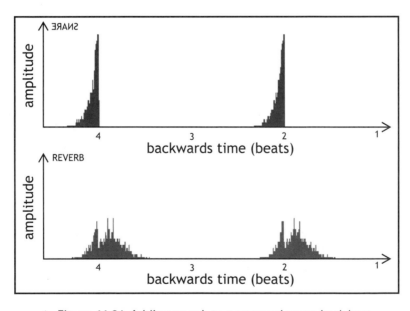

▲ *Figure 11.34 Adding reverb to a reversed snare back beat.*

The result is a kind of preverb. Sound energy swells up and into each snare hit, creating an unnatural, but musically effective, new snare sound. A short plate reverb was used for these illustrations, but there are no constraints on the type of reverb in this application. Musical anticipation, rhythmic

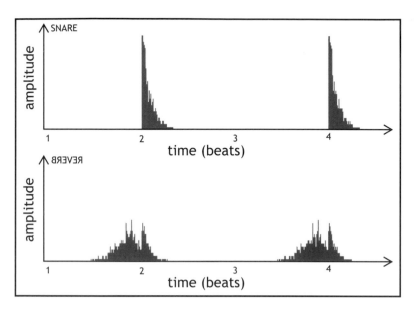

▲ *Figure 11.35 Creating reverse reverb on a snare back beat.*

syncopation, and the desire to fabricate an otherworldly sound motivate the sound engineer designing this effect.

This effect is somewhat simulated by reverb devices without resorting to backwards playback. Labeled "reverse reverb" or, more accurately, "non-linear reverb," the amplitude envelope of a reverberant decay is aggressively altered so that the reverb tail seems to go backwards in time. That is, the reverb starts off relatively quietly, becoming increasingly louder before abruptly cutting off. Unlike true reverse reverb, nonlinear reverb happens *after* the stimulating sound, not before. The unnatural, antidecaying shape to the reverb tail alludes to reverse reverb. This is a powerful way of lengthening the sustain of short sounds such as percussion, lifting them a bit up out of the mix so that they are easier to hear. It also provides dramatic ear candy without risk of mix-muddying, sustained reverberation.

Regenerative Reverb

Reverberation in a hall is initiated by the sounds coming from the musicians within. Reverberation in the recording studio is not so constrained. Figure 11.36 shows a reverb being fed by two sources. This might be any type of reverb: spring, plate, or digital. First, the typical signal routing using an aux send to a reverb is used. The track for which reverb is desired is sent to the input of the reverb (using effects send 1 in Figure 11.36). The reverb outputs

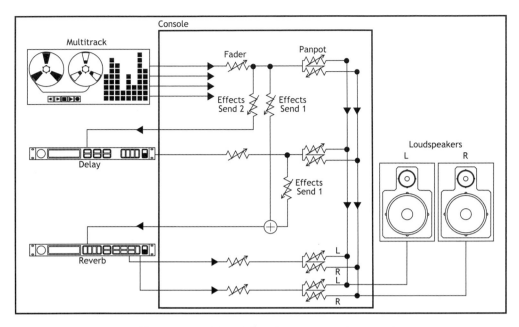

▲ *Figure 11.36 Signal flow for regenerative reverb.*

feed other inputs on the mixer, sending the reverberant signal into the stereo (or surround where applicable) mix.

For a regenerative reverb, this standard approach is augmented by a delay. The track for which this kind of reverb is desired is sent not only to the reverb, but also to a delay unit (shown using effects send 2 in Figure 11.36). The delay time is likely set to a musically-relevant time interval, such as a quarter note, a dotted eighth note, a quarter-note triplet. The output of this delay in turn feeds the same reverb used by the original audio track.

The delayed signal is likely lower in amplitude as it reaches the input of the reverb than the original, undelayed signal. The sonic result of this second, delayed feed to the reverb is a subtle extra pulse of reverberant energy, in time with the music. In addition, regeneration on the delay unit, routing its own output to its own input, further delays the already delayed signal. This feedback of the delay to itself creates gently decaying, musically-timed repetitions of the signal. This is combined with the direct, dry signal and fed to the same reverb device. The result does not sound like any physical space in existence. This elaborate reverb system creates an ethereal, swirling, enveloping wash of energy resonating underneath the track.

While any reverb type is effective for creating regenerative reverb, large hall programs from a digital reverb are most common. Clearly, such a dramatic effect would generally be applied only sparingly to a single element or two of a multitrack arrangement — not the entire mix. While it is a matter of taste, it is often the case that this pulsing, regenerative reverb is placed in the mix at a very low level. Acting almost subliminally, it offers a rich, ear-tingling sound that only exists in the music that comes from loudspeakers.

Dynamic Reverberant Systems

As the reverb-based synthesis approaches described above make clear, signals may be processed both before and after the reverberation device to alter and enhance the quality of the sound. Effects devices are placed before the reverb, as was done with the delay unit to create the regenerative reverb shown in Figure 11.36. Effects devices are placed after the reverb, as was done with the compressor and noise gate used to create the gated reverb shown in Figure 11.31. This figure also shows that multiple effects devices may be connected in series to further develop the effect.

This approach can grow still more complicated by introducing additional processors in parallel, both before and after the reverb device. A single such system is discussed here as an illustration (Figure 11.37), but an infinite number of options exist for the sound engineer to fabricate a reverb-based sound.

Consider having two parallel feeds into a reverb, each processed differently. For example, each of the two input options might have different EQ curves applied to the signal. Each equalizer is tuned to a different resonant frequency, or perhaps one is bright (high-frequency emphasis) while the other is warm (low-frequency emphasis). Preceding these two filters is an autopanner.

An autopanner (see Chapter 8) is a machine-controlled pan pot. It automatically adjusts the amplitude of a signal between two outputs so that, as the level is raised on one output, it is lowered by an equivalent amount in the other. It is two synchronized amplitude modulators moving with opposite polarity. Its typical application, used in a mixer, is to create the illusion of an audio track moving left to right and back again as the perceived localization follows the louder signal. In this dynamic reverberation system, the autopanner is applied in a different way.

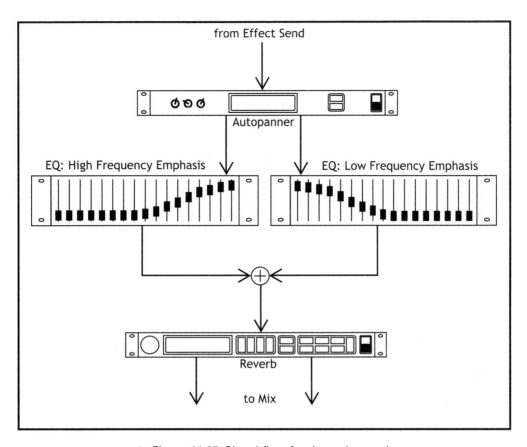

▲ *Figure 11.37 Signal flow for dynamic reverb.*

The effects send from the mixer feeds the autopanner. Each autopanner output feeds a different equalizer. Both equalizer outputs are combined to feed the input to the reverb. The reverb system is shown in Figure 11.37. The resulting reverb is a constantly changing sound, perhaps subtle, perhaps obvious. This dynamic reverberation takes on a life of its own, a soft pad of ever-changing, uncorrelated energy adding life and mystery to an audio track.

Convolution Creativity

No one requires that engineers must convolve their audio tracks with actual impulse responses. Why not convolve the vocal track with a snare drum sound? Or convolve the vocal with the piano track? This has no physical meaning. There is no space with an impulse response like a snare hit or a piano performance.

Just because it is physically meaningless does not mean it will not have creative value.

You are free to convolve anything with anything: background vocals with an insect buzzing, kick drum with glass breaking, a guitar solo with a dog barking, etc.

So the sound of a singer "in a snare" is rather puzzling to think about. But how does it sound? While such questions may bother those obsessed with reality, they are rather inspiring to those of us obsessed with unique and beautiful sounds.

When a recordist starts to use convolution as a signal-processing technique applied freely to any sound, convolved with any sound, they have begun to use convolution reverb as a synthesis device. In this way, engineers fabricate new sounds and textures using the convolution hardware/software as a sophisticated sound design tool.

11.5 Selected Discography

Artist: Roxy Music
Song: "Avalon"
Album: *Avalon*
Label: Warner Brothers Records
Year: 1982
Notes: This mix makes significant use of Chamber One at Avatar in New York City, a multistory stairwell. Note in particular the lush vocals.

Artist: Bryan Adams
Song: "Summer of '69"
Album: *Reckless*
Label: A&M Records
Year: 1984
Notes: More ear candy from Chamber One at Avatar. The snare hit at the top of the tune offers an exaggerated example, but the rhythm guitar has plenty too.

Artist: Michael Penn
Song: "Figment"
Album: *Resigned*
Label: Epic Records

Year: 1997

Notes: Dramatic change of scene driven by reverb shift between the A section ("Leave it for a while . . .") and the B section ("Before the day is done . . .").

Artist: Michael Penn
Song: "Out of My Hands"
Album: *Resigned*
Label: Epic Records
Year: 1997
Notes: Multiple simultaneous reverbs: Electric Bass is close-microphone track and dry. Drums enter with liveness and early reflections of a medium room. Acoustic guitar is also a close-microphone track and dry. Lead vocal is extremely wet, with reverb resembling a very large, dark hall program.

Artist: Paul Simon
Song: "Spirit Voices"
Album: *Rhythm of the Saints*
Label: Warner Brothers Records
Year: 1990
Notes: Reverb on a single note: The conga accent at 0:47 in the intro gets emphasis through a nonlinear reverb.

Basic Mix Approach

12

"The rubble over which you are stumbling
Just isn't that hard.
What did you do with all that grace, now?
What did you do with all my wine?
What makes you think that just' cause you dress bright,
it means that you shine?"

— "HALF HARVEST," MICHAEL PENN, *MARCH* (BMG MUSIC, 1989)

Recording engineers are musicians. Sometimes, recording engineers are excellent piano players, drummers, or singers. Always, recording engineers are musicians. Their instrument is the recording studio. The loudspeakers will play back a multitrack production that is artistically successful only if the engineer knows a good deal about music and how to play their instrument, which is the equipment of the studio. The challenge of coaxing art out of technology pervades the entire production process, but nowhere is it more apparent than at the mixdown session. The technical mastery and musical ability of the engineer are revealed in the mix session as every bit of equipment is applied to any and all elements of the multitrack production. This chapter offers a starting point in the highly nonlinear, unpredictable creation of a successful multitrack mix.

How do you write a song? How do you create a web page? The answer varies from person to person and from project to project. Creating a mix is a similar experience. That is part of the pleasure of the recording arts. No rote tasks are repeated daily in the creation of multitrack music. Engineers keep learning, creating, and exploring. There is no single right way to do it. There are no hard-and-fast rules to follow. There is, however, something of a standard approach to mixing pop music that is worth reviewing. It is not the only way. In some cases, it is not even the correct way. But it is a framework for study, a starting point from which any engineer can takeoff in their own direction.

12.1 Mix Challenge

Consider a pop/rock tune with the following somewhat typical set of challenges: drums, bass, rhythm guitar (doubled), lead guitar, clavinet, lead vocal, and background vocals. Where does the session start?

First, lay out the console. Using either a mixing console or a digital audio workstation, it helps to preset the signal flow with as much as can be anticipated. Mixing requires the engineer to be creative in how they shape and combine the various tracks and effects. Yet the engineer must hook everything up correctly too. The latter tends to interfere with the former, so it helps to do a chunk of the tedious and technical thinking ahead of time so that it does not interfere with the flow of inspiration while mixing.

12.2 Global Effects

All of the various signal-processing effects needed in a mix cannot be fully known before the mix is begun. But some basic effects are so frequent that they are very likely to be needed and can be patched up ahead of time. Many mixes will make use of a long reverb (hall-type program with a reverb time over two seconds); a short reverb (plate or small to medium room with a reverb time around one second); a spreader (explained below); and some delays (dotted eighth note, quarter note, or quarter-note triplet in time). Before digging into the details of the mix, the engineer will often launch the appropriate plug-ins or patch in the appropriate hardware to get these anticipated effects going. By setting them up ahead of time, these effects are available with minimal additional effort. The engineer can instantly send a bit of vocal, snare, and lead guitar to the same effect. The way to have all these effects handy is to use aux sends (see "Outboard Signal Flow" in Chapter 2). For example, the mix engineer might lay out the mixer so that aux 1 goes to the long reverb, aux 2 feeds to the short one, aux 3 sends a signal to the spreader — whatever is most comfortable for the engineer. The effects are returned to empty, available monitor faders on the mixer or digital audio workstation.

If the mixer is hardware based, the session may eventually run out of aux sends. The trick then is to use the track assignment network to get to additional sends. Sending a signal to track bus 1 is the way to record to track 1, of course. When mixing, not multitracking, the engineer can use the track bus as a way of accessing additional effects units.

Consider the production challenge of adding a delay to the electric guitar, using an additional delay unit not previously connected to the mixer. Making matters worse, all aux sends have been used. The electric guitar signal is sent to the stereo mix, of course. In order to add a new delay effect to the guitar, send it to track bus 1 as well as the mix bus. Patch track bus 1 send into the delay input. Patch the delay output into an available return to the mix bus on the console. In this way, the guitar goes to both the mix bus and the signal processor, using repurposed track busses as extra aux sends on the console.

12.3 Special Effects

What is the spreader? It is often desirable to take a mono signal and make it a little more stereolike. A standard effect in pop music, an engineer will spread a single track out by sending it through two short delays. Each is set to a different value somewhere between about 15–50 milliseconds (ms); if too short, it starts to flange/comb filter (see "Short Delay" in Chapter 9), and if too long, it pokes out as an audible echo. One delay return is panned left and the other panned right. The idea is that these quick delays add a kick of supportive energy to the mono track being processed, sort of like the early sound reflections that would be heard if the instrument were played in a real room. It is certainly more satisfying for a musician to practice in the kitchen than in the linen closet. The kitchen has cookies, which is nice. But it also has some shiny hard surfaces (not the cookies) that help musicians hear themselves. The strong sound reflections can make the performance sound better, stronger, more exciting. A linen closet absorbs all this, making for a dull place to practice. Then there is the whole "no cookies in the linen closet" rule enforced in most homes. So the *spreader* takes a single mono sound and sends it to two slightly different, short delays to simulate reflections coming from the left and right.

Any two delay lines can be used as a basis for the spreader effect. The effect is further refined by pitch shifting the delayed signals ever so slightly. That is, detune each delay by a nearly imperceptible amount, maybe 5–15 cents. Again, a stereo effect is the goal, so the spreader requires slightly different processing on the left and right sides. Just as a slightly different delay time was specified for each output, the engineer dials in a slightly different pitch shift as well (maybe the left side goes up 9 cents while the right side goes down 9 cents).

Introducing these slight pitch shifts to the short delays abandons any basis in acoustic reality that the spreader might have had. The effect is not just simulating early reflections anymore. Now the mix engineer is taking advantage of signal-processing equipment to create a widened stereo sound that only exists in loudspeaker music; it is not possible in the physical world. This sort of thinking is a source of real creative power in pop music mixing: Consider a physical effect and then manipulate it into something that is better than reality (good luck, and listen carefully).

This effect might be added to the lead vocal track, among others. The lead vocal track is most likely going to be panned straight up the middle. In order for the spreading effect to keep the vocal centered, it helps to do the following. Consider the delay portion of the spreader only. If one listens to the two panned short delays (the reader is encouraged to step through this in the studio), the stereo image pulls toward the shorter delay. For example, with a 30-ms delay panned left and a 20-ms delay panned right, a listener sitting on the median plane will hear the sound coming from the right, even as the unprocessed track sits panned dead center. Now consider just the pitch-shifted part of the spreading equation. The higher pitch tends to dominate the image. With a 9-cent pitch up to the left and 9-cent pitch down to the right, the image shifts ever so slightly left, toward the higher, brighter sound. The full spreader calls for both delay and pitch shift. Arrange it so that the two components balance each other out (delay pulls right while pitch pulls left). This way the main track stays centered. Experiment with different amounts of delay and pitch change. Each offers a unique signature to your mix. When overused, the vocal will sound too mechanized, too processed. When conservatively applied, the voice becomes bigger and more compelling; singers like this, and so does the record-buying public.

12.4 Kick Start

With the console laid out, the engineer can start mixing. The vocal is almost always the most important element of every pop song, no matter how many tracks there are. So most engineers start with . . . the *drums*. Starting with the vocal makes good sense, because every track should support it. But easily 99% of all pop mixes start with the drums.

The drums are often the most difficult instrument to get under control in the recording studio. It is an instrument with at least eight, generally more, separate instruments playing all at once in close proximity to each other (kick, snare, hi hat, two or three rack toms, a floor tom, a crash cymbal, a ride cymbal, and all the other various add-ons the drummer has managed to collect). It is hard to hear the problems and tweak the sounds of the drums without listening to them in isolation. So engineers tend to start with the drums so that this instrument can be addressed in isolation. Once the vocals and the rest of the rhythm section are going, it is difficult to dial in just the right amount of compression on the rack toms.

What does an engineer do with the drums? The kick and snare are generally the source of punch, power, and tempo for the entire tune — they have simply inspire awe. So it is natural to start with these tracks. Step one: Keep them dead center in the mix. The kick, snare, bass, and vocal are all so important to the mix that they almost always live center stage. The kick needs both a clear, crisp attack and a solid low-frequency punch. Equalization (EQ) and compression are the best tools for making the most of what was recorded. The obvious: EQ boost at around 3 kHz for more attack and EQ boost at about 60 Hz for more punch. Not so obvious: EQ cut with a narrow bandwidth around 200 Hz to get rid of some muddiness and reveal the low frequencies beneath (see Chapter 5).

Compression does two things for the kick. The first goal of compression on kick is to manipulate the attack of the waveform so that it sounds punchy and cuts through the rest of the mix. Chapter 6 describes the sort of low-threshold, medium-attack, high-ratio compression that sharpens the amplitude envelope of the sound. Second, compression controls the relative loudness of the individual kicks, making the slightly weaker kicks sound almost as strong as the more powerful ones. Drumming is physical work. Drummers understandably get tired in the fifth chorus of take 17. Compression helps them out.

Placing the compressor after the equalizer lets the engineer tweak the sound in some clever ways. The notch around 200 Hz keeps the compressor from reacting to that unwanted murkiness. As the engineer pushes up that low-frequency boost on the EQ, the compressor reacts. With an aggressive low frequency boost, the compressor is forced to yank down the signal hard. In this way, heightened low-frequency punchiness makes the drums sound larger than life.

12.5 Back Beat

The snare is next. As a starting point, it likely gets a similar treatment: EQ and compression. The buzz of the snares is broadband, from 2 kHz on up.

The engineer picks a desirable range: 8 kHz might sound to edgy or splashy, but 12 kHz starts to sound too delicate. Try 5 kHz. It is a subjective, musical decision. A low-frequency boost for punchiness is also common for snare. Typically the engineer looks higher in frequency than was done on the kick, maybe 80–100 Hz or so. The engineer might also look for some unpleasant frequency ranges to cut. Somewhere between 500 or 1,000 Hz lives a cluttered, boxy zing that does not help the snare sound and is only going to fight with the vocal and guitars anyway. One simply finds a narrowband to cut on the snare and the rest of the mix will go more smoothly.

The snare definitely benefits from the addition of a little ambience. In many mixes, it is not unusual to send it to the short reverb already set up and/or hope to find some natural ambience in the other drum tracks. The overhead microphones are a good source of supportive, natural snare sound. Any recorded ambience or room tracks should be listened to now. A gate across the room tracks keyed open by the close microphone on the snare can create a subtle touch of ambience on each snare hit (see "Keyed Gating" Chapter 7).

With the kick and snare punchy and nicely equalized, it is time for the engineer to raise the overheads and hear the kit fall into a single, powerful whole. The overheads have the best "view" of the kit and the snare often sounds phenomenal there. The mix engineer carefully blends the overhead tracks with the kick and snare tracks to make the overall drum performance congeal into a single, powerful event. It is tempting to add a gentle high-frequency boost across the overheads to keep the kit crisp. An engineer must listen carefully though, because if the tracks are already bright as recorded, additional high-frequency emphasis will lead to a harsh, un-pleasant drum sound. In fact a gentle and wide presence boost between 1 and 5 kHz can often be the magic dust that makes the drummer happy.

When the toms are on separate tracks, the engineer will again reach for those tried and true equalizers and compressors. Creativity is required, but nominally the engineer might EQ in a little bottom, and maybe some crisp attack around 6 kHz. One should consider EQing out some 200-Hz muddiness, as was done with the kick. Compress for attack and punch and the drum mix is complete, for now.

12.6 Get Down

With drums tentatively set, the bass guitar needs the engineer's attention. One often needs compression to balance the bass line. Some notes are louder than others, and some strings on the bass are quieter than others. Gentle compression (4 : 1 ratio or less) can even out these problems. A slow attack time adds punch to the bass in exactly the same way it did on the drums.

Release is tricky on bass guitar. Many compressors can release so fast that they follow the sound as it cycles through its low-frequency oscillations. That is, a low note at, say, 40 Hz, cycles so slowly (once every 25 ms) that the compressor can actually release during each individual cycle. This cycle-by-cycle reshaping amounts to a kind of harmonic distortion (see Chapter 4). Engineers typically slow the release down so that it does not distort the waveform in this way. The goal is for the compressor to ride the sound from note to note, not cycle to cycle.

The obvious EQ move is to add low end. But one must be careful as the track may already have a lot of low end. The bass player likely seeks it out. The tracking engineer likely emphasized it. The trick at mixdown is to get a good balance of low frequencies from 30 Hz through 300 Hz. The mix engineer must listen for a hump in the response — either too much or too little in a single low-frequency area. The engineer simply equalizes in the appropriate correction.

At this point, it makes sense for the engineer to glance back — with their ears — at the kick drum. If the kick sound is defined in the low end by a pleasing emphasis around 65 Hz, then one might need to make room for it in the bass guitar with a complementary, but gentle, cut. The trick is to find EQ settings on both the kick and the bass so that the punch and power of the kick does not disappear when the bass fader is brought up.

It is not unusual to add a touch of chorus to the bass (see "Medium Delay" in Chapter 9). This is most effective if the chorus effect does not touch the low frequencies. The bass provides important sonic and harmonic stability in the low frequencies. A chorus, with its associated motion and pitch bending, would undermine this. The solution: Place a filter on the send to the chorus and remove everything below about 250 Hz. The chorus effect works on the overtones of the bass sound, adding that desirable richness, without weakening the songs foundation at the low end.

12.7 Chug On

For this basic mix, the rhythm guitars are doubled. It is a rock-and-roll cliché to track the same rhythm guitar twice. The two tracks might be identical in every way except that the performance is slightly, humanly different. Panned apart, the result is a rich, wide, ear-tingling wall of sound. The effect is better still as the subtle differences between the two tracks are stretched slightly. Perhaps the second track is recorded with a different guitar, different amp, different microphones, different microphone placement, or some other slightly different sonic approach.

In mixdown, one makes the most of this doubling by panning them to opposite extremes: one goes hard left, the other hard right. It is essential to balance their levels so that the net result stays centered between the two speakers. A touch of compression might be necessary to control the loudness of the performance, but often electric guitars are recorded with the amp cranked to its physical limits, giving it amplitude compression effects already. Complementary EQ contours (boost one where the other is cut and vice versa) can add to the effect of the doubled, spread sound.

12.8 Key In

The clavinet completes the rhythm section in this basic mix. It probably needs compression to enhance its attack using much the same philosophy that was employed on the kick, snare, and bass guitar. Giving it a unique sound through EQ and effects ensures it gets noticed. Good starting points would be to add some flange or distortion (using a guitar foot pedal or an amp simulation plug-in) to make it a buzzy source of musical energy. Panning it midway off to one side makes effective use of the stereo sound stage. Depending on how the performances interact, a good approach is to pan it opposite the high hat, or the most active toms or the solo guitar in an effort to keep the spatial counterpoint most exciting. Add a short delay panned to the opposite side of the clavinet for a liver feeling.

With drums, bass, guitar, and clav carefully placed in the mix, a first draft of the rhythm section is complete. It is time at last to add the fun and important parts: vocal and lead guitar.

12.9 Speak Up

The vocal gets a good deal of the engineer's attention now. The voice must be present, intelligible, strong, and exciting. Presence and intelligibility live in the upper middle frequencies. Typically the engineer uses EQ to make sure the consonants of every word cut through that rich wall of rhythm guitars. A careful search from 1 kHz to maybe 5 kHz should reveal a region suitable for boosting that raises the vocal up and out of the clutter of the guitars and cymbals. The engineer might have to go back and modify the drum and guitar EQ settings to get this just right. Mixing requires this sort of iterative approach. The vocal highlights a problem in the guitars, so one goes back and fixes it. Trading off effects among the competing tracks, the engineer seeks to find a balance between crystal clear lyrics and perfectly crunchy guitars.

Strength in the vocal will come from panning it to the center, adding compression, and maybe boosting the upper lows (around 250 Hz). Compression controls the dynamics of the vocal performance so that it fits in the crowded, hyped-up mix that is screaming out of the loudspeakers. All of this compression and EQ track by track has so maximized the energy of the song that it will not forgive a weak vocal. Natural singing dynamics and expression are often too extreme to work, because either the quiet bits are too quiet or the loud screams are too loud, or both. Compressing the dynamic range of the vocal track makes it possible to turn the overall vocal level up. The soft words become more audible, but the loud words are pulled back by the compressor so that they do not over do it.

The vocal, a tiny point in the center, risks seeming a little small relative to the drums and guitars. The spreader effect is designed to combat this problem. Sending some vocal to the spreader helps it take on that much desired larger-than-life sound. As with a lot of mix moves, one may find it helpful to turn the effect up until it is clearly too loud and then back off until it is just audible. Too much spreader is a common mistake, weakening the vocal with a chorus-like sound. The goal is to make the vocal more convincing, adding a bit of width and support in a way that the untrained listener would not notice as an effect.

Additional polish and excitement comes from maybe a very high-frequency EQ boost (10 kHz or 12 kHz or higher) and some slick reverb. The high-frequency emphasis will highlight the breaths the singer takes, revealing more of the emotion in the performance. It is not unusual to add short reverb — try a plate — to the vocal to add midrange complexity and enhance the stereoness of the voice still further. The engineer likely adds a long reverb to give the vocal added depth and richness. Sending the vocal to an additional delay or two is another common mix move. The delay should be tuned to the song by setting it to a musically relevant delay time (maybe a quarter note). It is mixed in so as to be subtly supportive, but not exactly audible. The natural next step is to add some regeneration on the delay so that it gracefully repeats and fades. Taking it further still, the engineer might send the output of the delay to the long reverb too. Now the singer's every word is followed by a wash of sweet reverberant energy, pulsing in time with the music.

EQ, compression, delays, pitch shifting, and two kinds of reverb represent, believe it or not, a normal amount of vocal processing. It is going to require some experimentation to get it all under control. By turning up the various pieces of processing too loud and then backing off, engineers learn the role each effect plays. It certainly requires going back and forth among every piece of the long processing chain. Change the compression, turn up the delay, turn down the reverb, and back to the compressor again. With patience and practice, one finds even elaborate combinations of effects easy to control.

And that is just a basic patch. Why not add a bit of distortion to the vocal? Or flange the reverb? Or distort the flanged reverb? Anything goes.

The background vocals might get a similar treatment, but the various parts are typically panned out away from center, and the various effects can be made more prevalent in the mix. One cliché is to hit the spreader and the long reverb a little harder with background vocals to help give them more of that magic, pop sound. Intelligibility might be less important for background vocals. Repeated words or call and response lyrics can often be understood through context, freeing the engineer to pull out some presence in the middle frequencies and emphasize other timbral details in less competitive spectral spaces, if they wish.

12.10 Go Solo

The lead guitar can be thought of as replacing the lead vocal during the solo. Because they do not occur simultaneously, the solo guitar does not have to compete with the lead vocal for attention. The mix challenge is to get the lead guitar to soar above the rhythm section very much as the lead vocal was required to do. An EQ contour like that of the lead vocal is a good strategy: presence and low-end strength. Unlike the vocal, however, many styles of guitar require little to no compression. Electric guitars are naturally compressed at the amp when tracked at maximum volume. Additional reverb is also optional for guitars. The overall tone of the guitar might be fully set by the guitarist when the amp was setup and the settings were dialed in — that includes the reverb built into the amp.

Solo guitar might get sent to the spreader, and it might feed a short slapback delay. The slap delay might be somewhere between about 100–200 ms long. It adds excitement to the sound, adding a just perceptible echo reminiscent of live concerts and the sound of the music bouncing back off the rear wall. It can be effective to pan the solo about halfway off to one side and the slap a little to the other. If the singer is the guitarist, it might make more sense to keep the solo panned to center. Of course, one can always add a touch of phaser, flanger, or something from the long list of effects in the digital multi-effects unit. One can even add additional distortion. Such significant tone changes should probably be made with the guitarist present to get the blessing of an expert in the field.

12.11 Do It All

The entire stereo mix might get a touch of EQ and compression. As this can be done in mastering, it is wise to resist this temptation at first. With experience, one should feel free to put a restrained amount of stereo effects across the entire mix. The engineer is trying to make the mix sound the best it possibly can, after all. For EQ, usually a little push at the lows around 80 Hz and the highs around or above 10 kHz is the right sort of polish. Soft compression with a ratio of 2 : 1 or less, slow attack, and slow release can help make the mix coalesce into a more professional sound. As the entire mix is going through this equipment, it is essential to use very good sounding, low noise, low distortion effects devices.

That sums up the components of one approach to one mix. It is meant to demonstrate a way of thinking about the mix, not the step-by-step rules for mixing. Every song demands different approaches. The sometimes frustrating fact is that every multitrack production triggers so many ideas that the engineer does not know where to start. Let this chapter be a guide when the long list of options becomes paralyzing. The best mix engineers can stay in control of the many forces that conspire to reduce the sound quality of the mix, while reacting freely to the countless creative urges that might help the music transcend mediocrity and become real audio art.

Snare Drum FX 13

"Though it's hard to admit it's true
I've come to depend on you.
You, and your angelic shout,
Loud enough for two."
— "THAT IS WHY," BELLYBUTTON, *JELLYFISH* (CHARISMA RECORDS, 1990)

The role of the snare drum in popular recorded music cannot be overstated. It's backbeat drives the vast majority of all pop and rock tunes in western music. That energetic pulse on beats two and four is essential.

There is no such thing as a *correct* snare sound. The broad range of tone elements within this instrument offer the engineer great freedom in tailoring the sound. Evaluating a snare sound is not unlike criticizing modern, abstract art. Individual viewers have an instinctive, internal reaction to the art. Two people looking at the same abstract piece may have very different feelings about the work. The sound of the snare drum, abstract in its own right, often leads to similar differences of opinions. Heated debate may follow.

With this instrument, pretty much anything goes. Criticizing the snare sound in someone's mix is a lot like grading a class of kindergartners during Play Dough® hour. Anyone may believe some sculptures are better than others, but such preferences are personal and unique to their own experience of the piece of art. It is highly subjective, and each kid's parents can always comfort themselves with the *certain* knowledge that their child's Play Dough blob is best.

With Play Dough, there are just a couple rules: Keep it out of Andrew's hair, and do not eat it. Similarly, the snare drum follows just a few rules, which this chapter seeks to define. The book you are reading studies sound effects, but these effects are not limited to the mixdown session, when most effects are applied. The recording and overdub sessions present an opportunity to use signal processors too. Moreover, the kind of logic and creative thinking that informs an engineer's use of effects can be applied to microphone technique as well. The entire production life of the snare is therefore considered here.

13.1 Sound

To begin to make sense of the snare drum, it is useful to think of the snare drum as a *burst of noise*. The length of the burst ranges from a tiny impulse (cross stick) to a powerful wash of energy lasting more than a half note in duration.

That the snare can be classified as noise should not be a surprise. With the exception of that distinct, exaggerated ring used only for special effect, the snare is generally without pitch. It's a percussion instrument after all. In addition, unlike the rest of the drum kit, the snare drum has those wires (or gut) stretched across the bottom head — cleverly called "the snares" — to make sure that each hit of the snare is a buzzy, rattling mess of noise.

13.1.1 POSSIBILITIES

Like any good noise source, the snare can have acoustic energy throughout the audible range — lows, highs, and everything in between. Typical instruments used in pop, rock, and jazz have a diameter of 14 inches, however, 13-inch models occasionally make an appearance. Depths range from the 4-inch piccolo to more standard 5- to $6\frac{1}{2}$-inch snare depths, but may be as deep as 8 inches. Common materials for the instrument are wood (birch and maple are most common) and metal (frequently brass, bronze, steel, or aluminum). Factor in playing styles, stick selection, head selection, the specific location where the stick meets the drum, the tuning of the two heads, and a great variety of sounds becomes possible. The recording engineer is encouraged to work with drummers who understand the implications of all of these decisions, as it takes a musician's passion for his or her instrument to master these details and create a great sound.

For any given snare hit, the distribution of sound energy is rarely focused on a narrow range of frequencies or harmonic series. The overall sound is dissonant and noisy, spreading out spectrally during the decay. The result is a percussive thwack with a big spectral footprint. Few other instruments in the pop/rock genre offer such open-ended, noiselike possibilities. It is well known that the consonants of human vocals live around 2–5 kHz. The fundamental pitch of the piano's lowest note is around 30 Hz, and its highest note is around 4 kHz.

But the comfort of such standards does not exist with the snare drum. This is part of why it is so popular in pop/rock music. Spectrally, the snare is all over the map. Such a range makes it an ally in the anything-goes, rebel-for-the-sake-of-rebelling mission to be heard.

13.1.2 GOALS

The strategy for obtaining an effective snare sound begins with creative music and production decisions and is followed later by the more technical evaluation of engineering issues.

First the engineer, working with the drummer, must choose a certain shape, color, feeling, or sound that they personally wish to achieve. That target then influences the drummer's instrument selection, tuning, and playing. In parallel, the engineer makes recording decisions that further support that goal.

With snare, more than any other instrument in rock, there is opportunity for self-doubt. The risk is that someone might walk into the room and announce, "Dude. What are you thinking? I hate that snare sound. What is this, some sorta death-metal-polka album?"

Engineers must be prepared to ignore this. It is rarely important if this particular person does not like your idea of a snare sound. Find a snare sound that is acceptable to one person, and it will be pretty easy to find 10 listeners who vote against it. Odds are these snare critics are the ruffians who put the Play Dough in Andrew's hair back in kindergarten anyway. That said, there are a few people whose warnings should be heeded.

The drummer. Yes, this is one of those times when one should actually listen to the drummer. Drummers often have a very good idea of the sound they are going for and have a terrific amount of experience listening to this instrument. The snare drum is right up their alley. Remember, snare drums are round, just like pizzas, so the drummer is a real expert here.

The producer. Because the snare has such a strong influence on the overall sound of the tune, the producer often has strong snare-related desires.

The songwriter/composer. Many writers have specific sounds and feelings in mind when they build an arrangement. This is especially true for film scores. The engineer must not violate the songwriter's creative vision.

Ideally a recording project is a collaboration among all these and other talented musicians. Consider their input as a chance to hone in on the goal, narrowing the world of possibilities to a refined, much more specific goal. Consulting examples of released recordings or drum samples with the sonic character desired by all involved is a good way to manage this discussion, and this does not have to happen while the studio clock is running. This is a good rehearsal and preproduction activity.

Once a fairly specific goal is set, start making the decisions around which snare drum to use, how to tune it, which microphones to try, and so on. The tasty snare sound of the Wallflowers' "One Headlight," for example, could not be gotten from a 14-inch diameter, 8-inch deep snare tuned low for maximum punchiness. If the project consensus is "tight, with an edgy ring," then select and tune the drum accordingly. Clearly it is a mistake to wait until the mixdown session to target this sound.

13.2 Recording

The typical approach to tracking the snare drum dedicates several microphones to the task. In the end, the snare sound that reaches the home listener comes from two or three close microphones, a pair of overhead microphones, any number of more distant room microphones, plus supportive sounds generated through signal processing, most notably reverberation.

13.2.1 CLOSE MICROPHONES

It is not uncommon to put two or more microphones close to the snare drum. Most likely, a moving coil dynamic and a small diaphragm condenser will be tried, tucked out of the drummer's way, between the hi-hat and the first rack tom. The drummer must be able to comfortably whip the sticks between and among the drums and cymbals. It is unacceptable for the awkward presence of microphones to change the drummer's stick technique. It can be tricky indeed, but the engineer must find a close-microphone placement that does not interfere with the drummer's carefree, physical performance.

Moving Coil Dynamic

It is difficult to record a snare drum today without using the workhorse Shure SM57 or a similar moving coil dynamic cardioid microphone. Placed an inch or two above the top head of the drum, this track is likely to be a major part of the overall snare drum sound.

Of course the SM57 and its ilk are not the most accurate microphones in the studio. They are based on a moving coil design, after all. Cardioid moving coil microphones often have a strong lift in frequency response in the upper middle frequencies (around 3 or 4 kHz), but roll off toward higher frequencies, possessing a less than agile transient response. This is not as bad as one might think.

The apparent transient response weakness associated with all moving coil microphone designs is in fact a very helpful engineering asset. The relative laziness of the moving coil can be thought of as something of an acoustic compressor (see Chapter 6). By reacting slowly to a sudden increase in amplitude, the moving coil assembly acts mechanically as a compressor might act electrically. It reduces the amplitude of the peaks of a transient sound.

This is helpful for two major reasons. First, this reduction of peaks can help prevent the sort of distortion that comes from overloading the electronics that follow. The true spike of amplitude that leaps off a snare drum might easily distort the microphone preamplifier or overload the tape or converters in the multitrack recorder. The use of a moving coil dynamic microphone can be the perfect way to capture the sonic character of the instrument, blurring the details that might have caused problems.

The second advantage of the moving coil design is the signal-processing effect it creates. Moving coil dynamic microphones, with their natural lethargy, are often used for many of the creative reasons that make an engineer reach for a compressor. The sound of a clave, snare, kick, dumbek, and many other instruments is often much more compelling after the subtle reshaping of the transient that a moving coil microphone introduces.

Small Diaphragm Condenser

Just as the snare drum inspires many targets, sonically, the engineer must be prepared to chase many targets, technically. The percussive complexity of a snare drum is a perfect match for the transient detail of a small diaphragm condenser microphone. Place one coincident with the dynamic microphone, close to the snare drum, but out of the drummer's way.

The goal is to place the condenser microphone capsule in the exact same physical location as the dynamic microphone. In this way, the two signals can be mixed together freely without fear of comb filtering. Recall, that if the capsules are not the exact same distance from the snare drum, there will be a slight time of arrival difference between the two microphones. Mixing these two signals together will lead to coloration — possibly severe coloration — due to the frequency-dependent constructive and destructive interference that results (see "Comb Filter" in Chapter 9). Of course, it is impossible for two different microphones to occupy the same, single location. Place them as close together as possible, with their capsules the same distance from the top head of the snare drum.

It is not unusual to tape the small diaphragm condenser to the moving coil dynamic so that their capsules line up as closely as possible, and place them on a single stand. In this way the engineer can move the two microphones together as session experimentation dictates, in pursuit of the target snare sound.

With the moving coil and condenser microphones in a nearly identical, close placement on the snare drum, the engineer can listen critically to the different quality of each microphone individually. Most apparent will be the frequency response differences, with the condenser likely sounding a little brighter at the high end while the moving coil offers perhaps a presence peak in the upper mid-range. As important as it is, the engineer must listen beyond the spectral differences, to the character of the attack of the snare sound. It is generally the case that the moving coil dynamic microphone squashes the transient, possibly into a more exciting, more intense, easier to hear sound. With the two distinctly different sounds captured up close, the engineer can better achieve the snare goals previously planned. Either microphone individually, or perhaps some combination of the two, will give the engineer a good starting point for the snare sound.

The close-microphone snare sound is augmented by overhead microphones, room microphones, and signal processing.

13.2.2 OVERHEAD MICROPHONES

For the patient engineer, the search for the perfect approach to drum overhead microphone techniques is neverending: coincident (XY) cardioids, a Blumlein pair, spaced omnis, and MS (mid-side) approaches are worth considering. When it comes to overhead drum microphones, engineers are tempted to pull out all the stops on their stereo microphone techniques. It can help, in the heat of many a busy session, to simplify: abandon accuracy and pursue art.

Despite all of the many valid arguments for the overhead microphone approaches mentioned above, many engineers rely instead on a simple pair of spaced cardioids. This approach is more straightforward, well suited to the near panic of the basics session. There is no need for MS decoding. Avoid the gymnastics of setting up the Decca tree. Keep it simple by placing a pair of spaced cardioids up over the drum kit. This approach pays dividends in the forms of timbre, image, and control.

Timbral Benefits of Cardioids

For overhead microphone selection, think in terms of vocal microphones and focus on the snare. Large diaphragm cardioid condensers, with their mid-frequency presence peak, are perfectly suited to the snare drum challenge. These microphones are effective at grabbing hold of a detailed, buzzy, rattling snare sound. The sound of the snare a few feet away from the instrument almost always sounds better than the sound of the snare as viewed by the close microphone. Moreover, the final mix has the luxury of combining overheads and close microphones (and more) to fabricate the ultimate snare sound. In this context, engineers appreciate the contribution of cardioid overheads.

With the mid-frequency emphasis of a vocal microphone, the sound from the overhead microphones will be easy to use. The snare will rise up out of the mix.

Of course, overhead microphones capture the whole kit, not just the snare drum. In fact, it might be tempting for the less-experienced recordist to equate overhead microphones with cymbal microphones. It is customary to place several microphones close to the snare, kick, toms, etc. Logic might suggest the cymbals also deserve close microphones, placed above the kit. Should the overheads then be called the cymbal microphones?

This simply is not the case. It is true, the overheads are needed to capture the sound of the cymbals. Obvious to everyone who plays in a band, cymbals are loud and rich in high-frequency sound energy. Sonically, cymbals are hard to miss. It is not generally necessary to dedicate microphones just to cymbals or to seek out especially bright microphones to capture the cymbals.

Hang large diaphragm cardioid condensers over the kit, with snare drum as the first priority, and more than enough sound energy from the cymbals will also be recorded.

With cardioid overheads, an engineer's creativity and experience are rewarded. Just as producers and engineers audition different microphones for different singers, it also makes sense to search for the "right" cardioid for the overheads. Focusing on the sound of the snare within the overhead sound, swap microphones for the most flattering mid-frequency lift. When the production goals call for it, one can reach for the 1-kHz edge of a Neumann U47. Other sessions might be better served by the 6- to 8-kHz forwardness of those Neumann U87s. Song by song, match the microphone with the snare drum in a way that best suits the music and the overhead microphone selection process becomes an easier, intuitive process.

Overhead Image

Spaced cardioids might trouble the realist. Is there a less accurate, more problematic approach to stereo microphone placement than spaced cardioids? Probably not. Spaced omnis offer higher timbral accuracy and enhanced envelopment. Coincident microphone approaches, such as XY and MS, offer solid mono compatibility. Spaced cardioids seem to fail on both fronts. A pair of cardioids two to three feet apart over a drum kit are the basis for a larger-than-large, better-than-the-real-thing drum sound.

Again, even as the overhead microphones pick up the entire drum kit, the snare is the focus. With a pair of spaced cardioids, the audio engineer has the chance to augment the signal from the close microphone on the snare with the uncorrelated, mono-incompatible sound of the overheads. The close microphone offers punch and power and a center-localized snare sound. The overheads add to it a wide burst of energy that expands outward from the snare drum to the left and right speakers on each and every snare hit. This is a major part of what makes pop music fun. The exaggerated, unrealistic stereo image of the snare sound that lives within the spaced cardioid overheads combines with the timbral detail and well-centered stability of the close microphone to create the sort of snare sound that does not exist in real life. It is a studio-only creation.

Directional Overheads

Placing unidirectional microphones over the kit rewards the engineer for microphone placement tweaks. To change the sound — the timbre, the ambience, the balance among various drums and cymbals — one simply moves or rotates the cardioid microphones. The sonically narrow point of view of a cardioid enables the recordist to focus on and even exaggerate specific elements of the kit. The rejection capabilities of the cardioid enable the deemphasizing of the unappealing parts of the drums and any unwelcome aspects of the room. The off-axis coloration of any given cardioid makes it an equalizer whose settings shift as the microphone is rotated into a new orientation.

Abandoning realism and reaching for cardioid overheads makes the recording engineer more productive during the tracking session. One can modify the drum sounds in no time through even small movements of the overheads.

13.2.3 ROOM MICROPHONES

With the exception of the drummer, most people listen to snare drum from a distance. The sound farther away from the drum can sound more natural and more exciting as it is a blend of the direct sound from the drum with all of the reflected energy from the room. While the snare starts to sound better farther away from the kit, it is frequently the case that the kick drum sounds weak when recorded by room microphones. One approach is to build a tunnel out of gobos and blankets around the microphones in front of the kick drum. This makes for interesting sounds at the close microphones and it attenuates the amount of kick drum in the room sound. The room tracks become easier to tuck into the mix. Room tracks are also fodder for a range of signal-processing approaches discussed next.

13.3 Signal Processing

With a goal agreed to by all relevant parties, and the most appropriate sound achieved by the drummer out in the live room, and a combination of close and distant microphones recorded onto the multitrack, the engineer can at last consider what many others mistakenly believe to be the starting point when recording the snare drum: equalization (EQ) (see Chapter 5).

13.3.1 EQUALIZATION

Less-experienced engineers relying too much on EQ may mangle a snare sound with radical equalization curves, trying to make a chocolate mousse out of a crème brûlée. More mature engineers form a plan ahead of time — choosing the mousse, not the brûlée — and start achieving that sound at the drum, with the drummer, using microphone strategies that support the goal. This enables the engineer to calmly fine-tune the exact flavor of the snare with additional processing. The engineer applies EQ with elegant finesse, not brute force.

It is common, with EQ on a snare drum, to emphasize some region of high frequencies (around 10 kHz). This is logical as few productions benefit from a dull snare. If this was already addressed through the selection and tuning of the drum, and selection and placement of the microphones, any further brightening needed will be less severe. Perhaps EQ is unnecessary.

Alternatively, consider the less obvious equalization move: subtractive EQ. Notching out a narrowband around 1 kHz (± an octave or so) can open up the sound of the snare and give it a more distinct sound.

Subtractive EQ has the welcome additional benefit of making room spectrally for things like vocals and guitar. It is difficult to overstate the importance of vocals and guitars in the history of pop music. Minimize the masking (see Chapter 3) of these important instruments caused by the snare, and the production is likely to be more successful. Cutting rather than boosting is an excellent way to shape the character of the snare sound while improving other elements of the overall mix.

Meantime, watch out for the low end. Too much energy in the bottom octaves (below 300 Hz) in search of "punch" is an all too common error in snare drum recording. Remember that through close-microphone recording using directional microphones, a fair amount of proximity effect is likely.

Place any directional microphone in the near field of an instrument, and the frequency content of the signal will lift up at the low-frequency end of the spectrum. This is an effect loved by radio disc jockeys and jazz crooners. Their voices sound fuller and larger than life. But proximity effect is not just for vocals.

In fact, because of proximity effect, one should beware of deep snare drums and give serious consideration to piccolo and other shallow drums. Many rock drummers play big, deep snare drums for maximum low-frequency umph. Combined with too much proximity effect, the result can be a snare sound with too much low end. Spectrally, the sound may not be balanced, lacking the upper mid rattle and high-frequency sizzle that are often a part of the snare sound. What works in live venues does not always work in the studio.

A small drum with proximity effect can be a fun combination of high-frequency crack with low-frequency punch and power. This means that the snare sound, realized through studio recording and loudspeaker playback, can be a sound that does not really exist when we listen in a room to an actual drum kit. Low-frequency boosts must be made with care, with a keen awareness of proximity effect at the close microphones.

Of course subtractive EQ and low-frequency avoidance is not as much fun as equalizing in an ear-grabbing boost. Try going for some upper middle frequency "crack" (3–7 kHz) instead of low end "punch."

Meanwhile, the snare can have plenty of punch without boosting. Consider again some subtractive EQ, this time around 400 Hz, but pay attention to the low end of the snare sound below 400 Hz. Removing a narrow notch of lower middle frequencies usually reveals plenty of clear, tight low end in the frequencies just below. Engineers sometimes challenge themselves to find a controlled amount of low-frequency power without boosting.

To help make the point, consider that a typical three-verse song might have more than 130 snare hits in the backbeat, plus any number of snare hits in the fills and rolls. If each hit of the snare overpowers everything in sight, even just for that instant, the vocals and guitars and everything else will become musically weak; the energy of the tune may dissipate.

Said another way, too strong a snare weakens the entire mix. A spectrally well-placed snare, on the other hand, which is appropriate to the feeling of the tune, remaining balanced and controlled, can fit in well with other elements of a complicated multitrack arrangement.

13.3.2 ENVELOPE

More useful than EQ at helping the snare cut through a stereo pop mix is compression (see Chapter 6). With percussion it is common to think of compression as a tool for smoothing out the overall dynamics of a performance for a more consistent performance in the mix. This type of compression effect seeks to turn down the loud notes and turn up the quiet notes.

As the chapter dedicated to the topic makes clear, there is so much more to compression and limiting than this. Compression can be used to change the amplitude envelope of the waveform. The envelope describes the "shape" of the sound, how gradually or abruptly the sound begins and ends, and what happens in between.

Drums might generally be expected to have a sharp attack and immediate decay. That is, the envelope resembles a spike or impulse.

Envelope: Attack

Low-threshold, medium-attack, high-ratio compression can alter the shape of the beginning of the sound, further enhancing it's natural attack. It may take a little practice, so explore this effect on a calm project, perhaps working off the clock when the client is not in attendance. Be sure the compressor's attack time setting is not too fast, set the ratio at 4 : 1, preferably higher, and gradually pull the threshold down. This type of compression gives the snare a snappier attack. The result is a snare better equipped to poke out of a crowded mix and be noticed.

In the same vein, microphone selection can take care of the front end of the snare envelope instead of, or in addition to, compression. Condenser microphones are generally physically better equipped than most dynamic microphones to follow sharp, quick transients. Use a combination of dynamic and condenser microphones with this in mind, mixing in a little condenser for attack and a little dynamic for color.

Envelope: Length

Besides the attack, another variable to massage when recording and mixing the snare drum is at the other end of the envelope: the decay.

If this burst of noise can be persuaded to last a little longer by lengthening its decay time, it will achieve more prevalence in the mix. Recall that sounds shorter than about 200 ms are more difficult to hear than the same signal at the same level lasting longer than 200 ms (see Chapter 3). Compression to the rescue again, but this time it is through the knob labeled "release."

A fast release pulls up the amplitude of the snare sound even as it decays. Dial in a fast enough release time, and the compressor can raise the volume of the snare almost as quickly as it decays. The result is sustain instead of decay. Stretched longer, perhaps longer than 200 ms with the help of a little reverb, this lengthened snare sound can be placed in the mix at a lower fader setting and still be audible. It is long enough to get noticed, so it does not have to be turned up as loud. Filling in the mix with other tracks is easier when the snare performs its musical role without excessive level.

13.3.3 IMAGE

An important variable beyond spectrum and envelope is the stereo or surround image of the snare sound. Engineers are encouraged to master the power of simple stereo approaches to the snare drum pretty quickly, because it gets complicated fast in a world of multichannel, surround-sound audio.

The snare's image is really created by the pair of overhead microphones placed on the drum kit. Compared to close-microphone techniques, there are a couple of advantages to capturing the sound of the snare through overhead microphones.

First, they capture the snare from a more rational distance. When fans hear music live, they do not usually have their ears just a couple inches away from the snare drum. There is no real-life reference for the sound of the drum at the close microphone; it is just not natural. In addition, as discussed above, the close-microphone sound will include possibly detrimental proximity effect.

The close microphone presents an incomplete or skewed picture of the snare: too focused on the sound of the top head, without an appropriate amount of the sound components that radiate from the bottom head and drum shell. The snare drum is a complicated, omnidirectional sound source. While the drummer strikes the top head, sound radiates up, down, and outward in all directions from the entire drum. The sound at the overhead microphones is more consistent with a typical listening distance, integrating elements of the snare sound from all directions as the room reflects sound from the floor, ceiling, and walls into the overhead microphones.

Beyond the advantages of using more distant microphones to capture a more complete timbre of the drum and its room reflections, a pair of overhead microphones gives the snare its stereophonic image. Placing a single microphone up close to the snare will create only a narrow, small, monophonic snare sound. Such a tiny image will have a difficult time keeping up with a wall of electric guitars and rich layers of sweet vocals.

The solution is to make the snare image a little (or a lot!) bigger. A stereo pair of microphones above the drum kit can be all it takes to make the snare more substantial and exciting again.

Proper placement of the microphones and careful treatment of the signals is required to keep the snare sound balanced relative to the other elements of the drum kit with a stereophonic image that is centered in between the loudspeakers for maximum rock-and-roll effect.

Listen carefully for amplitude consistency when tracking the overheads, as a snare that pulls to one side usually goes from "cool" to "annoying" pretty quickly. Also maintain phase consistency between the two overhead mikes. "Phasey" overheads lead to snare hits that seem to drift behind the listening position. The snare needs to help hold the band together and drive the music forward. It must be fully front and center to do this effectively.

In addition, any such phase differences suggest you will have some serious mono-compatibility problems. Identify this problem by listening to the pair of overhead signals, level matched, in mono. Slight phase differences will cause the snare to change tone, lose power, and drop in loudness. For large phase differences, these changes can be substantial. The goal is to have the snare sound reach each overhead microphone at the same instant. Time-of-arrival differences lead to comb filtering (see Chapter 9). Once matched, the overhead microphones present a strong, stable, convincing image of the snare drum.

This stereo overhead microphone technique is taken an important step further through the use of ambient/room microphones. These more distant pairs of microphones can further exaggerate the stereo width of the snare sound, but they are tricky to use.

As microphones are placed further from the drum set, they pick up relatively more room sound and less direct, or close, drum sound. As these ambient microphone pairs are moved still further away from the kit, the drum tracks quickly become a messy wash of cymbals ringing and a room rumbling. The musical role of the drums — keeping time and enhancing the tune's rhythmic feel — is diminished, as the actual drum hits become difficult to distinguish, lost in a roar of drum-induced noise.

This problem is alleviated through the use of noise gates (see Chapter 7). Two noise gates can be patched across the room tracks to gate out the problematic wash of noise that fills the room between snare hits. The trick is getting the gates to open the ambient tracks musically at each snare hit. The gates need key inputs to accomplish this; stereo linking capability is extremely helpful. Patch the close-microphone snare track into the key input of the gates. A little tweaking of the gates' threshold, attack, hold, and release settings enables the engineer to synthesize a "gated room" sound. Every snare hit is more powerful than can be captured with close and overhead microphones alone. The snare becomes a wide explosion of adrenaline-inducing noise, a phenomenon known to please music fans.

Gated rooms (and gated reverb, for that matter) need not conjure up the possibly bad memories of the hyped, synth-pop 1980's music. Gated sounds were "discovered" and made popular in this distinct decade of music. Like the fashion and hairstyles of the time, the effect seems inappropriate today. It is important to note that the close-microphone track can be strengthened substantially by even a small amount of gated room sound, tucked almost subliminally into the mix. Used in this way, gated ambience still has relevance in the production contemporary music styles.

13.3.4 ALL OF THE ABOVE

Naturally, the snare responds well to creative combinations and variations of any and all of the approaches discussed above.

Compressing the drum tracks for enhanced attack and stretched duration is not limited to the close snare track. This effect can be applied to room tracks, gated room tracks, and overhead tracks. Effects from subtle to aggressive are welcome.

A valuable alternative to natural room tracks is synthesized ambience — reverb (see Chapter 11). A short reverb, with a reverb time less than a second, adds duration to each snare sound, making it easier to hear. The spectral content of the reverb, from the metallic presence of a plate to the warm richness of a digital room emulation, alters the perceived spectral content of the snare sound accordingly. Reverb presents a good alternative to EQ for adjusting the spectral color of the snare sound. Compress and gate stereo and surround reverb returns (plate reverb is a common choice, but anything goes) to add width, distance, envelopment, duration, and caffeine to the snare.

Engineers should not be afraid to use studio devices that may not have been intended for drums. For example, it is not unusual at mixdown to send some of the snare track through a guitar amp. Be careful to adjust for impedance and keep careful control over levels. Use a reamping device or a passive direct injection (DI) box "backwards," sending a very low-amplitude, low-impedance signal from the console through the DI and into the guitar amplifier. The passive DI will happily boost the amplitude and impedance of the signal into the sort of electrical signal that the guitar amp expects to see coming from a guitar.

At the amp, use radical EQ, distortion, spring reverb (generally only available in guitar amps), wah-wah pedals, and stereo ambient microphone techniques on the guitar amp to capture a wholly new snare sound. Make this the snare sound that pushes the mix along, or layer it in more subtly with the less-processed snare tracks.

Similarly, guitar amp simulators can create stereo, distorted noises from a snare track with a lot less headache. Think of these effects as hybrid distortion/compressor/equalizer combinations that welcome creative snare alterations.

13.4 Summary

A snare sound in its raw form — a burst of noise — represents the jumping off point for almost any kind of sound. The brief, broadband noise of the snare drum is a piece of granite awaiting the engineer's chisel to be shaped. Target a specific musical goal for the snare sound, shape it along the dimensions of spectrum, envelope and image, and the vague opportunity of the snare drum becomes a refined source of production success.

13.5 Selected Discography

Artist: Led Zeppelin
Song: "D'yer Mak'er"
Album: *Houses of the Holy*
Label: Atlantic Records
Year: 1973
Notes: A touchstone for many engineers and drummers. Recipe: well-tuned drums, hit harder than most humans dare. Examples from this drummer abound, but this track is a favorite with exciting ambience through distant microphone placement.

Artist: Nine Inch Nails
Song: "Closer"
Album: *The Downward Spiral*
Label: Nothing/Interscope
Year: 1994
Notes: Through gating and/or sampling, this snare is literally a burst of noise.

Artist: Red Hot Chili Peppers
Song: "The Power of Equality"
Album: *Blood Sugar Sex Magik*
Label: Warner Brothers Records
Year: 1991
Notes: Snare pulls right, with close microphone/ambient microphone differences left versus right.

Artist: Bruce Springsteen
Song: "Born in the USA"
Album: *Born in the USA*
Label: Columbia Records
Year: 1984
Notes: One of the first larger-than-life snare sounds in pop music. This is a studio-only creation, not available in the real world.

Artist: Tears for Fears
Song: "Woman in Chains"
Album: *The Seeds of Love*
Label: Polygram Records, Inc.
Year: 1989
Notes: Snare plus reverb for very long duration in the intro.

Artist: U2
Song: "Daddy's Gonna Pay for Your Crashed Car"
Album: *Zooropa*
Label: Island Records
Year: 1993
Notes: Proof that there are no rules. This sounds like a gated guitar opening on each snare hit.

Artist: Spin Doctors
Song: "Two Princes"
Album: *Pocket Full of Kryptonite*

Label: Epic Records
Year: 1991
Notes: Glorious room sound, gated. This sound is available only at Avatar Recording Studios in New York, NY.

Artist: Stevie Ray Vaughn and Double Trouble
Song: "The Sky Is Crying"
Album: *The Sky Is Crying*
Label: Epic Records
Year: 1991
Notes: Classic blues, with that lazy snare falling a hair late on the backbeat. Textbook application of plate reverb.

Artist: The Wallflowers
Song: "One Headlight"
Album: *Bringing Down the Horse*
Label: Interscope
Year: 1996
Notes: Tight, great ring, seriously compressed. No cymbals but hi-hat for the entire tune, yet there is no lack of energy from the drums in any chorus.

Artist: XTC
Song: "Here Comes President Kill Again"
Album: *Oranges and Lemons*
Label: Geffen
Year: 1989
Notes: Three different snare sounds by the first chorus.

Piano

14

"She's so swishy in her satin and tat
In her frock coat and bipperty-bopperty hat.
Oh God, I could do better than that."
— "QUEEN BITCH," DAVID BOWIE, *HUNKY DORY* (RCA RECORDS, 1971)

The piano is nothing short of a miracle. Most weigh more than 1,000 pounds and hold in check more than 10 tons of string tension. Its very existence is an impressive accomplishment for the instrument maker. Able to reach about as low and as high as any note that can be written into a musical score, the piano very nearly possesses the full musical range needed by all of western European music. It is an industrial-strength machine with extraordinary artistic sensibilities.

For the recording engineer, the piano is something of an enigma. Possessing broad spectral abilities, extraordinary dynamic range, complex and dynamic harmonic structure, percussive sharp attack, and signature decay, the piano represents a range of sonic capabilities that fills the audio window (see Chapter 3). An instrument almost without limits, it presents the engineer with open-ended opportunities and seemingly unanswerable questions. This chapter studies some common production approaches to the piano, illustrating not only some valid signal-processing techniques specifically for the piano, but also a way of thinking through an instrument that engineers at all levels might find helpful on this and other instruments.

14.1 Mother Instrument

If the piano were invented today, it probably would not catch on. Priced more like an automobile than a musical instrument, possessing no musical instrument digital interface (MIDI) jacks, requiring constant (and laborious) tuning, and unavailable in a rack version, the mother instrument asks much of us. In the hands of mere average musicians, it is difficult to play expressively; there is no mod wheel, no volume pedal, and no vibrato.

While talents from Germany's Beethoven to Peanuts' Schroeder embraced the piano and helped it earn its place as one of music's most important instruments, some of the issues described above have begun to take their toll. Courtesy of the convincing value and convenience of MIDI-triggered sample playback devices, the word "piano" often describes a set of samples or synth patches, not an instrument full of wood, wire, steel, and felt. Fair enough. Few musicians, your author included, are willing to hoist a piano into the back of the wagon for the next general business gig.

Engineers who have not had many opportunities to record a piano and really spend some time with one need not worry. Though it is a daunting instrument to record and mix, it pays rich rewards. This chapter studies some ap-proaches to this enigmatic instrument so that any recording engineer can learn to make accurate, beautiful, unique, weird, or ____ (insert any preference here) recordings quickly, leveraging what may be limited (and pricey) time in the increasingly rare piano-equipped studio.

14.2 Defining the Piano

Before the microphones are selected and placed around the piano, a more important issue should be resolved first. What is the piano to sound like? A piano in a room presents an opportunity to record any of a number of very different sounding instruments. The engineer's approach to recording and mixing the instrument can dramatically affect the sound of the in-strument. As moving microphones is more time consuming than scrolling through patches, it is wise to begin with a rather specific goal in mind.

It is essential to view the piano within the context of the song. Production, composition, and arranging decisions must be made regarding the feeling to be conveyed, the piano's contribution to the story of the song. Engineering goals for its placement in the mix should be envisioned. These strategic questions should be asked of all pieces of the arrangement of any pop song before the tactics of the multitrack recording session are detailed. The piano requires special attention because it is so very flexible.

It is the very definition of a full spectrum instrument. It is essentially capable of playing throughout the entire musical range. Its harmonic content can easily fill the top, bottom, and middle frequencies of the entire mix. Ah, such options. But such options require some decision making. If the song is not solo piano, then the piano must leave room for the other players. That means, first, limiting the musical part itself so that it interacts with and complements the other instruments musically. Second, it requires restraining and shaping the sound of the piano so that it finds its place in the ensemble sonically.

14.3 Effects

The musical issues are left to the players, composers, and arrangers. It is the engineer's responsibility to look more closely at the sonic issues. It is useful to consider the piano's place in the mix along three broad categories: spectrum, image, and envelope. These are not wholly independent variables as adjusting one property likely changes another. In search of some semblance of order in the heat of a busy recording session, it helps to break the decision into these useful components.

14.3.1 SPECTRUM

Spectrally, the piano can pretty much cover the entire audible range, from rich and warm low end, to a shimmering high end, and everything in between. Grand pianos thunder below 30 Hz while offering harmonics above 10,000 Hz. While most of it is in the hands of the instrument maker and the performer, the engineer should at first seek to achieve a balanced use of spectrum. Except as a special effect, it is generally undesirable to allow any narrow range of frequencies to overpower any other range of frequencies in the overall sound. All of the frequency elements that make up the unique timbral signature of the piano need to be persuaded to play fair with the rest of the spectrum.

Microphone selection and placement do the most to determine this, but equalization (EQ) (see Chapter 5), gently applied, gives the engineer finer control. In general, wide-bandwidth (low Q) parametric EQ is gently ($\pm$ 6 dB or less) applied.

With the full spectral palette of the piano available and under control, the engineer can greatly improve a complicated mix by selecting specific frequency regions to highlight and/or others to deemphasize. Again, the primary tools for accomplishing this are microphone selection, microphone placement, and EQ. When two instruments play in similar frequency ranges continually, at the same time, there is the chance to rather aggressively filter one of them. It would be essential to avoid the low-end build-up that might occur when a gorgeously rich, left-hand oriented piano part competes with a six-string chorused bass over a kick, tom groove. Such a thick arrangement demands that something in the bottom two octaves (20–80 Hz) be sacrificed. The piano tolerates this quite well. If the kick, toms, and bass are unrelenting, and the piano rarely if ever plays without them, then something of an illusion can be created. Reduce the low-frequency content of the piano (perhaps using a shelving EQ attenuating 6–12 dB below approximately 80 Hz) to make room in the mix for these other low-frequency tracks. Distracted by the now clarified low end of the bass, kick, and toms, listeners may not notice the low-frequency reduction to the piano sound. The listener's imagination — built on experience with the instrument, plus maybe a little wishful thinking — fills in the missing spectrum. The net result is that the piano sounds full and natural yet no longer competes with bass, kick, and toms.

At the high-frequency end of things, if the session ran out of time and failed to track to the intended tambourine or shaker performance that was meant to be added before the drummer left town, then the mix engineer might be free to emphasize a tasty part of the piano sound anywhere in the 6- to 12- kHz range; the delicate high end of a ribbon microphone or a small diaphragm condenser might be in order. Freed of competition in that high-frequency range, the piano welcomes a high-end highlight.

If piano and acoustic guitar compete, playing similar musical parts, try a high-pass filter or low shelf cut on the acoustic guitar to get it out of the way of the piano from about 200–400 Hz and below. The bright, metallic sound of the acoustic guitar is distinct from the round, rich warmth of the piano. Similar parts play together yet each instrument is identified and enjoyed easily within the mix through unique spectral processing.

Engineers who work frequently on music that includes synths and samplers have no doubt experienced the horror of *the synth patch that ate the mix*. Many patches — strings and pads are most often guilty — are mixes in and of themselves. They are nicely balanced from 20 Hz to 20 kHz and fill the entire stereo image with their lush swirliness. This creates a great sounding patch, when enjoyed all alone. In the context of a mix, however, these sounds present such broad spectral content that they are unmixable.

The piano commits a similar sin if the audio engineer is not careful. There is a certain microphone manufacturer based in Evanston, IL, that makes a microphone often accused of being appropriate for any application. This includes the piano. Throwing a $100 microphone on a multithousand dollar instrument might feel logically inconsistent. Expensive pianos sound great through expensive microphones, to be sure. However, the less expensive microphone may create a tone color that sounds just right. The presence emphasis of a Shure SM57 can pull the piano forward in the mix just as it does the Marshall stack.

WARNING: Applying EQ to a piano will at times feel a bit like building sand castles. Just as the sound takes shape, it shifts and moves. A lot of hard work seems to disappear and it is time to start over. The spectral content of the piano changes over time. This is part of the beauty, and the curse, of the instrument. A single note or chord, allowed to sustain indefinitely, will have harmonics waft in and out slowly, over time. No other instrument offers this property to this degree. A complex set of harmonics, changing over time, coming and going, some getting softer while others get briefly louder, is all part of the pianoness of the piano. This leaves the engineer applying spectral processing to a moving target. Finding the right boost and cut and the right frequency with the right bandwidth takes a bit of luck, a bit of faith, and a strong engineering sense of what is needed. Listen patiently and expect to be frustrated at first.

Of course the piano demands more engineering resources than microphones and equalizers. Audio engineers achieve new heights sonically on the piano through careful use of stereo placement.

14.3.2 IMAGE

The stereophonic image of the piano sound is probably the best weapon for bringing clarity and unity to a mix. To discuss image for the piano, it is important to understand what left and right mean.

For a drum kit, left might mean the drummer's left. However, as the drummer typically faces the audience, the drummer's left is the audience's right. Left/right confusion sneaks in. Left and right for drums is sorted out by choosing the drummer's perspective or the audience's perspective.

Left and right are straightforward concepts for an orchestra — the conductor's and audience's definitions of left and right are the same and translate well to loudspeakers. But what is left and what is right on the piano? Listeners might be sitting at the piano bench (player's perspective). Listeners might be in the audience looking directly at the open lid with the piano player to the left (audience's perspective) More fun still, listeners could be looking straight down on a piano with the lid removed entirely (Heaven's perspective). With the piano, these are all valid recording and mixing approaches. There are no strict rules to follow, only creative options to pursue.

Carefully defining the width and placement of the image will help bring order to a mix. The center is always the most challenging. In particular, sorting out the competition for the middle frequencies can be made easier by thinning out the traffic in the middle of the stereo field. Two approaches to consider are to isolate or to contrast.

Isolation is achieved by panning midrange-hogging instruments (even stereo tracks of them) to different locations. Left-right separation reduces the spectral masking (see Chapter 3). The lead vocal and solos typically go to the center, so it can be helpful to pan the rhythm section to locations hard left and right. Laws of pop mixing (which of course do occasionally need breaking) require leaving the vocal unquestionably front and center. It follows that one could then put the guitars on one side and the piano on the other (Figure 14.1).

On the other hand, contrast is created by making one instrument spacious and enveloping with a large stereo image, perhaps wider than the loudspeakers (using fully panned stereo or doubled tracks, wide and enveloping stereo reverb, stereo phaser/chorus effects, and so on) while the competing instrument is a more precise point in space (mono, dry, compressed) (Figure 14.2). The piano is capable of both imaging extremes. Record and mix the piano toward a goal from the start — a pinpoint piano or a panoramic piano — and all elements of even a complicated mix can be better enjoyed.

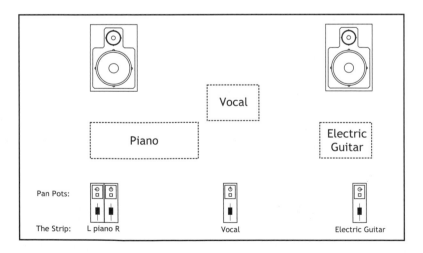

▲ *Figure 14.1 Separating parts through isolation panning.*

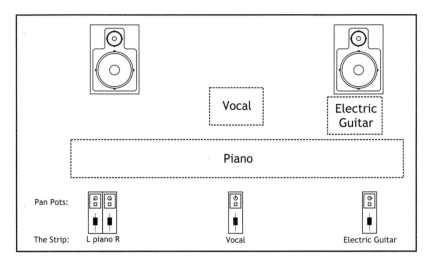

▲ *Figure 14.2 Separating parts through contrasting images — wide piano.*

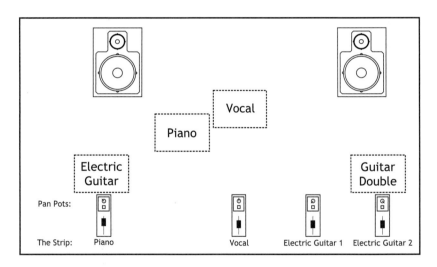

▲ *Figure 14.3 Separating parts through contrasting images — pinpoint piano.*

If it is late at night and no one can tell left from right, do not be afraid to track a mono piano. Often the essence of the piano can be captured by a single well-placed microphone (Figure 14.3). Alternatively, a low-fidelity-by-design sort of sound might be the right piano sound for the tune. Open the lid and place a microphone right in the middle of the soundboard, starting about two or three feet above it, and a surprisingly effective piano sound can be found. The single microphone approach can benefit from the use of unusual microphones, noisy and distorted use of recording media, heavy compression, etc. File under special effects, but know the piano rewards this type of approach.

14.3.3 ENVELOPE

Lastly, consider the dynamics this instrument can offer. Again, musical dynamics (e.g., *mf* decrescendo to *ppp*) is left to the musical judgment of the players, composers, and producers. The amplitude envelope is the creative variable here: the attack, sustain, and decay of the instrument (Figure 14.4). All that time foraging through the parameters of the synths and samplers and all those hours squinting at the screen of the digital audio workstation, offers an unexpected payoff on the piano. Creating, looping, editing, and simply staring at sampled waveforms gives an engineer a useful way to think about sound. The shape of the sound wave (called the amplitude envelope, and often just the envelope) shows the recordist a bit about how the sound might fit into the mix. Instruments fall somewhere on a range between clave (very short attack with essentially no sustain and nearly instant decay — the analog meter barely twitches) and rhythm electric guitar (every engineer who has recorded this can admire how the meter just sits at one level during the entire song). Sharp, spiky attacks will poke out of the mix naturally and need not be especially loud in the mix (this is particularly true for sounds panned far to one side). Smooth and continuous sounds need careful placement to remain audible without covering other elements of the mix. The piano, ever the challenge, can cover quite a variety of shapes: percussion to pad.

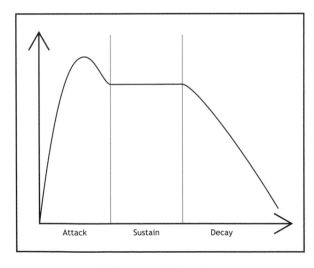

▲ *Figure 14.4 Envelope.*

Consider the attack side of the envelope. The very striking of the hammer against the strings in the piano is often emphasized through microphone placement. The sharp attack enables the piano to cut through a wall of guitars; the contrast in envelopes (piano versus guitars) is the trick here. As discussed in Chapter 6, there is more than one way to sharpen the attack. The piano all but begs engineers to patch in that most wonderful studio tool, the compressor. Heavy compression (low threshold with 10 : 1 ratio or higher) with a medium to slow attack and slow release is all that is needed. The piano note sounds. Several milliseconds later (depending on the attack time of the compressor), the compressor kicks in and yanks the signal down in level, changing the signal's envelope. Through this compressor action, the attack is sharpened (as described in Figure 14.5). Perhaps the attack was entirely fabricated — a smooth and gentle envelope turned mean. Engineers should explore a range of attack times, ratios, and thresholds in search of a variety of sounds caused by alteration of the amplitude envelope.

At the other end of the envelope lives still more opportunity for signal processing: decay. The piano, score, and performance will determine the

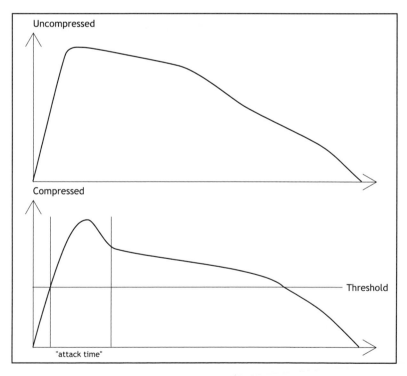

▲ *Figure 14.5 Sharpening the attack through compression.*

character of the sustain of each note. Sixteenth-note funk parts are expected to have a different envelope than whole-note, textural accompaniment. The release parameter of the compressor offers the engineer a chance to modify the sustain. Keeping the release time medium to fast can have the effect of lengthening the decay time. One may tweak it into a natural or a bizarre sound.

The piano can be converted into an almost bell-like sound through use of a very quick attack setting, fast release time setting, and very high (limiting) ratio. Each note of the piano possesses a soft, almost nonexistent attack, followed by a very long, seemingly infinite decay. This radical use of limiting pairs nicely with the use of an unusual old microphone.

Finally, consider a more classical approach to recording the piano. Instead of using close microphone techniques on the piano, distant pairs of microphones are preferred. This aesthetic values the sound of the piano in a hall. Perhaps it is better described as the sound of the piano *and* the hall. Distant microphones record a greater proportion of the early reflections and reverberation in the hall and less of the direct sound of the piano itself. Among other things, this approach captures a radically different envelope, unlike any of the close microphone approaches discussed above. There is no chance to hear the hammers hitting the strings with this approach. The decay of the instrument is augmented by the decay of the hall's reverb. In the home studio, engineers approximate this through the addition of the studio's best sounding hall reverb, built on a not-too-aggressive close microphone approach.

The piano tolerates all extremes of recording approaches. One can place microphones very near the hammers, and apply a long hall reverb patch and use severe compression. Distortion and wah pedals are sometimes appropriate, as well. The piano welcomes radical approaches, but demands careful control. First target the sound most appropriate to the tune and to the mix, and follow any path necessary to get there.

14.4 Outside the Box Behavior

An instrument as important as the piano has inspired many imitators.

14.4.1 "ACOUSTIC" PIANO

There is no such thing as an "acoustic" piano. Pianos are acoustic *by definition.* If it is a box full of silicon and solder rather than wood and wire, then, and only then, does it need the extra adjective (e.g., a synth piano or a digital piano). Such are the rants of hopelessly romantic piano lovers like your author, but there is an important point to be made here. That heavy piece of wood and steel may be difficult to move, but its sound is easy to shape. A producer or engineer may want an "acoustic" piano sound that evokes the feeling and perhaps the image of an honest to goodness piano in an honest to goodness room. Alternatively, they may want a sound whose personality is recognizable as a piano but whose overall sound has been aggressively manipulated to suit some creative desires. Discussed above were the many techniques for recording and mixing the piano toward a range of options. To be complete, we also consider the less expensive version.

14.4.2 DIGITAL PIANO

The alternative to the real-life piano is the *digital* or *MIDI piano*: the sampled/synthesized piano patches that live in synth racks. While often owing some of their heritage to an actual piano somewhere in their past, these sounds are in fact pretty far removed from "acoustic" reality. In addition to being subjected to all of the processes discussed above for true piano, it is useful to explore some additional spectral, image, and envelope opportunities offered by these pianolike devices.

For engineers whose synthesis programming skills are sharp, there is opportunity to really take the piano out, spectrally. With several types of effects built into the synth architecture of many digital pianos, it is easy and rewarding to apply sweeping resonant filters to the pianolike sound. Layered with strings or other sounds, the piano sinks into a deeper texture.

The stereophonic image of a digital piano warrants special attention. Probably the most obvious peculiarity/feature of the digital piano is the panning distribution of the notes. Each note of the typical piano patch occupies a location in the stereo field that very closely follows the position of each individual (ivory or, more likely, plastic) key. This is standard, dare we call it traditional, practice for digital pianos. With the popularity of synths and the impracticality of pianos, there is some risk that the true sonic image of a piano is being replaced with this synthesized/sampled nonreality. Surely the engineers who recorded the original piano sounds that form the basis of the sample playback piano synthesizers noticed that the notes of a piano come from the entire piano, not from such precise points in space. The sound of the individual notes certainly does not come from the black keys and white keys themselves. The sound does not laser beam its way from the string to the ears. It is wonderfully more vague than that. It is the soundboard, all 15 square feet or so stretching out before the pianist, which radiates the sound. Both single notes and thick chordal voicings swell out from the soundboard (above and below it) swirling and bouncing throughout the structure of the piano and the room, finally reaching the listeners as a complex sound field in which the localization of individual notes is anything but discrete.

So what is an engineer to make then of this ridiculous note-by-note, left-to-right, continuum-panning paradigm that pervades the digital pianos? Leverage it. It is an opportunity. Real pianos cannot do this. Digital pianos can. The humble panning effect is used on pretty much every element of every multitrack production, but in much simpler, broader ways: lead vocal, snare, kick, and bass panned center; acoustic guitar panned right; horn section panned left, etc. Digital pianos upgrade panning to a note-by-note detail that few other instruments can offer. The clever composer takes advantage of it. The creative engineer might try to show it off. Digital pianos (and most any synthesizer or sampler output) represent an opportunity for audio engineers to utilize the convenience of sample playback to make more intricate use of the stereo field. Like hard-panned guitar doublings, or ping-ponging vocal delays, digital piano panning can be a great use of the two speakers one has to work with.

WARNING: Do not pretend it is a piano (an "acoustic" piano). It simply is not. It cannot be. In fact, when arpeggios and glissandos pan across the stereo landscape, a digital piano is identified. If the creative goal is a real piano, this panning standard will undermine the illusion, and can only be used as a special effect, which likely is not valid. If the musical desires of the production team include this note-by-note stereophonic image, then a piano should not be used; record a MIDI piano instead. They are two different instruments. An engineer should consciously choose the one that meets the needs of the music and the mix.

Should a still more interesting stereo image effect be desired, one may further exploit the power of the sound module. Chords need not sit still as a low-to-high, left-to-right image. LFOs (low-frequency oscillators, like those found in the modulation section of most delay lines, see Chapter 9) or randomizing features can give the MIDI piano a more varied sonic image. It may be necessary to narrow the total spread of the pan pots to exercise some control over a radical, motion-filled image and confine the total space of the pianolike sound to a small region within the stereo field. Imagine frenetic guitar arpeggios on the right, radical-hyper-swirling-virtual piano on the left. That is easier to make sense of than having both instruments move and dance across the entire left to right spectrum. Confining them makes it easier to enjoy their motion. When musicians use synth/sample playback units for pop-music parts, anything goes, so an engineer must at times seek out the more unusual effects opportunities these devices make possible.

On the envelope front, digital pianos offer another signal processing opportunity: reverse piano. Sneaking in with a gradual attack, sustaining with a rich piano tone, and ending abruptly, the reverse piano gives the arranger and the engineer a lot to work with. Flip those samples so that they play backwards, use a sequencer to get the part right, and an attention grabbing element drops into the arrangement — one that would be very difficult, if not impossible, to create with a piano.

14.5 Summary

The piano retains its significance today, despite it price, heft, and mechanical complexity. It is still the instrument on which most of us compose. It is the instrument in most every classroom at most every music school. It has an important place in the gear intense, pop-music world. The piano on its own represents a vast range of sounds, an infinite bank of patches that respond well to any kind of sound effect an engineer can imagine.

Automated Mix

15

"Life's the same
I'm moving in stereo.
Life's the same
Except for my shoes.
Life's the same,
You're shaking like tremolo.
Life's the same,
It's all inside you."
— "MOVING IN STEREO," THE CARS, *THE CARS* (ELEKTRA/ASYLUM RECORDS, 1978)

No study of recording studio effects would be complete without the careful consideration of mix automation. The power of studio effects cannot be fully realized without it. This chapter is a tutorial on mix automation, appropriate for any automation system, analog or digital, on a console or in a digital audio workstation.

15.1 Unautomated Mixing

So what is so automatic about automation? An automation system plays back an engineer's mix moves, however elaborate, automatically. That is all automation can do. It repeats the mix done by someone else. It is nothing without the engineer. In order to see what can be done with automation, it makes sense to take a look at what can be done without automation. Mixing without automation is called *manual mixing*. The following is pretty typical:

- Intro: all vocals cut, extra reverb on the strings, fade organ in. Bass enters at bar four.
- Verse 1: guitar, drums, and lead vocal in, keep background vocals out, less reverb on the strings, pan organ left, and make room for horns on the right at bar 12.
- Chorus 1: lead vocal double comes in, six background vocal tracks up and perfectly blended (three part harmony, doubled), gated room sound added to snare, strings out, acoustic guitar in.

That is a lot for one engineer to do all at once, so the assistant engineer, the studio manager, and the Chinese food delivery person help out. There is a lot to remember, so notes are scribbled on the track sheet, the console, and the Chinese food menu. It is hard to listen critically while doing so much, so in truth the engineer does not hear the mix objectively until the master is played back later.

Oops, the horns were too loud on verse 3? Try again. Do everything the same way as the last mix, but get the horns right on the third verse. This is not easy. It takes several more passes to get close to that last good version, the whole time the mix team is trying to remember the horns and not forget anything else.

When everyone has run out of the ability to remember another thing, it is time to start mixing in pieces: "Well, we finally got the intro right. Now, let's move on to the first verse . . ." The song is mixed section by section and later edited together into what sounds like a single pass.

That is manual mixing, which can be summarized as:

- *All hands on deck.* Only an octopus with golden ears could make all these moves at once, so the production pulls in the help of others in the control room.
- *Extensive documentation.* Make creative but informative notes about everything that changes in the course of the mix, putting tick marks next to the faders for their key levels at each part of the song, sticking red tape on the pan pot that gets twisted left in every verse, sticking a post-it on the reverb that gets cut in the bridge, etc.
- *Trial and error.* Print several passes to the master deck. Then have a listening session with the band and choose the best one.
- *Cut and paste.* Be prepared to edit the good pieces from several good mixes into a single best mix.

Manual mixes often become an intense process of choreographing the contortions of the various helpers grabbing knobs and faders on cue, while speed-reading the notes, scratches, and scribbles all over the studio. This can be an adrenaline-filled experience, pulling the engineer into the musical performance on the multitrack; the engineer starts to feel like part of the band. It is important that the excitement not cloud the engineer's judgment. Sometimes these engineering thrill rides are not fun for anyone except the engineer. Sometimes the music suffers. The thrill of mastering the complicated logistics in a manual mix can mask any opinion the engineer has about the music. One must remember that those who listen to the mix later will not, for the most part, have any idea what occurred in the studio. Listeners will react to the sound of the art, not the complexity of the craft.

Automation to the rescue. Because it can control any number of faders, automation can be several sets of hands doing several mix moves at once. Because it "remembers" different settings and fader positions by storing them, automation can make all the crazy documentation unnecessary. Automation makes trial and error and cut and paste obsolete, turning each mix pass into a controlled, repeatable fine-tuning of the mix.

15.2 Automated Mixing

There are degrees of automation capability. In the world of digital workplaces (digital audio workstations, digital hard disk recorders with built-in mixing capability, or digital consoles), it is often possible to automate nearly every knob, switch, slider, or parameter in the mixer. This is more difficult to do with an analog work surface where it is quite likely that only the faders and cut buttons are automated. Somewhere in between is musical instrument digital interface (MIDI) automation, where simple note on/note off and other performance gestures are used to drive a mixer instead of a synthesizer or sequencer.

15.2.1 FADERS AND CUTS AUTOMATION

Some consoles only automate the faders and the cut buttons. It is possible to spend more than $200,000 and only get faders and cuts automation. That might seem disappointing at first. There is a lot more to a mix than faders moving and mute buttons cutting tracks in and out. What about pan positions, equalization (EQ), reverb time? Of course these other settings are important to a mix. But a key to successful automated mixing is being honest with oneself in assessing whether or not these settings really need complex changes throughout a mix. Keeping the mix moves simple helps keep the mind free enough and calm enough to think creatively and listen carefully. It is possible to have too much of a good thing, even in rock and roll. By the way, it is safe to say that at least 80% of all the hit pop and rock records made in the 1980s and 1990s were mixed with fader and cut automation only. Wonderfully elaborate and complicated mixes can be built with this relatively limited amount of automation capability.

Alternative Signal Path

The horn arranger has the creative mandate to build colors and feelings through the controlled use of different kinds of horns. Which horn plays which part of the chord? How will the chords be connected to each other? The arranger answers these sorts of questions creatively to produce a horn chart. Multitrack mixes are also arranged; they are arranged by the mix engineer. The engineer decides the sound and texture of each track, using signal processing to tweak or mangle the sound as desired. The mix engineer decides which part plays when, using the cut buttons.

The mixer/arranger often wishes to push the sonic development of the song further by changing the signal-processing structure of a given track for a special part of the song. For example, it might be desirable for the vocal to take on a less aggressive persona during the bridge. This can be accomplished through the use of a different EQ contour, less compression, more reverb, and a touch of chorus on the reverb tail — signal-processing details that were not a part of the vocal sound during the rest of the tune.

Not surprisingly, automation is the mixer's/arranger's tool. What perhaps is surprising, however, is the discovery that such elaborate changes to different elements of a mix can be achieved through simple faders and cuts automation. Often it is not necessary to automate the equalizer as it transitions from its primary sound to the less radical tone desired in the bridge. Nor is it necessary to automate the rest of the signal-processing components as the compression decreases, the reverb increases, and the chorused reverb appears. All that is needed is a parallel vocal channel, getting the same original vocal track, but sending it through the signal-processing chain required for the bridge. During the bridge, automated cuts simply mute the aggressively treated vocal sound that is open for the rest of the tune, and turn on the parallel, sweet, and gentle one. The action of two mute switches, turning one vocal patch off while the other is turned on, affects this significant change to the mix.

The sonic result feels like an elaborate mix move and, hopefully, a compelling musical statement. But on the console, it is created through the use of one additional channel using different effects, and a couple of mute and unmute commands of the automation system.

Automated Send

A variation on the theme above is the automated send. It may not be necessary to create an entirely different effects structure to accomplish a creative twist in the mix. For example, it might be desirable to have extra reverb on the acoustic guitar during the intro, but back off once the band kicks in. In this case, the acoustic guitar track is routed to an additional fader that sends the guitar to the reverb only, not the mix bus. Most consoles can do this. The input into the additional channel is a mult (short for "multiple," the mult is a copy of the signal created simply by splitting the signal at the patch bay) of the guitar sound. The aux send patched to the reverb is turned up, but this channel's output is specifically not assigned to the mix bus. The cut button on this extra guitar channel amounts then to a reverb on/off button; the fader is a reverb send level. Pull the fader down for less reverb; push it up for more. This automated send offers the engineer a way to layer in areas of more or less effects, again using only straightforward faders and cuts automation.

The opportunities for automated sends are limited only by the engineer's imagination — and good taste. Add a triplet echo to key words through some automated cuts on a send to a delay. Consider the delay accent on the promising line, "My baby's got a new pair of ear plugs . . . ear plugs . . . ear plugs . . . ear plugs." The automated send comes from a module with a mult of the vocal, feeding a delay. It is muted the entire tune until it unmutes for the words, "ear plugs," sending it to the delay, only to mute again for perhaps the rest of the tune. Alternatively, the engineer can ride a fader-sending signal only to reverb and gently add width and depth to some background vocals during the chorus. In all of these cases, the mix engineer is creating very sophisticated layers to their mix using faders and cuts automation only.

MIDI Equivalents

Depending on the extent of the desired sonic change, it may be possible to use MIDI messages sent to the effects units to achieve some mix goals. The engineer creates and saves two different reverbs on the same effects device and uses program change commands to switch to the "verse reverb" and back to the "main reverb." These reverbs might be the same exact patch with a single parameter, like reverb time, changed.

Some of the more clever effects units allow not just program changes, but also the changing of many parameters while the effect is running. That is, using a MIDI controller, the reverb time might be shortened slightly without any audio artifacts; the reverb smoothly transitions into the smaller sound.

Once again, detail and complexity are added to the character of the mix through the use of very simple commands: program changes or mod wheel motions.

Automated Comps

Ever done a session in which the lead vocal ends up scattered across several tracks? Actually, the real question is, has anyone ever done a session when it was not? Take one: sounds really good, especially the last verse and chorus — the singer really dug in there. Save that track and do another take. Take two: Everything is sounding good and the missing parts of track one are all now covered, except a couple of words drift flat. Now record those problematic words, but use a third track as punching in risks erasing a portion of the keeper part. Naturally these tracks are to be "comped" into a single track: Bounce the best part from each of these three source tracks onto a fourth track, feeling no pressure when punching in and out at each transition. If a punch is missed, the engineer just tries again. The original magic moments are safe on other tracks. As shown in Figure 15.1, the process has created a best, composite take from the many different options tracked.

Automate the cuts associated with creating that composite track for a more relaxed experience. Better yet, do not bounce the tracks to a new one. Using automated cuts, create a virtual comp that plays back the best take at all times, even those single words over on track 17 that were resung to correct pitch. Used wisely (i.e., not constantly), this can enable the project to enjoy a single, convincing, consistent, and powerful performance. Faders and cuts automation is all that is needed.

Printed Mix Moves

Sometimes, and this might be once every dozen mixes or so, the engineer gets really inspired and throws together a mix move too complicated for the automation system, and too magic to let go. The solution, tracks permitting, is to record the effect. In the interest of creating a distinct sound for the vocal on the last chorus, one might feel inspired to run it through an old guitar amp with tremolo, using a wah-wah pedal, sticking the amp in the shower (the water is off!). Wait, there's more. It is too tempting to use that strange looking ribbon microphone that the tech says, "Doesn't

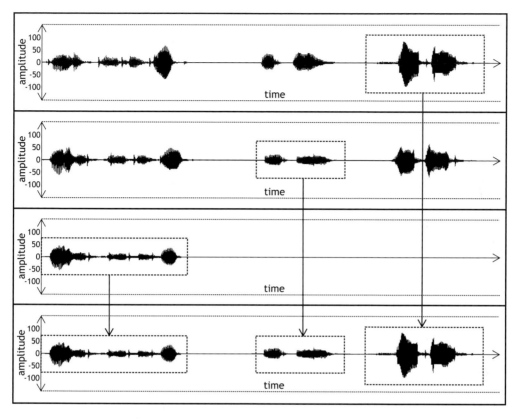

▲ *Figure 15.1 Compositing a vocal from three takes.*

work," but that the assistant engineer says, "Sounds freakin' wicked." Without really knowing what "freakin' wicked" means, one tries it and it sounds good. While the singer plays the wah-wah pedal and the engineer pans it in time with the music, this crazy effect is printed to multitrack. Not only will it be impossible to recreate the amp tone, microphone placement, etc. of this sound tomorrow, there is the possibility that the microphone will quit working altogether. This mix move is not saved in automation, it is saved to multitrack. Simple automated mute commands will bring it into the mix anytime.

15.2.2 EVERYTHING AUTOMATION

Of course, when the audio or the mixer live in the digital world, automation as a technology becomes a lot like a word processor or video game. If software controls the music, there is almost no limit to what the automation system can do. Digital- or computer-based mixers offer this sort of opportunity. The alternative signal path explored in faders and cuts automation is still useful, but it lives in software not hardware. To have an alternative vocal sound, just program the changes in: automate the equalizer, compressor, reverb, and chorus. There is no need for an entirely different signal path for the vocal to have this altered state, just a different kind of signal processing. In addition, the automated sends discussed above become trivial to set up if the echo send level control itself can be automated. That is, one need not dedicate an entire channel to automating an effect. For more ambience on the intro, ride the automated echo send level up on the intro. Nice and easy. Everything automation is a real ally to the engineer's creativity. The mix engineer can do more sonically because the automation system can do more. Automate the EQ or the compressor for de-essing. Automate a mid-frequency sweep for wah-wah. The imagination is the limit.

The discussion of automation approaches for faders and cuts automation above reveals that it sometimes takes an elaborate signal flow structure just to create a small effect. The engineer often has trouble feeling inspired if every sonic idea ("Wouldn't it sound great if...") is followed by an analysis of the automation manual ("How the heck is that automated?"). With everything automation, sound engineers do not have to restrain their creative impulses to enhance or manipulate a sound. The mix moves can be programmed almost as soon as the engineer thinks of the type of sound effect wanted.

The possibilities are endless, but the risk is that it's too much of a good thing. Too many options can paralyze the mixing engineer: "Let's see . . . I'll push the lead vocal fader up on the word, 'baby,' roll-off a bit of high end on the hi-hat, cut the snare track between each hit, pan the piano and guitar back and forth in triplet time, add a nasally EQ to the bass, and shorten the delay on the digeridoo for the first two bars of the first verse. Then, I'll . . ."

Clearly the music could suffer. There is a real temptation to explore mix moves because the gear can do it, not because the music needs it. With everything automation, the recording studio has lost a sort of "reality check" that the more limited automation system imposed. It takes restraint and maturity. Most engineers are at least occasionally seduced by the equipment in this way, no matter how committed they are to the music. The equipment will take over or just plain interfere with the process if left unchecked. The trick is to know when to explore the automation possibilities and when to take a step back and let the music be.

15.2.3 SNAPSHOT AUTOMATION

Everything automation is made less intimidating through what is known as *snapshot automation*. While it may be musically desirable to have the bridge of the tune live in another sonic world (e.g., different fader, pan, EQ, and effects settings), often it is not necessary to continuously modify all the controls with subtle shifts from one setting to the other. Snapshot automation can store the configuration of every parameter of the console at one instant (snapshot A), and then store all the settings wanted for the bridge of the tune (snapshot B). The snapshot automation system enables the mixer toggle from one setting to the other, smoothly, all at once. The mix engineer does not have to program the changes in fader by fader, pan pot by pan pot, effects parameter by effects parameter, and so on. Just store the best sounding chorus setup and the best sounding bridge setup, and let the snapshot automation system connect the dots.

The clever snapshot system gives the user some control over how the change from one snapshot to another is made. Adjustable cross-fade or morph times, perhaps with adjustable slopes, can make the snapping process itself more musical. A very complicated set of mix moves accomplished with very little tedious automation programming results. This often more than adequately serves the musical needs of the mix.

15.3 Mix Modes

While there are as many different automation systems as there are brands of consoles and workstations, they have enough in common that the reader can be prepared to use any automation system through the following orientation.

15.3.1 WRITE OR READ

Either the engineer is performing the mix moves, or the automation system is. Terms like "write" or "read" are common ways of making this rather important distinction. Think of it as *automation record* and *automation play*. When a mix move has been designed and rehearsed, the engineer enters automation write mode and records the move in. Automation read mode makes the automation system reproduce the mix move.

Write and read can usually, and sometimes dangerously, be done globally across the entire console or workstation. Alternatively, write and read may be applied in a more focused way, a few channels at a time, fader by fader, cut button by cut button.

Like any piece of software, there are some quirks that seem strange at first but become more natural with experience. For example, some automation systems need you to write at least a rough version of the whole mix for all faders and mutes on the entire console before one is allowed to go back and do those smaller tweaks on a single fader within a small part of the tune. Sometimes something as trivial as the start time of the mix is an unmovable anchor for the automation system. Getting expressive ideas into a computer is never easy. The artist's approach to creating computer graphics is at least a little different from how they draw on paper. Similarly, entering a mix into the computer requires some navigation through menus and mastering of peculiar syntax that slightly alters the routine from just manual mixing. Once these quirks are mastered, one writes and reads mix automation moves as naturally as one records a track, rewinds and does another pass, plays it back, moves onto another track, and so on. Expect at least a short learning curve.

15.3.2 WRITE: ABSOLUTE OR RELATIVE

When recording mix moves, there are usually two broad approaches, often described by words like *absolute mode* and *relative mode*. Absolute mode tells the system that the engineer wants the automation to store the exact mix parameters currently being adjusted. Any previously written automation moves at this point of the song are completely forgotten, erased and replaced by the new mix moves.

Relative (a.k.a. update or trim) mode lets the engineer revise an existing mix through nudging and tweaking. If the mix engineer likes where the vocal is sitting in the mix, but wishes to push a couple of key words up a little louder, a relative mode might be a good approach. Entering write mode in this way does not make the automation system disregard the mix information already recorded; rather, it updates it based on the new relative moves. Push the fader up and it adds that increase to the move already there. These relative mode trims can be accumulated through additional automation passes to create a very complicated set of fader rides in a simple and intuitive way.

Using Figure 15.2, consider this pretty typical scenario. The first pass sets the general level of the vocal in the mix. Absolute mode writes this into the

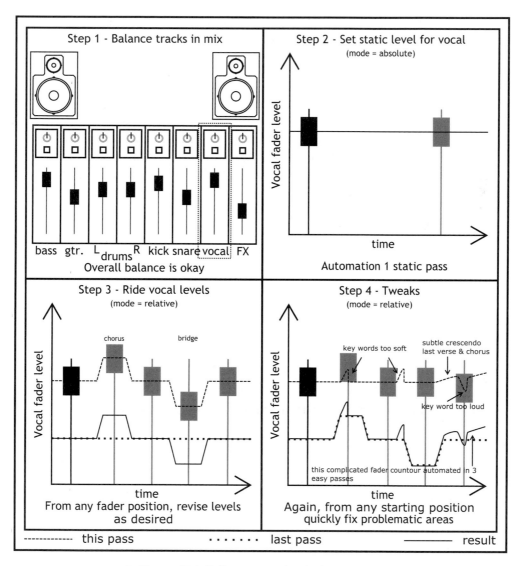

Step 1 - Balance tracks in mix

bass gtr. L drums R kick snare vocal FX

Overall balance is okay

Step 2 - Set static level for vocal
(mode = absolute)

Vocal fader level

time

Automation 1 static pass

Step 3 - Ride vocal levels
(mode = relative)

Vocal fader level

chorus

bridge

time

From any fader position, revise levels
as desired

Step 4 - Tweaks
(mode = relative)

Vocal fader level

key words too soft

subtle crescendo
last verse & chorus

key word too loud

this complicated fader countour automated in 3
easy passes

time

Again, from any starting position
quickly fix problematic areas

---------- this pass   . . . . last pass   ————— result

▲ *Figure 15.2 Fader automation in four easy steps.*

computer as a starting point. The second pass might address some overall
arrangement issues: Turn the vocal up from this basic level at every chorus,
and have it be a little lower at the bridge. Relative mode would be a good
way to do this. With all other faders safely in read mode, enter write mode
on the vocal with the fader in any position, push it up in the choruses, pull
it back down to the original starting point in the verses, and lower it in the

bridge. Because the mix is being trimmed in relative mode, the automation system looks only at the *changes* in the fader's position, not the actual, absolute position of the fader itself. On the third pass the engineer might then modify those already changing fader settings on a word-by-word basis, reducing the words that poke out of the mix and raising the words that are getting lost. Relative mode would be the appropriate choice again. As the mix plays, there is no need to find and match fader levels already written, just write automation moves in relative mode and make these quick fader rides to revise and refine the evolving fader moves already stored in automation.

15.4 Automation Strategies: Organizing the Infinite Options

It is essential to bring some amount of order to the mix approach. The ability to do everything makes it difficult to do anything. Here is a common way to break it down into bite-size increments that grow into an elaborate, musical mix.

15.4.1 PHASE I: BALANCE

Consider the various stages of building a mix. First, one lays out the rough balances that start to make musical sense. Listening to the entire song, one finds fader positions and pan pot settings that enable the song to stand on its own. As discussed in Chapter 8, the pop music vocal and the snare sit pretty loud in the mix, dead center. Kick and bass are also in the center, likely at a lower level. The other pieces of the arrangement fill in underneath and around these critical tracks. Guitars and keys must not mask the vocals or harm the intelligibility of the lyrics. The musical role of each and every track must be understood, and the mix balanced accordingly, so that elements playing together blend or achieve counterpoint as needed. Work hard to find a balance that is fun to listen to, supporting the music while revealing the complexity and subtlety of the song. Importantly, the engineer does not yet automate the mix.

15.4.2 PHASE II: SIGNAL PROCESSING

With the fine tuning that comes from the addition of various forms of signal processing, the engineer builds up a mix that tastefully highlights every element of the pop arrangement that needs it, while achieving subtle blending and layering for the behind-the-scenes, supporting tracks. Various forms of distortion, EQ, compression, limiting, expansion, gating, tremolo, flanging, chorus, echo, pitch shift, reverb, and any number of other effects are introduced, tested, rejected, adjusted, and refined. The sonic traits that will ultimately define this mix are developed now. Every effect introduction requires the engineer to check the balance again. Effects on the snare likely affect its apparent loudness. Effects on the snare are likely to influence its relative balance versus the guitars and the vocal. The engineer constantly and iteratively works to keep the multitrack arrangement balanced as effects are added and deleted.

Please keep in mind that, so far, the whole mix is static. That is, the mix engineer has got a pretty decent sounding tune coming out of the loudspeakers without doing any clever fader moves and effects twiddles. These two steps (phases I and II) actually do the most to determine the overall sound of the mix. The engineer works hardest here, and still has not entered automation yet. As they are so important, these first two phases should take the most time and consume the most creative energy. In fact, because automation is so darn fun, this is rarely the case.

15.4.3 PHASE III: CUTS

With the mix very near where it needs to be sonically (i.e., the producer, engineer, vocalist, and the rest of the band pretty much approve of the sound), the engineer will at last begin to automate the mix. The first step is to apply the appropriate cuts to channels that either are not being used or that were decided against for performance or arrangement reasons. This amounts to making the mix arrangement official.

If the producer does not want horns in the first chorus, the engineer automates the mutes on all of the horn tracks accordingly. If, in the engineer's judgment, the bridge sounds better using doubled harmony vocals, the appropriate mutes and unmutes are automated. If the singer likes the second chorus from the third take of the lead vocal, the engineer "comps" it in using cuts automation.

Now with these cuts happening automatically on cue, the producer and engineer can listen carefully to how the song feels to make sure those are the right decisions. It comes as a surprise at first, but just the process of diving for the cut buttons to mute and unmute tracks at the appropriate times is enough to interfere with one's hearing, psychologically. That is, it is hard for the engineer to form a certain opinion of the mix idea while also remembering all these manual moves. It is much more effective to plug the cuts into automation, sit back, and listen to the song unburdened by any other activities or distractions. Hands folded, eyes closed, the engineer can join the producer and the artist as they decide how much they like the production so far. Does the arrangement make musical sense? Does it grow musically? Does it sag and feel empty in the bridge? Is the detail of the snare lost when the doubled electric guitars enter each chorus? Is the piano in tune with the bass? All these important questions can best be answered when one is just listening and not pressing buttons.

15.4.4 PHASE IV: RIDES

To complete the multitrack arrangement of the song, the cuts described above are followed by some general fader rides. This is where the engineer does things like push the vocal up in the choruses, pull the piano down during the guitar solo, and such. Generally, these are pretty subtle rides. These fader moves are aimed at the musical interpretation of the mix, trying to make the song feel right, whatever that means. A little ride here and another one there helps shape the energy level and mood of the mix.

15.4.5 PHASE V: TWEAKS AND SPECIAL EFFECTS

Only after the musically compelling and well-organized automated mix built carefully through the four phases described above is complete, should the mix engineer attempt the more elaborate, the silly, and the downright crazy moves that are now so tempting. At this point the engineer can add quarter-note delays to keywords of the lead vocal. Now is the time for the mixer to bump up the lost notes of the solo. At last, the engineer can experiment freely with some more elaborate effects to dress up the bridge. With the fundamental elements of the mix being faithfully replayed by the automation, the mix engineer's mind is free to explore the complicated stuff: "In the last chorus, let's run the returns of the long piano reverb through a distortion pedal. Then run that through a noise gate being keyed open by a swing eighth-note delayed snare hit. Then, if we have any more patch cables, it would sound cool if we . . ." Anything goes at this phase of the automated mix.

15.5 Playing the Instrument

It is not enough, after all this, just to know the theory of how to operate the automation system. The mix engineer has to develop some performance ability on the automated mixing console.

15.5.1 PRACTICE

Truly musical mixes come only after the engineer is comfortable with the automated mixer, mentally and physically. A goal of automation is to make the console more like a musical instrument — a device on which all good engineers can perform. It follows then that engineers need to *practice*. Like practicing scales, the mix engineer needs to have down a set of typical moves and be able to do them quickly, under pressure, without thinking. The "scales" of mixing to be practiced are things like the quick cuts associated with comping, appropriate level rides on a vocal track or horn solo through a performance, musical fades for the end of a tune, quick fader moves to attenuate an unwanted squeak or highlight a subtle phrase, setting up an automated send, etc.

Building on this set of often used basic moves, the engineer also develops the techniques needed to attempt more unusual moves. Automation, just like piano, requires practice, practice, practice. It may sound a little silly, but it is also recommended that engineers find some calm time to simply noodle around with the automation system. Think of practicing a musical instrument. Scales, arpeggios, etudes, and prepared pieces are part of that discipline. But all musicians depart, during practice sessions, and just jam. Engineers should do the equivalent on the mixer. Experiment with tracks when the client is not there. Design and explore elaborate automation moves just to see if it can be pulled off. Practicing with the mixing console in this way does more than make the engineer a better mixer. The engineer also develops the ability to work quickly and to improvise on the mixing console. Musicians and producers notice this ability in an engineer.

15.5.2 USER INTERFACE

The techniques developed for mixing are intimately related to the type of mixer used, be it an analog work surface, a digital console, a digital audio workstation, or some combination thereof. Some interfaces welcome a pretty natural approach, while others require special techniques. To get an understanding of how to automate all controls (faders, pan pots, aux sends, equalizers, compressors, and so on) it is helpful first to look closely at faders alone. They serve as an excellent example that provides insight into automating the rest of the console.

The faders on an automated console take instructions from two possible sources: the engineer or the automation system. When the automation system is in control, one of two rather unusual things happens. One possibility is that the faders start moving on their own, without being touched. This can be rather creepy, especially when one is all alone with the console, late at night, lacking sleep, overloading on caffeine. Perhaps more bizarre is when the mix is played back by the automation system, with all of those carefully engineered fader rides happening, yet the faders are *not* moving. It is difficult to say which ghost is preferred within the machine.

Moving Faders

The performance techniques — the physical, dexterity part — are slightly different based on this sole criterion: Do the faders move? Consider moving fader automation first, as it is a little more intuitive. When the console has the ability to move the faders, there is less chance for confusion. Moving faders offer the engineer that much-desired feature of WYSIWYG — what you see is what you get. That is, one can see the mix moves during playback. Guitars too loud in the bridge? One glance at the guitar faders reminds the engineer — oops, I forgot to pull them down.

Not only is there excellent visual feedback about the mix, there is also a nice physical feature. Most moving fader systems will automatically leave read mode and enter write mode whenever a fader is touched. Moreover, they can return to read mode as soon as the engineer lets go of the fader. This gives the engineer a real opportunity to perform. Could not hear that word? Wind the mix back to a point a couple of bars before the word. Play the mix and simply nudge the fader up on that word. When the engineer touches the fader and pushes it up, the automation system instantly starts recording the mix move. When the engineer lets go, the fader returns naturally to the old level and the automated mix playback resumes. It could not be more intuitive. Automation systems without moving faders can achieve the same mix move, but without motorized faders it is less intuitive how the change gets programmed in.

The ease of use as well as the immediate visual, aural, and physical gratification that moving fader systems offer makes them the most desirable way to work. Most world-class studios have moving faders. All digital audio workstations offer a click and drag equivalent. And, thankfully, many less expensive automation systems have moving faders too. But moving fader systems are not without drawbacks. Perhaps most obvious is cost. Consoles are already expensive devices full of seemingly countless components. Putting a reliable, accurate, consistent, small, and quiet motor on every fader is not cheap. If they are cheap, then one has to worry about their reliability, accuracy, consistency, size, and noise. Those world-class studios build this into their big-ticket studio fees. And they probably have spare motors in the tech room. The solution is to have high revenue that justifies this, or to shop very carefully. Such a financial commitment to automation is not always the best choice when it might be more productive to buy another compressor rather than maintain a set of servomotors. There is an additional potential drawback to moving faders, and this one is a little scary. If the automation system can move the faders when the engineer tells it to, then it might also be possible for the automation to move the faders when the engineer did not (mean to) tell it to.

Trade shows and studio tours often show-off demos of the moving fader systems on large-format consoles. In the demo, the faders move together, not for a musical mix, but for some sort of visual effect — sort of a Radio City Rockettes sort of entertainment. The faders arrange themselves into a sine wave that moves left to right and then right to left across the console. First slow, then fast. Those are nifty demonstrations. But imagine spending a not unreasonable four or more hours finding the balance for a tune (phase I) and, just before entering automation, someone accidentally hits the "demo" button. Buzz, click, hum. The faders start doing the wave like they are at the Super Bowl. A careful balance is destroyed by the pre-programmed dance of the faders. It sounds silly, but it really happens. Worse, it is not just the demo that wipes away hours of hard work.

Imagine the following situation: The engineer has found a decent static balance for the tune when, on accident, the multitrack winds back too far, into the previous song. If the engineer is not careful, locating to the previous song can cause the faders to race to the appropriate levels from *that* mix. The automation has seen that time code address before; it knows what to do. The balance for this tune is gone, baby, gone. The solution is to know how to turn the motors off. If fader motors are disabled until after the careful balance is safely stored in the automation memory, then those important fader levels will not be lost. Even after doing this, your author has to admit, he still feels a little rush of terror every time he turns the motors on.

VCA Faders

If the faders cannot be moved by the automation system, how can they be automated? Good question. The solution relies on a voltage-controlled amplifier (VCA) (see Chapter 8). VCAs are faders whose amplifiers boost or attenuate a signal based not on the position of a fader control, but on the value of a control voltage. The result is that the engineer and the automation system share control of the fader level. The engineer uses the slider on the console to adjust the control voltage, while the automation computer uses software. Either control voltage then determines the fader level, the engineer's or the automation system's. But it is tricky mixing when the faders are not moving. What one sees definitely is not what one gets.

No problem. With some adjustment to technique and some practice, most engineers find VCA automation perfectly easy to use. As with true audio faders, the first automated pass is done in an absolute mode. All subsequent passes, however, are generally done in a relative mode. A typical series of mix automation passes is shown in Figure 15.2. Once the static level is written in absolute mode, additional adjustments are made with the VCA in relative mode.

Broadly, two approaches to VCA automation have evolved. The first approach is to return all faders to their zero position after the first pass. This way, the engineer can cut or boost the level, and the scale next to the fader quantifies just how far it has been moved. If the fader is at −12 dB, the engineer has reduced the level 12 decibels from the last stored fader level. There is little confusion as to how extreme the fader move on any automation pass has become, and one can always return to the level of the last pass by returning to the 0 dB marking. This is a good way to keep track of all changes within a pass on a nonmoving fader system.

On the down side, however, it does remove entirely the visual representation of all the fader levels relative to each other. That is, if all faders are set to zero after the first automation pass, then the careful, overall balance built before automation — where the relative levels of each and every track in the multitrack production were thoughtfully and iteratively coaxed into position — lives in the computer's memory only, and the engineer cannot see it. Some engineers find this discomforting. Another option is to do even the later relative mode passes with the faders left where they originally were in the rough balance. That rough balance is usually the level for most of the faders most of the time. Leave the faders at that position for the start of any relative trims and then the actual, physical position of the fader will stay at or very near the fader level that is programmed into the automation. It is a little more confusing to enter write mode at −8 dB, nudge it up briefly to −5 dB, and then remember to return to −8 dB before finishing the revision. With practice, it becomes more natural.

Software Faders

When the faders live on a computer screen, they can move without motors. So much of the benefit of moving fader automation is preserved in digital audio workstations without the expense and maintenance disadvantages. Perhaps the biggest frustration is that, without an external hardware controller with motorized faders, one has to click the fader, not touch the fader. For old-school engineers who were weaned on large-format analog consoles, clicking on faders is an acquired taste. For anyone who is reasonably computer savvy, this is not a problem. Power-users, who are quick on the word processor, efficient surfing the Internet, and agile playing computer games, probably have enough mouse dexterity to find mixing by clicking on a screen perfectly comfortable.

In the software domain, all the other controls on the computer screen can usually be automated as well. The approach is the same, so an engineer's knowledge of write/read and absolute/relative modes is easily applied to pan pots, echo sends, reverb parameters, etc.

Yet another automation feature appears when one can interact with it on screen: graphic editing of mix parameters. There are times when one might not need to perform an additional mix pass just to fix a mix problem. Wish the trombone solo were just a little louder? Click on the graphical representation of the level and drag it up a bit. Listen to it to confirm it is right. Accidentally uncut the track containing the organ solo a hair late and clipped off a bit of the first note? No need to repeat passes and hope through trial and error to eventually time the uncut move at the right instant. Just type in a slightly earlier unmute time in the computer to trim it until it sounds right. These screens, menus, lists, and pictures are often a powerful way to program a mix.

15.6 Caveats

Mix automation gives engineers the power to achieve more. Sloppily handled, that power turns on the engineer, leading almost certainly to lower-quality mixes.

15.6.1 MASTER THE GEAR BEFORE IT MASTERS YOU

Most engineers do this sort of work because they love music. The focus during mixdown is to make the most of the music already recorded, using as much gear as it takes and employing every trick ever seen or heard. It is not easy work. The proverbial magic dust that gets sprinkled on a single vocal track might realistically include two stages of compression; a "secret recipe" of EQ; the subtle addition of rhythmic delays to create a pulsing, highly customized reverb; and fader rides that do not just change from verse to chorus, but word to word, and sometimes syllable by syllable. It is easy to spend several hours of mix time on the vocal alone. The quantity and quality of the signal processors used — on just the vocal — might take a year's salary to acquire. It is not easy to dive that deep into the detail of the vocal sound and still keep track of the musicality of the mix. Inexperienced engineers find it frustrating when the bass player walks in while they are working on the vocal. Likely, the bass player is focused on the bass. The engineer has been focusing on the vocal. The unfortunate engineer "glances" at the bass sound to find a twangy bit of sonic meatloaf, overpowered by every other track in the mix, including the shaker.

Mixing requires an engineer to see the forest *and* the trees. The gear pulls all engineers into a microscopic level of resolution. The music asks them to zoom back out and listen to the whole. Experienced mixers can handle this sort of conflict because they have exercised themselves in this discipline and because they know the gear backwards and forwards. New and intermediate engineers should endeavor to do the same. Recognize that the automation system itself is a very seductive amount of signal-processing capability, and it is often so elaborate as to be at least a little intimidating. Overcome this through overpreparation. Great engineers become so comfortable with the syntax, techniques, tricks, and limitations of the automation system that they can freely move between big-picture musical issues and highly focused automation moves. They can listen to the bass, and the snare, and the kazoo, even as they refine the vocal.

15.6.2 RESIST TEMPTATION

Avoid the common trap of entering automation too soon. This is tricky, and it takes most engineers a number of painful experiences to get over this temptation. Automation employed carefully (and sometimes that means sparingly) is a lifesaver. When misdirected, automation will fall somewhere on a spectrum from distracting to crippling. That is, at best one is faced with the need to constantly undo or redo moves as the mix unfolds. At worst, the automation system so takes over and interferes with the engineer's ability to make changes that, musically, the mix unravels. The solution is to activate the automation system as late as possible in the course of the mixdown session. It is easier to explore mix ideas when the automation does not have to be rewritten. For example, the producer might like to hear the background vocals panned left, without the doubled tracks, but with the addition of some heavily flanged reverb panned right. Trying this out is no problem preautomation. The engineer just turns the knobs and pushes the faders until it sounds right. If the mix is already automated, the engineer will have to revise the automation data governing the fader levels, pan positions, and cut buttons on every track involved. That will slow the session down. It may take so long to get the idea sounding good that everyone gives up on it. A good mix idea is suppressed by the interference of the automation system.

Only the most complicated mix ideas need automation. Explore those late in the game after the basic mix arrangement is automated. The creative process associated with most mix ideas will be much more successful if they are explored without automation.

15.6.3 SAVE OFTEN

The performance gestures that make a mix soar live in software. They must be treated with the same level of paranoia that other software documents inspire: save often, make safety backups, and be prepared for frustration and irretrievable loss when working during a thunderstorm with its associated power spikes and flickers.

15.7 Summary

Through organization, study, and practice, automation can become an easy-to-use asset in the studio. It does not make manual mixing easier or faster. On the contrary, automation enables mixes to become much more elaborate. Mix sessions can sometimes be much, much slower.

Automation does not make everyone a better mix engineer. If the manual mixes were pretty dull, the automated mixes will be too. It is only in the hands of a talented engineer that automation makes mixes more interesting and sophisticated. Automation makes the rather uninteresting device known as a mixing console into more of a musical instrument. It stores the engineer's performance. It frees them to improvise. Many engineers do not think of automation as a helpful feature, they think of it as a required tool that empowers them to orchestrate all of the effects associated with making recorded music.

Bibliography

Ando, Yoichi, 1998, *Architectural Acoustics*, Springer-Verlag, New York.

Atagi, J., Ando, Y., and Ueda, Y., 1996, "*The Autocorrelation Function of a Sound Field in an Existing room as a Time-Variant System*," Abstract, *J. Acoust. Soc. Am.*, vol. 100, pp. 2838.

Bacon, Tony, Ed., 1981, *Rock Hardware*, Crown Publishers, New York.

Ballou, Glen, Ed., 2005, *Handbook for Sound Engineers*, 3rd Ed., Focal Press, Oxford, England.

Barron, M., and Marshall, A., 1981, "Spatial Impression Due to early Lateral Reflections in Concert Halls: the Derivation of a Physical Measure," *J. Acoust. Soc. Am.*, vol. 77, pp. 211–232.

Benade, Arthur H., 1990, *Fundamentals of Musical Acoustics*, Dover Publishing, Inc., New York.

Beranek, Leo, 1996, *Concert and Opera Halls, How They Sound*, Acoustical Society of America, Woodbury, NY.

Beranek, Leo, 2004, *Concert Halls and Opera Houses, Music, Acoustics, and Architecture*, 2nd Ed., Springer-Verlag, New York.

Békésy, Georg Von, 1960, *Experiments in Hearing*, McGraw-Hill New York.

Blauert, Jens, 1997, *Spatial Hearing*, MIT Press, Cambridge, MA.

Blesser, Barry, 2001 October, "An Interdisciplinary Synthesis of Reverberation Viewpoints," *J. Audio Eng. Soc.*, vol. 49, pp. 867–903.

Bradley, J.S., and Soulodre, G.A., 1995, "The Influence of Late Arriving Energy on Spatial Impression," *J. Acoust. Soc. Am.*, vol. 97, pp. 2263–2271.

Eargle, John, 1996, *Handbook of Recording Engineering*, 3rd Ed., Chapman & Hall, New York.

Elmore, William C., and Mark A. Heald, 1969, *Physics of Waves*, Dover Publications, Inc., New York.

Gardner, William G., 1998, "Reverberation Algorithms," in *Applications of Digital Signal Processing to Audio and Acoustics*, Kahrs, M. and Brandenburg, K., Eds., Kluwer Academic Publishers, Boston, MA.

Griesinger, David, 1989, "Practical Processors and Programs for Digital Reverberation," *Proc. Audio Eng. Soc. 7th Int. Conf.*, Toronto, Ontario, Canada, pp. 187–195.

Griesinger, David, 1995, "How Loud is My Reverberation?," *Proc. Audio Eng. Soc. Conv.* Preprint 3943.

Hammond, Laurens, 1941, "Electrical Musical Instrument", U.S. patent 2,230,836.

Hartmann, William, 1998, *Signals, Sound, and Sensation*, Springer-Verlag, New York.

Hidaka, T., Beranek, L.L., and Okano, T., 1995, "Interaural Cross-Correlation, Lateral Fraction, and Low- and High-Frequency Sound Levels as Measures of Acoustical Quality in Concert Halls," *J. Acoust. Soc. Am.*, vol. 98, pp. 988–1007.

Hunt, Frederick Vinton, 1992, *Origins in Acoustics*, Acoustical Society of America, New York.

Keet, M.V., 1968, "The Influence of Early Lateral Reflections on the Spatial Impression," *Proc. 6th Intern. Congr. Acoust.*, Tokyo, Japan, Paper E-2-4.

Kinsler, Lawrence, and Austin Frey, et al., 1982, *Fundamentals of Acoustics*, 3rd Ed., John Wiley & Sons, New York.

Kuhl, Walter, 1960, "Acoustic Reverberation Arrangements," U.S. patent 2,923,369.

Kuhl, Walter, 1973, "Reverberation Device," U.S. patent 3,719,905.

Kuttruff, Heinrich, 1991, *Room Acoustics*, 3rd Ed., E & FN Spon, London.

Martin, Geoff, et al., 1996, "*Sound Source Localization in a Five-Channel Surround Sound Reproduction System*," *Proc. Audio Eng. Soc. Conv.* Preprint 4994.

Mitchell, Doug, 1996, "*Rear and Side Pairwise Evaluations of the Precedence Effect*," *Proc. Audio Eng. Soc. Conv.* Preprint 4995.

Moore, Brian C. J., 2003, *An Introduction to the Psychology of Hearing*, 5th Ed., Academic Press, London.

Moorer, J.A., 1979, "About This Reverberation Business," *Computer Music Journal*, vol. 3, pp. 3255–3264.

Moulton, David, 2001, *Total Audio*, Kiq Producitons, Los Angeles, CA.

Moylan, William, 2002, *The Art of Recording*, Focal Press, Oxford, England.

Moylan, Williiam, 2006, *Understanding and Crafting the Mix*, Focal Press, Oxford, England.

Nicomachus the Pythagorean, Flora R. Levin, Translator, 1994, *The Manual of Harmonics*, Phanes Press, Michegan.

Newton, Sir Isaac, 1952, *Opticks*, Dover Publications, Inc., New York.

Oppenheim, A.V, and Schafer, R.W., 1989, *Discrete-Time Signal Processing*, Prentice Hall, Englewood Cliffs, New Jersey.

Pickles, James O., 1988, *An Introduction to the Physiology of Hearing*, Academic Press, London.

Rumsey, Francis, 2002 September, "Spatial Quality Evaluation for Reproduced Sound: Terminology, Meaning, and a Scene-Based Paradigm," *J. Audio Eng. Soc.*, vol. 50, pp. 651–666.

Sabine, Wallace Clement, 1992, *Collected Papers on Acoustics*, Peninsula Publishing, Los Altos, CA.

Schroeder, Manfred R., 1962 July, "Natural–Sounding Artificial Reverberation," *J. Audio Eng. Soc.*, vol. 10, pp. 219–223.

Schroeder, Manfred R., 1996 May, "The 'Schroeder Frequency' Revisited," *J. Acoust. Soc. Am.*, vol. 99, pp. 3240–3241.

Woram, John, 1989, *Sound Recording Handbook*, Howard W. Sams & Co, Indianapolis, IN.

Zwicker, E., and Fastl, H., 1999, *Psychoacoustics, Facts and Models*, 2nd Ed., Springer-Verlag, Berlin.

Index